Mathematical Physics Studies

More information about this series at http://www.springer.com/series/6316

Gaëtan Borot · Alice Guionnet
Karol K. Kozlowski

Asymptotic Expansion of a Partition Function Related to the Sinh-model

Springer

Gaëtan Borot
Max Planck Institut für Mathematik
Bonn
Germany

Alice Guionnet
Department of Mathematics
MIT
Cambridge, MA
USA

Karol K. Kozlowski
ENS de Lyon
Laboratoire de Physique-UMR 5672 du CNRS
Lyon
France

ISSN 0921-3767 ISSN 2352-3905 (electronic)
Mathematical Physics Studies
ISBN 978-3-319-81499-5 ISBN 978-3-319-33379-3 (eBook)
DOI 10.1007/978-3-319-33379-3

Printed on acid-free paper

This Springer imprint is published by Springer Nature
The registered company is Springer International Publishing AG Switzerland

An opening discussion

The present work develops techniques enabling one to carry out the large-N asymptotic analysis of a class of multiple integrals that arise as representations for the correlation functions in quantum integrable models solvable by the quantum separation of variables. We shall refer to the general class of such integrals as the sinh-model:

$$\mathfrak{z}_N[W] = \int_{\mathbb{R}^N} \prod_{a<b}^{N} \{\sinh[\pi\omega_1(y_a - y_b)]\sinh[\pi\omega_2(y_a - y_b)]\}^{\beta} \cdot \prod_{a=1}^{N} \mathrm{e}^{-W(y_a)} \cdot \mathrm{d}^N\mathbf{y}.$$

When $\beta = 1$ and for specific choices of the constants $\omega_1, \omega_2 > 0$ and of the confining potential W, $\mathfrak{z}_N$ represents norms or arises as a fundamental building block of certain classes of correlation functions in quantum integrable models that are solvable by the quantum separation of variable method. This method takes its roots in the works of Gutzwiller [1, 2] on the quantum Toda chain and has been developed in the mid-1980s by Sklyanin [3, 4] as a way of circumventing certain limitations inherent to the algebraic Bethe Ansatz. Expressions for the norms or correlation functions for various models solvable by the quantum separation of variables method have been established, *e.g.* in the works [5–12]. The expressions obtained there are either directly of the form given above or are amenable to this form (with, possibly, a change of the integration contour from $\mathbb{R}^N$ to $\mathscr{C}^N$, with $\mathscr{C}$ a curve in $\mathbb{C}$) upon elementary manipulations. Furthermore, a degeneration of $\mathfrak{z}_N[W]$ arises as a multiple integral representation for the partition function of the six-vertex model subject to domain wall boundary conditions [13]. In the context of quantum integrable systems, the number N of integrals defining $\mathfrak{z}_N$ is related to the number of sites in a model (*e.g.* in the case of the compact or non-compact XXZ chains or the lattice regularisations of the sinh or sine-Gordon models) or to the number of particles (*e.g.* in the case of the quantum Toda chain). From the point of view of applications, one is mainly interested in the thermodynamic limit of the model, which is attained by sending N to $+\infty$. For instance, in the case of an integrable lattice discretisation of some quantum field theory, one obtains in this way an exact and non-perturbative description of a quantum field theory in 1 + 1 dimensions and in finite volume. This limit, at the level of $\mathfrak{z}_N[W]$, translates itself in the need to extract the large-N asymptotic expansion of $\ln \mathfrak{z}_N[W]$ up to $o(1)$. It is, in fact, the constant term in the expansion of $\ln(\mathfrak{z}_N[W']/\mathfrak{z}_N[W])$ with W' some deformation of W that gives rise to the correlation functions of the underlying quantum field theory in finite volume. These applications to physics constitute the first motivation for our analysis. From the purely mathematical side, the motivation of our works stems from the desire to understand better the structure of the large-N asymptotic expansion of multiple integrals whose analysis demands surpassing the scheme developed to deal with β-ensembles.

As we shall argue in Section 2.1, it is possible to understand the large-N asymptotic analysis of the multiple integral $\mathfrak{z}_N[W]$ from the one of the rescaled multiple integral

$$Z_N[V_N] = \int_{\mathbb{R}^N} \prod_{a<b}^{N} \{\sinh[\pi\omega_1 T_N(\lambda_a - \lambda_b)]\sinh[\pi\omega_2 T_N(\lambda_a - \lambda_b)]\}^{\beta} \cdot \prod_{a=1}^{N} \mathrm{e}^{-NT_N V_N(\lambda_a)} \cdot \mathrm{d}^N\boldsymbol{\lambda}.$$

There T_N is a sequence going to infinity with N whose form is fixed by the behaviour of $W(x)$ at large x, and: $V_N(\xi) = T_N^{-1} \cdot W(T_N\xi)$.

The main task of the book is to develop an effective method of asymptotic analysis of the rescaled multiple integral $Z_N[V]$ in the case when $T_N = N^{\alpha}$, $0<\alpha<1/6$ and V is a given N-independent strictly convex smooth function satisfying to a few additional technical hypothesis.

Prior to discussing in more detail the results obtained in this work, we would like to provide an overview of the developments that took place, over the years, in the field of large-N asymptotic analysis of N-fold multiple integrals, as well as some motivations underlying the study of these integrals in a more general context than the focus of this book. This discussion serves as an introduction to various ideas that appeared fruitful in such an asymptotic analysis, that we place in a more general context than the focus of this book. More importantly, it will put these techniques in contrast with what happens in the case of the sinh-model under study. In particular, we will point out the technical aspects which complicate the large-N asymptotic analysis of $\mathfrak{z}_N[W]$ and thus highlight the features and techniques that are new in our analysis. Finally, such an organisation will permit us to emphasise the main differences occurring in the structure of the large-N asymptotic expansion of integrals related to the sinh-model as compared to the β-ensemble like multiple integrals.

The book is organised as follows. Chapter 1 is the introduction where we give an overview of the various methods used and results obtained with respect to extracting the large number of integration asymptotics of integrals occurring in random matrix theory. Since we heavily rely on tools from potential theory, large deviations, Schwinger–Dyson equations and Riemann–Hilbert techniques, which are often known separately in several communities but scarcely combined together, we thought it would be useful to give a detailed introduction for readers with various backgrounds. We shall as well provide a non-exhaustive review of various kinds of N-fold multiple integrals that have occurred throughout the literature. Finally, we shall briefly outline the context in which multiple integrals such as $\mathfrak{z}_N[W]$ arise within the framework of the quantum separation of variables method approach to the analysis of quantum integrable models. In Chapter 2, we state and describe the results obtained in this book. In Chapter 3 appears the *first part of the proof*: we carry out the *asymptotic analysis of the system of Schwinger–Dyson equations* subordinated to the sinh-model. It relies on results concerning the inversion of the master operator related with our problem. It is a singular integral operator whose inversion enables one to construct an N-dependent equilibrium measure. The *second part of the proof* is precisely the *construction of this inverse operator*: it is carried out in Chapter 4 by solving, for N large enough, an auxiliary 2×2 Riemann–Hilbert problem. The inverse operator itself and its main properties are described in Section 4.3. The *third part of the proof* consists in obtaining *fine information on the large-N behaviour of the inverse operator*: Chapter 5 is devoted to deriving uniform large-N local behaviour for the inverse operator. Chapter 6 deals with the asymptotic analysis of one- and two-fold integrals of interest to the problem. In Section 6.1 we build on the results established so far to carry out the large-N asymptotic analysis of single integrals involving the inverse operator. Finally, in Section 6.3 we establish the large-N asymptotic expansion of certain twofold integrals, a result that is needed so as to obtain the final answer for the expansion of the partition function. The book contains five appendices. In Appendix A we remind some useful results of functional analysis. In Appendix B we establish the asymptotics for the leading order of $\ln \mathfrak{z}_N[W]$ by adapting known large deviation techniques. In Appendix C we derive some properties of the N-dependent equilibrium measures of interest to the analysis. Then, in Appendix D, we derive an exact expression for the partition function $Z_N[V_G]$ when $\beta = 1$ and V_G is a quadratic potential. We also obtain there the large-N asymptotics of $Z_N[V_G]$ up to $o(1)$. This result is instrumental in deriving the asymptotic expansion of $Z_N[V]$ for more general potential, since the Gaussian partition function always appears as a factor of the latter. Finally, Appendix E recapitulates all the symbols that appear throughout the book. Some basic notations are also collected in Section 1.6.

References

1. Gutzwiller, M.C.: The quantum mechanical Toda lattice. Ann. Phys. **124**, 347–387 (1980)
2. Gutzwiller, M.C.: The quantum mechanical Toda lattice II. Ann. Phys. **133**, 304–331 (1981)
3. Sklyanin, E.K.: The quantum Toda chain. Lect. Notes in Phys. **226**, 196–233 (1985)
4. Sklyanin, E.K.: Functional Bethe ansatz. In: Kupershmidt, B. (ed.) Integrable and Superintegrable Theories, p. 833. World Scientific, Singapore (1990)
5. Babelon, O.: On the quantum inverse problem for the closed Toda chain. J. Phys. A **37**, 303–316 (2004). arXiv:hep-th/0304052

6. Derkachov, S.E., Korchemsky, G.P., Manashov, A.N.: Noncompact Heisenberg spin magnets from high-energy QCD: I. Baxter Q-operator and separation of variables. Nucl. Phys. B **617**, 375–440 (2001). arXiv:hep-th/0107193
7. Derkachov, S.E., Korchemsky, G.P., Manashov, A.N.: Separation of variables for the quantum SL(2,R) spin chain. JHEP 0307, 047 (2003). arXiv:hep-th/0210216
8. Grosjean, N., Maillet, J.-M., Niccoli, G.: On the form factors of local operators in the lattice sine-Gordon model. J. Stat. Mech.: Theor. Exp. P10006 (2012). arXiv:hep-th/1204.6307
9. Kozlowski, K.K.: Unitarity of the SoV transform for the Toda chain. Comm. Math. Phys. **334** (1), 223–273 (2015). arXiv:hep-th/1306.4967
10. Kozlowski, K.K.: Aspects of the inverse problem for the Toda chain. J. Math. Phys. **54**, 121902 (2013). arXiv:hep-th/1307.4052
11. Sklyanin, E.K.: Bispectrality for the quantum open Toda chain. J. Phys. A: Math. Theor. **46**, 382001 (2013). arXiv:hep-th/1306.0454
12. Wallach, N.R.: Real reductive groups II. Pure and applied mathematics, vol. 132-II. Academic Press, inc. (1992)
13. Kazama, Y., Komatsu, S., Nishimura, T.: A new integral representation for the scalar products of Bethe states for the XXX spin chain. J. High. Energ. Phys. **2013**, 13 (2013). arXiv:hep-th/1304.5011

Acknowledgements

The work of GB is supported by the Max-Planck Gesellschaft and the Simons Foundation. The work of AG is supported by the Simons Foundation and the NSF award DMS-1307704. KKK is supported by the Centre National de Recherche Scientifique. His work has been partly financed by the Burgundy region PARI 2013 and 2014 FABER grants 'Structures et asymptotiques d'intégrales multiples'. KKK also enjoys support from the ANR 'DIADEMS' SIMI 1 2010-BLAN-0120-02.

Contents

Abstract

This book develops a method to carry out the large-N asymptotic analysis of a class of N-dimensional integrals arising in the context of the so-called quantum separation of variables method. We push further the ideas developed in the context of random matrices of size N, but in the present problem, two scales $1/N^{\alpha}$ and $1/N$ naturally occur. In our case, the equilibrium measure is N^{α}-dependent and characterised by means of the solution to a 2×2 Riemann–Hilbert problem, whose large-N behaviour is analysed in detail. Combining these results with techniques of concentration of measures and an asymptotic analysis of the Schwinger–Dyson equations at the distributional level, we obtain the large-N behaviour of the free energy explicitly up to $o(1)$. The use of distributional Schwinger–Dyson is a novelty that allows us treating sufficiently differentiable interactions and the mixing of scales $1/N^{\alpha}$ and $1/N$, thus waiving the analyticity assumptions often used in random matrix theory.

Chapter 1
Introduction

1.1 Beta Ensembles with Varying Weights

One of the simplest and yet non-trivial example of an N-fold multiple integral that we are interested in is provided by the partition function of a β-ensemble with varying weights:

$$\begin{aligned}\mathcal{Z}_N^{(\beta)}[V] &= \int\limits_{\mathbb{R}^N} \prod_{a<b}^{N} |\lambda_a - \lambda_b|^{\beta} \cdot \prod_{a=1}^{N} \mathrm{e}^{-NV(\lambda_a)} \cdot \mathrm{d}^N \boldsymbol{\lambda} \\ &= \int\limits_{\mathbb{R}^N} \exp\Big\{ \sum_{a<b} \beta \ln|\lambda_a - \lambda_b| - N \sum_{a=1}^{N} V(\lambda_a) \Big\} \cdot \mathrm{d}^N \boldsymbol{\lambda} \,. \qquad (1.1.1)\end{aligned}$$

$\beta > 0$ is a positive parameter and V is a potential growing sufficiently fast at infinity for the integral (1.1.1) to be convergent. $\mathcal{Z}_N^{(\beta)}[V]$ can be interpreted as the partition function of the statistical-mechanical system of N particles at temperature β^{-1}, that interact through a two-body repulsive logarithmic interaction and are placed on the real line in an overall confining potential V. This logarithmic interaction is the Green function for the Laplacian in $\mathbb{R}^2$ equipped with its canonical metric. By "varying weights" we mean that the potential V is preceded by a factor of N, such that the logarithmic repulsion can typically be balanced by the effect of V for λ_a remaining in a bounded in N interval. This is an important feature of the model that we shall comment further on. We shall however start the discussion by explaining the origin of β-ensembles.

The partition function (1.1.1) can be interpreted as the result of integrating over the spectrum of certain random matrices whose distribution is invariant under one of the classical groups. Consider the real vector spaces:

G. Borot et al., *Asymptotic Expansion of a Partition Function Related to the Sinh-model*, Mathematical Physics Studies,
DOI 10.1007/978-3-319-33379-3_1

$$\mathscr{H}_{N,\beta} = \begin{cases} \beta = 1 & : \text{ real symmetric} \\ \beta = 2 & : \text{ complex hermitian} \\ \beta = 4 & : \text{ quaternionic self-dual} \end{cases} \qquad N \times N \text{ matrices.}$$

We denote $\mathrm{d}M$, the product of the Lebesgue measures for the linearly independent real coefficients of such matrices. The Lie groups:

$$\mathscr{G}_{N,\beta} = \begin{cases} \beta = 1 & : \text{ real orthogonal} \\ \beta = 2 & : \text{ complex unitary} \\ \beta = 4 & : \text{ quaternionic unitary} \end{cases} \qquad N \times N \text{ matrices}$$

act on $\mathscr{H}_{N,\beta}$ by conjugation. If M is a random matrix in $\mathscr{H}_{N,\beta}$ drawn from the distribution[1] $C_{N;V}\mathrm{e}^{-N\mathrm{tr}[V(M)]} \cdot \mathrm{d}M$, $C_{N;V}$ being the normalisation constant, the induced distribution $\mathbb{P}_N^{(\beta)}$ of eigenvalues must be of the form:

$$p_{N;V}^{(\beta)}(\boldsymbol{\lambda}) \cdot \mathrm{d}^N\boldsymbol{\lambda} \qquad \text{with} \qquad p_{N;V}^{(\beta)}(\boldsymbol{\lambda}) = \frac{1}{\mathcal{Z}_N^{(\beta)}[V]} \prod_{a<b}^{N} |\lambda_a - \lambda_b|^\beta \prod_{a=1}^{N} \left\{\mathrm{e}^{-NV(\lambda_a)}\right\}.$$

Hence, in this context, the partition function $\mathcal{Z}_N^{(\beta)}[V]$ corresponds to the normalisation constant of the induced distribution of eigenvalues. The three cases $\beta \in \{1, 2, 4\}$ are very special, since they feature a determinantal or Pfaffian structure that is unknown for general β. This additional structure allows one to reduce the computation of $\mathcal{Z}_N^{(\beta)}[V]$ to one of a family of orthogonal or skew-orthogonal polynomials [1].

For general $\beta > 0$ and polynomial V, the partition function (1.1.1) can also be interpreted as the integral over the spectrum of a family of random tri-diagonal matrices [2, 3], whose entries are independent and have a well-tailored distribution depending on V. As there is no symmetry group acting here, this class of random matrices is very different from the invariant ensembles. It is in nature closer to stochastic Schrödinger operators.

The β-ensembles have been extensively studied for more than 20 years, see *e.g.* the books [1, 4–6], for two main reasons, that we shall develop below. From the probabilistic perspective, the statistical-mechanics interpretation of β-ensembles makes $\mathcal{Z}_N^{(\beta)}[V]$ and its associated probability distribution a good playground for testing the local universality of the distribution of repulsive particles [7]. From the perspective of geometry and physics, the interest in the $\beta = 2$ case—*viz.* random hermitian matrices—has been fostered, since the pioneering works of Brézin–Itzykson–Parisi–Zuber [8], by the insight it provides into two-dimensional quantum gravity and the enumerative geometry of surfaces. This interest was eased by the algebraic miracles that make the case $\beta = 2$ quite tractable from the computational point of view, and also raised by the desire to understand the geometry (related to the integrable structure associated with the orthogonal polynomials) behind these miracles.

[1] Such distributions are indeed invariant under conjugation by $\mathscr{G}_{N,\beta}$.

1.1.1 Local Fluctuations and Universality

The physical idea behind universality is that the logarithmic repulsion dictates the local behaviour of the particles.[2] The universality classes should only depend on β and the local environment of the chosen position on $\mathbb{R}$. Typically one expects that, when $N \to +\infty$, the particles will localise on some union of segments $\cup_k [a_k \,;\, b_k]$ and that, up to a $O(N^{-1})$ precision, the pth particle will localise around a "classical" position γ_p^{cl}.

To be precise, we introduce the k-point density functions $\rho_N^{(k)}(x_1, \ldots, x_k)$. These are symmetric functions of k real variables characterized (if they exist) by the property that, for any sequence of pairwise disjoint intervals $(A_i)_{i=1}^k$:

$$\mathbb{P}_N\Big[\exists i_1, \ldots, i_k \in \{1, \ldots, N\} \;:\; \lambda_{i_j} \in A_j\Big] = \int_{A_1} \cdots \int_{A_k} \rho^{(k)}(x_1, \ldots, x_k) \prod_{i=1}^k \mathrm{d}x_i \,. \tag{1.1.2}$$

$\rho_N^{(k)}$ fails to be a density probability function, because of the unusual normalization:

$$\int_{\mathbb{R}^k} \rho_N^{(k)}(x_1, \ldots, x_k) \prod_{i=1}^k \mathrm{d}x_i = \frac{N!}{(N-k)!}$$

which follows from (1.1.2) by taking a partition of $\mathbb{R}$ into k pairwise disjoint intervals and symmetrising the integration range. In particular, $N^{-1}\rho_N^{(1)}(x)$ is the local mean density of particles.

For instance, if $\beta = 2$ and we look at intervals of size $1/N$ around a point $x_0 \in \mathbb{R}$ where the mean density of particles is smooth and positive—*i.e.* in the bulk—one expects the distribution of the eigenvalues to converge to the determinantal process of the sine kernel. This means that, if $\lim_{N\to\infty} N^{-1}\rho_N^{(1)}(x_0) > 0$, we expect that

$$\lim_{N\to\infty} \frac{\rho_N^{(k)}\Big(\Big\{x_0 + \xi_i/\rho_N^{(1)}(x_0)\Big\}_{i=1}^k\Big)}{\big[\rho_N^{(1)}(x_0)\big]^k} = \rho_{\sin}^{(k)}(\xi_1, \ldots, \xi_k)$$

where $\rho_{\sin}^{(k)}$ are given by:

$$\rho_{\sin}^{(k)}(\xi_1, \ldots, \xi_k) = \det_{1 \le i,j \le k}\big[K_{\sin}(\xi_i, \xi_j)\big], \qquad K_{\sin}(x, y) = \frac{\sin \pi(x-y)}{\pi(x-y)} \,. \tag{1.1.3}$$

Still for $\beta = 2$ and if we look at intervals of size $1/N^{2/3}$ around a point x_0 where the mean density vanishes like a square root, one rather expects to observe the

[2] By local, we mean "looking at intervals shrinking with N so that these contain typically only a finite number of particles in the $N \to \infty$ limit".

determinantal process of the Airy kernel:

$$\rho_{\mathrm{Ai}}^{(k)}(\xi_1,\ldots,\xi_k) = \det_{1\leq i,j\leq k}\big[K_{\mathrm{Ai}}(x_i,x_j)\big], \qquad K_{\mathrm{Ai}}(x,y) = \frac{\mathrm{Ai}(x)\mathrm{Ai}'(y)-\mathrm{Ai}'(x)\mathrm{Ai}(y)}{x-y}\,. \tag{1.1.4}$$

To reformulate, if the condition:

$$\lim_{x\to x_0}\frac{\lim_{N\to\infty}N^{-1}\rho_N^{(1)}(x)}{\sqrt{|x_0-x|}} = A > 0 \tag{1.1.5}$$

holds, we expect that,

$$\lim_{N\to\infty} N^{-k/6}\rho_N^{(k)}\Big(\Big\{x_0+(\pi A)^{2/3}N^{-2/3}\xi_i\Big\}_{i=1}^{k}\Big) = (\pi A)^{-2k/3}\,\rho_{\mathrm{Ai}}^{(k)}(\xi_1,\ldots,\xi_k)\,.$$

Without being too precise, let us say that (1.1.5) is the generic behaviour at the edge of the spectrum of random matrices of large sizes.

The expression for the potentially universal distribution of particles for other shapes of large-N local mean density of particles, and other values of β are known [9, 10]—although their understanding is currently much more developed for $\beta \in \{1,2,4\}$. The main theme in universality problems is therefore to prove that given models exhibit these distributions for the local behaviour of particles in the large-N limit. As a matter of fact, the precise mode of convergence to the universal laws that one can obtain mathematically is not always optimal from a physical point of view, namely it may hold only once integrated against a class of test functions, or only after integration on intervals of size $N^{-1+\eta}$ for η arbitrarily small and independent of N. We refer to the original works cited below to see which mode of convergence they establish.

First results of local universality in the bulk where obtained by Shcherbina and Pastur [11] at $\beta = 2$. Then, at $\beta = 2$ and for polynomial V, Deift, Kriecherbauer, McLaughlin, Venakides and Zhou [12] established the local universality in the bulk within the Riemann–Hilbert approach to orthogonal polynomials with orthogonality weight $\mathrm{e}^{-NV(x)}$ on the real line. These results were then extended by Deift and Gioev to $\beta \in \{1,2,4\}$ for the bulk [13] and then for the generic edge [14] universality. The bulk and generic edge universality for general $\beta > 0$ were recently established by various methods and under weaker assumptions. Bourgade, Erdös and Yau built on relaxation methods so as to establish the bulk [15, 16] and the generic edge [17] universality in the presence of generic C^k potentials. Krishnapur, Rider and Virág [3] proved both universalities by means of stochastic operator methods and in the presence of convex polynomial potentials. Finally, the bulk universality was also established on the basis of measure transport techniques by Shcherbina [18] in the presence of real-analytic potentials while universality both at the bulk and generic edge was derived by Bekerman, Figalli and Guionnet [19] for C^k potentials with k large enough.

1.1.2 Enumerative Geometry and N-fold Integrals

The motivation to study the all-order asymptotic expansion of $\ln \mathcal{Z}_N^{(\beta)}[V]$ when $N \to \infty$ initially came from physics and the study of two-dimensional quantum gravity, especially in the case $\beta = 2$ corresponding to hermitian matrices. In the landmark article [8], Brézin, Itzykson, Parisi and Zuber have argued that, for potentials given by formal series:

$$V(x) = \frac{1}{u}\Big(\frac{x^2}{2} + \sum_{j \geq 3} \frac{t_j}{j}\, x^j\Big)$$

the free energy $\ln \mathcal{Z}_N^{(\beta)}[V]$ has the formal expansion:

$$\ln\Bigg(\frac{\mathcal{Z}_N^{(2)}[V]}{\mathcal{Z}_N^{(2)}[V|_{t_\bullet=0}]}\Bigg) \overset{\text{formal}}{=} \sum_{g \geq 0} N^{2-2g}\, \mathcal{F}^{(g)} \tag{1.1.6}$$

which is to be understood as an equality between formal power series in $\{t_j\}$, and we use the notation $V|_{t_\bullet=0}(x) = \frac{x^2}{2u}$. The coefficients $\mathcal{F}^{(g)}$ correspond to a weighted enumeration of "maps", *i.e.* equivalence classes of graphs $\mathcal{G}$ embedded in a topological, connected, compact, oriented surface $\mathcal{S}$ of genus g such that all connected components $\mathcal{C}_i$ of $\mathcal{S} \setminus \mathcal{G}$ are homeomorphic to disks. Each $\mathcal{C}_i$ which is bordered by j edges of $\mathcal{G}$ is counted with a local weight $-t_j$, each vertex in $\mathcal{G}$ is counted with a local weight u, and the overall weight of $(\mathcal{G}, \mathcal{S})$ is computed as the product of all local weights, divided by the number of automorphisms of $(\mathcal{G}, \mathcal{S})$. For instance, choosing $t_3 \neq 0$ and $t_{j>3} = 0$, $\mathcal{F}^{(g)}$ enumerates triangulations of an oriented surface of genus g. More generally, $\ln \mathcal{Z}_N^{(\beta)}[V]$ with $\beta \neq 2$ gives rise to enumerations of graphs embedded in possibly non-orientable surfaces [20, 21]. Then, the expansion (1.1.6) also contains half-integer g's. Although the mathematical nature of the expansions that had been obtained in [8] and in the many subsequent works in physics was often not precised, these handlings can be set in the appropriate framework of providing an equality between formal power series in u and the parameters $\{t_j\}_{j \geq 3}$ in [22]. Indeed, the fact of subtracting the free energy of the quadratic potential $V|_{t_\bullet}(x) = x^2/2u$ in the right-hand side of (1.1.6) turns the formula into a well-defined equality between formal series; a combinatorial argument based on the computation of the Euler characteristics $2 - 2g$ shows that the coefficient of a given monomial $u^k \prod_\ell t_\ell^{n_\ell}$ is given by a sum over finitely many genera g. For a restricted class of potentials[3] for which the integral (1.1.6) is convergent, the formal power series also corresponds to the $N \to \infty$ asymptotic expansion of $\mathcal{Z}_N^{(2)}[V]$. It is because of such a combinatorial interpretation that these expansions are called "topological expansions", this independently of their formal or asymptotic nature.

---

[3] Roughly speaking, when there exists $1/N$ asymptotic expansion, its coefficients are the same as the formal expansion.

The random hermitian matrix model ($\beta = 2$) for finite N was also interpreted as a well-defined discretised model of two-dimensional quantum gravity. For fixed $\{t_j\}$, there is a finite value $u = u_c$ at which the model develops a critical point: the coefficients $\mathcal{F}^{(g)}$ exhibit as a singularity of the type $(u_c - u)^{(2-\gamma_{\rm str})(1-g)}$ with a critical exponent $-1/2 \leq \gamma_{\rm str} < 0$ depending on the universality class. One of the consequences of the appearance of such singularities is that the average or the variance of the number of faces in a map of fixed genus diverges when $u \to u_c$. This allows one to interpret the $u \to u_c$ limit as a continuum limit. In taking such a limit, it becomes particularly interesting to tune the u-parameter in an N-dependent way such that $u = u_c - N^{-1+\gamma_{\rm str}/2} \cdot \tilde{u}$, hence making each term $N^{2-2g}\mathcal{F}^{(g)}$ of order 1 when $N \to \infty$. In such a double scaling limit, the expansion (1.1.6) does not make sense any more. However, it is expected, and it can be proved in certain cases, that the double scaling limit of the appropriately rescaled partition function $\mathcal{Z}_N^{(2)}[V]$ exists. This limit was proposed as a way of defining the partition function of two-dimensional quantum gravity with coupling constant $\tilde{u}$. We refer to the review [23] for more details. The investigation of these double scaling limits is similar in spirit to the investigation of universality that we already mentioned, with the only difference being that a continuation to complex-valued $\tilde{u}$ does have an interest from the physics point of view, whereas it is often excluded from mathematical study of universality given the difficulty to address it with probabilitistic techniques.

Finally, the all-order expansion (be it formal or asymptotic) of N-fold integrals in the β-ensembles and generalisations thereof have numerous applications at the interface of algebraic geometry and theoretical physics. The key point is that the coefficients of the all-order expansions have an interesting geometric interpretation, and the study of matrix models in the large N limit can give some insight into topological strings, gauge theories, *etc*. Describing the exponentially small in N contributions to $\ln \mathcal{Z}_N^{(\beta)}[V]$ has also an interest of its own. It is particularly interesting [24] as a path towards understanding the possible non-perturbative completion(s) of the perturbative physical theories. As an illustration closer to the scope of this book, we shall give in Section 1.3 a non-exhaustive list of N-fold integrals which have a physical or geometrical interpretation.

1.2 The Large-N Expansion of $\mathcal{Z}_N^{(\beta)}$

1.2.1 Leading Order of $\mathcal{Z}_N^{(\beta)}$: The Equilibrium Measure and Large Deviations

Given a sufficiently regular potential V growing at $x \to \pm\infty$ faster than $\big(\beta + \epsilon\big) \ln |x|$ for some $\epsilon > 0$, the leading asymptotic behaviour of the partition function $\mathcal{Z}_N^{(\beta)}[V]$ takes the form:

$$\ln \mathcal{Z}_N^{(\beta)}[V] = -N^2\Big(\mathcal{E}^{(\beta)}[\mu_{\mathrm{eq}}] + \mathrm{o}(1)\Big) \tag{1.2.1}$$

with

$$\mathcal{E}^{(\beta)}[\mu] = \int V(x)\,\mathrm{d}\mu(x) - \beta \int_{x<y} \ln|x-y|\,\mathrm{d}\mu(x)\mathrm{d}\mu(y)\,.$$

In these leading asymptotics, the functional $\mathcal{E}^{(\beta)}$ is evaluated at the so-called equilibrium measure μ_{eq}, the probability measure on $\mathbb{R}$ that minimises the functional $\mathcal{E}^{(\beta)}$. The notion of equilibrium measure arises in numerous other branches of mathematical physics, for instance the study of zeroes of families of polynomials or the one of the characterisation of the thermodynamic behaviour at finite temperature of quantum integrable models [25]. The minimiser μ_{eq} can be characterised within the framework of potential theory [26]. One can show that the equilibrium measure associated with the functional $\mathcal{E}^{(\beta)}$ exists and is unique. We stress that μ_{eq} is characterised by (1.2.3) and thus depends on β only via a rescaling of the potential, hence imposing the same dependence in the leading order term of the expansion (1.2.1).

Let us explain, on a heuristic level, the mechanism which gives rise to (1.2.1). For this purpose, observe that the integrand of $\mathcal{Z}_N^{(\beta)}[V]$ can be recast as

$$\exp\Big\{ - N^2\mathcal{E}^{(\beta)}[L_N^{(\boldsymbol{\lambda})}]\Big\} \qquad \text{where} \qquad L_N^{(\boldsymbol{\lambda})} = \frac{1}{N}\sum_{a=1}^{N}\delta_{\lambda_a} \tag{1.2.2}$$

is the so-called empirical measure while δ_x refers to the Dirac mass at x. For finite but large N, $\boldsymbol{\lambda} \mapsto \mathcal{E}^{(\beta)}[L_N^{(\boldsymbol{\lambda})}]$ with $\lambda_1 < \cdots < \lambda_N$ attains its minimum at a point $\boldsymbol{\gamma}_{\mathrm{eq}} = (\gamma_{\mathrm{eq};1}, \ldots, \gamma_{\mathrm{eq};N})$ whose coordinates $\gamma_{\mathrm{eq};1} < \cdots < \gamma_{\mathrm{eq};N}$ are bounded, uniformly in N, from above and below. This minimum results from a balance between the repulsion of the integration variables induced by the logarithmic interaction and the confining nature of the potential V since the entropy is negligible.[4] It seems reasonable that the main contribution to the integral, namely the one not including exponentially small corrections, will issue from a small neighbourhood of the point $\boldsymbol{\gamma}_{\mathrm{eq}}$ (or those issuing from permutations of its coordinates) and hence yield, to the leading order in N, $\ln \mathcal{Z}_N^{(\beta)}[V] = -N^2\big(\mathcal{E}^{(\beta)}[L_N^{(\boldsymbol{\gamma}_{\mathrm{eq}})}] + \mathrm{o}(1)\big)$. As a matter of fact, the $\gamma_{\mathrm{eq};a}$ are distributed in such a way that they densify on some compact subset of $\mathbb{R}$ and in such a way that, in fact, $L_N^{(\boldsymbol{\gamma}_{\mathrm{eq}})}$ converges, in some appropriate sense, to the probability measure μ_{eq}.

This reasoning thus indicates that the leading asymptotics of $\ln \mathcal{Z}_N^{(\beta)}[V]$ issue from a saddle-point like estimation of the integral (1.1.1). This statement can be made precise within the framework of large deviations. Ben Arous and Guionnet [27]

[4] The Lebesgue measure does not participate to the setting of this equilibrium: the aforementioned terms induce a $\mathrm{e}^{\mathrm{O}(N^2)}$ behaviour in the light of (1.2.1), while on compact subsets of $\mathbb{R}^N$, the Lebesgue measure produces at most a $\mathrm{O}(\mathrm{e}^{cN})$ contribution, with c depending on the size of the compact set.

showed that the distribution of $L_N^{(\lambda)}$ under the sequence $\mathbb{P}_N^{(\beta)}$ of probability measures associated with $\mathcal{Z}_N^{(\beta)}[V]$ satisfies a large deviation principle with good rate function $\mathcal{E}^{(\beta)}[\mu]$ at speed N^2. This means that, for any open set Ω and any closed set F of the space of probability measures endowed with the weak topology, we have:

$$\liminf_{N\to\infty} N^{-2} \ln \mathbb{P}_N^{(\beta)}[L_N^{(\lambda)} \in \Omega] \geq -\inf_{\mu\in\Omega} \mathcal{E}^{(\beta)}[\mu]$$
$$\limsup_{N\to\infty} N^{-2} \ln \mathbb{P}_N^{(\beta)}[L_N^{(\lambda)} \in F] \leq -\inf_{\mu\in F} \mathcal{E}^{(\beta)}[\mu] .$$

Saying that $\mathcal{E}^{(\beta)}$ is a good rate functional means that its level sets $(\mathcal{E}^{(\beta)})^{-1}([0;M])$ are compact for any $M \geq 0$. As a direct consequence of this large deviation principle, the random measure $L_N^{(\lambda)}$ converges almost surely and in expectation, in the weak topology, towards the (deterministic) equilibrium measure μ_{eq}.

The properties of the equilibrium measure μ_{eq} have been extensively studied [26, 28, 29]. Its uniqueness follows from the strict convexity of $\mathcal{E}^{(\beta)}$. Indeed, given two probability measures μ_0, μ_1 and $t \in [0, 1]$, one has

$$\mathcal{E}^{(\beta)}[(1-t)\mu_0 + t\mu_1] = (1-t)\mathcal{E}^{(\beta)}[\mu_0] + t\mathcal{E}^{(\beta)}[\mu_1] - \beta\, Q[\mu_1 - \mu_0]\cdot t(1-t)$$

where, for any signed finite measure ν of zero mass, one has:

$$Q[\nu] = -\int_{x<y} \mathrm{d}\nu(x)\mathrm{d}\nu(y)\, \ln|x-y| = \int_0^{\infty} \frac{\mathrm{d}k}{2k}\, \big|\mathcal{F}[\nu](k)\big|^2$$

which implies that $Q[\nu] \geq 0$. Furthermore, it is clear that equality holds if and only if $\nu = 0$. The latter does ensure the strictly convexity of $\mathcal{E}^{(\beta)}$.

As a solution of a minimisation problem, μ_{eq} must satisfy an "Euler–Lagrange equation". This condition states the existence of a constant C_{eq} such that:

$$V_{\mathrm{eff}}(x) = V(x) - C_{\mathrm{eq}} - \beta \int \ln|x-y|\mathrm{d}\mu_{\mathrm{eq}}(y), \qquad \begin{cases} V_{\mathrm{eff}}(x) = 0 & \mu_{\mathrm{eq}}\text{–almost everywhere} \\ V_{\mathrm{eff}}(x) \geq 0 & \mu\text{–almost everywhere} \end{cases} \tag{1.2.3}$$

where the second condition holds for any probability measure μ on $\mathbb{R}$ such that $\mathcal{E}^{(\beta)}[\mu] < +\infty$, *cf.* [3, Theorem 6.126]. The inequality comes from the fact that one minimises over positive measures, and the constant C_{eq} is a Lagrange multiplier for the constraint that the total mass should be 1. $V_{\mathrm{eff}}(x)$ is the effective potential felt by a particle; it takes into account V and the repulsion it feels from all other particles distributed according to μ_{eq}. The characterisation (1.2.3) expresses that the effective potential is constant and minimal on the support of μ_{eq}. The constant C_{eq} is chosen in such a way that this minimum is zero. It thus appears reasonable to expect that, in the large N limit, the particles should mostly likely accumulate in $\mathrm{supp}[\mu_{\mathrm{eq}}]$. More precisely, [4, 30, 31] proves a large deviation principle for the position of individual particles at speed N with good rate function V_{eff}. This means that, for any open subset Ω and closed subset F of $\mathbb{R}$, we have:

$$\liminf_{N\to\infty} N^{-1} \ln \mathbb{P}_N^{(\beta)}\big[\exists a \in \{1,\dots,N\},\ \lambda_a \in \Omega\big] \geq -\inf_{x\in\Omega} V_{\text{eff}}(x)$$
$$\limsup_{N\to\infty} N^{-1} \ln \mathbb{P}_N^{(\beta)}\big[\exists a \in \{1,\dots,N\},\ \lambda_a \in F\big] \leq -\inf_{x\in F} V_{\text{eff}}(x)\ .$$

One can prove that if V is C^k for $k \geq 2$, then μ_{eq} is Lebesgue continuous with a C^{k-2} density. Besides, if V is real-analytic, the density is the square root of an analytic function what, in its turn, implies that its support consists of a finite number of segments, called *cuts*. Critical points of the model occur when the topology of the support becomes unstable with respect to small perturbations of the potential. Namely when there exist arbitrarily small perturbations of the potential which result in a support of the equilibrium measure in which one of the component has split in two, or where a new cut has appeared. When this is not the case, one says that the potential is *off-critical*. For V real-analytic on $\mathbb{R}$, a necessary condition for off-criticality is that μ_{eq} vanishes exactly like a square root at the edges of the support:

$$\lim_{x\to a\in\partial\text{supp}\mu_{\text{eq}}} \frac{1}{\sqrt{|x-a|}}\,\frac{\mathrm{d}\mu_{\text{eq}}}{\mathrm{d}x} > 0$$

and this is the "generic" behaviour. When μ_{eq} vanishes like $|x-a|^{k+\frac{1}{2}}$ with $k > 0$, a small island of particles around a can separate from the rest and form a new cut under certain small perturbations of the potential.

The simplest example of an equilibrium measure is provided by the one subordinate to a quadratic potential $V_G(x) = x^2$. This equilibrium measure is given by the famous Wigner semi-circle distribution [32]:

$$\mathrm{d}\mu_{\text{eq}}(x) = \frac{\mathrm{d}x}{\beta\pi}\cdot(\beta - 4x^2)^{\frac{1}{2}}\cdot \mathbf{1}_{[-\frac{\sqrt{\beta}}{2}\,;\,\frac{\sqrt{\beta}}{2}]}(x)$$

which has only one cut. Although there is no easy characterisation of the set of potentials V leading to one-cut equilibrium measures, strictly convex V do belong to this set [33]. Indeed, since for any y the function $x \mapsto -\ln|x-y|$ is strictly convex, integrating it over y against the positive measure μ_{eq} still gives a convex function. As a result, if V is strictly convex, the effective potential (1.2.3) is a *fortiori* strictly convex. This imposes that the minimum of V_{eff} is attained on a connected set, therefore the support of μ_{eq} is a segment.

A remarkable feature of β-ensembles is that, looking at the case of equality in (1.2.3), the density of μ_{eq} can be built in terms of the solution to a *scalar* Riemann–Hilbert problem for a piecewise holomorphic function having jumps on the support of μ_{eq}. If one assumes the support to be known, such Riemann–Hilbert problems can be solved explicitly leading to a *one-fold* integral representation for the density of μ_{eq}. These manipulations originate in the work of Carleman [34], and some aspects have also been treated in the book of Tricomi [35]. In the case where the support consists of single segment $[a; b]$ and for V at least C^2, one gets that the density of the equilibrium measure reads:

$$\mathrm{d}\mu_{\text{eq}}(x) = \mathrm{d}x \cdot \mathbf{1}_{[a;b]}(x) \cdot \int_a^b \frac{\mathrm{d}\xi}{\pi\beta} \cdot \frac{V'(x) - V'(\xi)}{x - \xi} \cdot \left\{ \frac{(b-x)(x-a)}{(b-\xi)(\xi-a)} \right\}^{\frac{1}{2}} . \tag{1.2.4}$$

The above representation still contains two unknown parameters of the minimisation problem: the endpoints a, b of the support of μ_{eq}. These are determined by imposing additional non-linear consistency relations. In the one-cut case discussed above, the conditions on the endpoints a and b are:

$$\int_a^b \frac{\mathrm{d}\xi}{\pi\beta} \cdot \frac{V'(\xi)}{\{(b-\xi)(\xi-a)\}^{\frac{1}{2}}} = 0, \qquad \frac{a+b}{2} \int_a^b \frac{\mathrm{d}\xi}{\pi\beta} \cdot \frac{\xi\, V'(\xi)\,\mathrm{d}\xi}{\{(b-\xi)(\xi-a)\}^{\frac{1}{2}}} = 1 . \tag{1.2.5}$$

The situation, although more involved as regards explicit expressions, is morally the same in the multi-cut case where one has to determine all the endpoints of the support. We stress that the very existence of a *one-fold* integral representation with a *fully explicit* integrand tremendously simplifies the analysis, be it in what concerns the description of the properties of μ_{eq}, or any handling that actually involves the equilibrium measure. The one-cut case is computationally easier to deal with than the multi-cut case: for instance when V is polynomial, the conditions (1.2.5) determining the endpoints a and b are algebraic. As we will explain later, there is another, more important difference between the one-cut case and the multi-cut case, that pertains to the nature of the O(1) corrections to the large-N behaviour of the partition function $\mathcal{Z}_N^{(\beta)}[V]$.

1.2.2 Asymptotic Expansion of the Free Energy: From Selberg Integral to General Potentials

For very special potentials, the partition function $\mathcal{Z}_N^{(\beta)}[V]$ can be exactly evaluated in terms of a N-fold product. The quadratic potential $V_G(x) = x^2$ is one of these special cases and the associated partition function is related to the Selberg integral [36], from where it follows:

$$\mathcal{Z}_N^{(\beta)}[V_G] = (2\pi)^{N/2} \cdot (2N)^{-\frac{N}{2}(1-\frac{\beta}{2}+\frac{\beta}{2}N)} \cdot \prod_{m=1}^{N} \frac{\Gamma\left(1+\frac{m\beta}{2}\right)}{\Gamma\left(1+\frac{\beta}{2}\right)} .$$

With such an explicit representation, standard one-dimensional analysis methods lead to the large-N asymptotics of the partition function:

$$\ln \mathcal{Z}_N^{(\beta)}[V_G] = \left\{ \frac{\beta}{4} \cdot \ln\left(\frac{\beta}{4}\right) - \frac{3\beta}{8} \right\} \cdot N^2 + \frac{\beta}{2} \cdot N \ln N$$
$$+ \left\{ \left(\frac{1}{2} + \frac{\beta}{4}\right) \cdot \ln\left(\frac{\beta}{4e}\right) + \frac{\beta}{2} \cdot \ln 2 + \ln(2\pi) - \ln \Gamma\left(1 + \frac{\beta}{2}\right) \right\} \cdot N$$
$$+ \frac{1}{12}\left(3 + \frac{\beta}{2} + \frac{2}{\beta}\right) \cdot \ln N + \chi'\left(0\,;\, \frac{2}{\beta}, 1\right) + \frac{\ln(2\pi)}{2} + o(1)\,. \tag{1.2.6}$$

The function $\chi\,(s\,;\, b_1, b_2)$ is the meromorphic continuation in s of the function defined for $\mathrm{Re}\, s > 2$ by the formula:

$$\chi\,(s\,;\, b_1, b_2) = \sum_{\substack{m_1, m_2 \geq 0 \\ (m_1, m_2) \neq (0,0)}} \frac{1}{(m_1 b_1 + m_2 b_2)^s}\,.$$

Note that, when $\beta = 2$, the constant term in (1.2.6) can be recast in terms of the Riemann zeta function as:

$$\chi'(0\,;\, 1, 1) = \zeta'(-1) - \frac{\ln(2\pi)}{2}\,.$$

The $o(1)$ remainder admits an asymptotic expansion in $1/N$ whose coefficients are expressed as linear combinations with rational coefficients of Bernoulli numbers and $2/\beta$.

For generic potentials V, there is no chance to obtain a simple closed formula for $\mathcal{Z}_N^{(\beta)}[V]$. Nevertheless, the cases that are computable in closed form do play a role in the asymptotic analysis of the more general $\mathcal{Z}_N^{(\beta)}[V]$ beyond the leading order. Indeed, most of the methods of asymptotic analysis rely, in their final step, on an interpolation between the potential of interest V, and a potential of reference V_0 for which the partition function can be exactly computed. The strategy for obtaining the leading corrections is to conduct, first, a study of the large-N corrections to the macroscopic distribution of eigenvalues, and in particular to the fluctuations of the linear statistics:

$$\mathbb{E}_N^V\left[\sum_{i=1}^{N} f(\lambda_i) - N \int f(x) \cdot \mathrm{d}\mu_{\mathrm{eq}}(x)\right] = N \cdot \mathbb{E}_N^V\left[\int f(x) \cdot \mathrm{d}(L_N^{(\boldsymbol{\lambda})} - \mu_{\mathrm{eq}})(x)\right]$$
$$\equiv N \int_{\mathbb{R}^N} p_{N;V}^{(\beta)}(\boldsymbol{\lambda}) \left\{ \int f(x) \cdot \mathrm{d}(L_N^{(\boldsymbol{\lambda})} - \mu_{\mathrm{eq}})(x) \right\} \cdot \mathrm{d}^N \boldsymbol{\lambda} \tag{1.2.7}$$

for a sufficiently large class of test functions f. The subscript V indicates that we are considering the sequence of probability measures for which $\mathcal{Z}_N^{(\beta)}[V]$ is the partition function. Assume that one is able to establish the large-N behaviour of (1.2.7) for a one parameter t family $\{V_t\}_{t\in[0;1]}$ of potentials, and this uniformly in $t \in [0,1]$ and up to a $O(N^{-\kappa-1})$, $\kappa > 0$ and fixed, remainder. Then one can build on the basic formula:

$$\ln\left(\frac{\mathcal{Z}_N^{(\beta)}[V_1]}{\mathcal{Z}_N^{(\beta)}[V_0]}\right) = -N^2 \cdot \int_0^1 \mathbb{E}_N^{V_t}\left[\int \partial_t V_t(x) \cdot \mathrm{d}L_N^{(\lambda)}(x)\right] \cdot \mathrm{d}t \tag{1.2.8}$$

so as to obtain the asymptotic behaviour of the left-hand side up to a $O(N^{-\kappa})$ remainder. If by some other means one can access to the large-N asymptotics of $\ln \mathcal{Z}_N^{(\beta)}[V_0]$ up to $O(N^{-\kappa})$ remainder, then one can deduce the expansion of $\mathcal{Z}_N^{(\beta)}[V_1]$ to the same order. The terms of order $N \ln N$, $\ln N$, and the transcendental constant term in the asymptotic expansion of the partition function usually do not arise from the fluctuations of linear statistics, but rather from an "integration constant" or from some additional singularities present in the confining potential. The comparison to some known, Selberg-like integral often seems the only way of accessing to these terms, especially in what concerns the highly non-trivial constant terms in such asymptotic expansions. Note that when $\beta = 2$ one can build on orthogonal polynomial techniques to access to the $N \ln N$ and $\ln N$ terms. Also, recently, some progress in an alternative approach to computing the logarithmic terms has been achieved in [37].

1.2.3 Asymptotic Expansion of the Correlators via Schwinger–Dyson Equations

As already mentioned, for a general potential, going beyond the leading order demands taking into account the effect of fluctuations of the integration variables around their large-N equilibrium distribution. The most effective way of doing so consists in studying the large-N expansion of the multi-point expectation values of test functions versus the probability measure induced by $\mathcal{Z}_N^{(\beta)}[V]$, *i.e.* the quantities:

$$\mathbb{E}_N^V\left[\int f(x_1,\ldots,x_n) \cdot \prod_{i=1}^n \mathrm{d}L_N^{(\lambda)}(x_i)\right],$$

sometimes called n-linear statistics. Indeed the access to the sufficiently uniform in the potential large-N expansion of the 1-linear statistic allows one to deduce the asymptotic expansion of $\mathcal{Z}_N^{(\beta)}[V]$ by means of (1.2.8). As we shall explain in the following, these linear statistics satisfy a tower of equations which allow one to express the n-linear statistics in terms of k-linear statistics with $k \leq n+1$. These equations are usually called the Schwinger–Dyson equations and, sometimes, also referred to as "loop equations".

We are first going to outline the structure and overall strategy of the large-N analysis of the Schwinger–Dyson equations on the example of a polynomial potential. The latter simplifies some expressions but also allows one to make a connection with the problem of enumerating certain maps. Then, we shall discuss the case of a general potential which will be closer, in spirit, with the techniques developed in the core of the book.

1.2.3.1 Moments and Stein's Method

For the purpose of this subsection, we shall assume that the potential is a polynomial of even degree $V_{\mathrm{pol}}(x) = \frac{\beta}{2}\{\frac{x^2}{2} + \sum_{j=1}^{2d} \frac{t_j x^j}{j}\}$ with $t_{2d} > 0$ so that $\mathbb{P}_N^{(\beta)}$ is well-defined. (1.1.6) implies that for any integer number p, the pth moment

$$m_N(p) := \mathbb{E}_N^{V_{\mathrm{pol}}}\left[\frac{1}{N}\sum_{a=1}^{N} \lambda_a^p\right] = -\frac{2p}{\beta N^2}\,\partial_{t_p} \ln \mathcal{Z}_N^{(\beta)}\big[V_{\mathrm{pol}}\big]|_{t_\bullet=0}\,. \tag{1.2.9}$$

has a power series expansion in $(t_j)_{j=1}^{2d}$, whose coefficients enumerate maps (see Section 1.1.2) with one marked face of degree p. To explain what convergent matrix integrals and their asymptotic expansions on the one hand, and generating series of maps and their topological decomposition in (inverse) powers of N on the other hand, have to do with each other, we shall use Schwinger–Dyson equations.

Probably, the simplest example of a Schwinger–Dyson equation can be provided by focusing on the standard Gaussian law γ on $\mathbb{R}$. An integration by parts shows that γ satisfies

$$\int x f(x)\,\mathrm{d}\gamma(x) = \int f'(x)\,\mathrm{d}\gamma(x)\,. \tag{1.2.10}$$

for a sufficiently big class of test functions f. In fact, γ is the unique probability measure on $\mathbb{R}$ which satisfies (1.2.10). Recall that the pth-moment of the Gaussian law has the combinatorial interpretation of counting the number of pairings of p ordered points. One can, in fact, build on the above Schwinger–Dyson equation so as to deduce such a combinatorial interpretation by checking that the moments satisfy the same recurrence relation than the enumeration of pairings.

The strategy to extract the large-N behaviour of moments (1.2.9) relies first on the derivation of the system of Schwinger–Dyson equations they satisfy. As for the Gaussian law, it is obtained by an integration by parts. To write this equation down, we denote by $\overline{m}_N$ the quenched moments, *i.e.* the real-valued random variable:

$$\overline{m}_N(p) = \frac{1}{N}\sum_{a=1}^{N} \lambda_a^p\,.$$

We shall as well adopt the convention that $\overline{m}_N(p) = 0$ when $p < 0$. Then, integration by parts readily yields

$$\mathbb{E}_N^{V_{\text{pol}}}\left[\sum_{l=0}^{p-1} \overline{m}_N(l)\,\overline{m}_N(p-1-l) + \frac{1}{N}\left(\frac{2}{\beta}-1\right)p\,\overline{m}_N(p-1) - \overline{m}_N(p+1) - \sum_{j=1}^{2d} t_j\,\overline{m}_N(p+j-1)\right] = 0\,. \tag{1.2.11}$$

Compared to (1.2.10), this equation depends on the dimension parameter N. Note that (1.2.11) not only involves the observables $m_N = \mathbb{E}_N^{V_{\text{pol}}}[\overline{m}_N]$, but also the covariance of $\{\overline{m}_N(p)\}_{p\geq 0}$. To circumvent this fact, let us assume first that these moments self-average, so that this covariance is negligible. Let us assume as well that the expectations $m_N(p)$ are bounded by some C^p for some C independent of N, this for all $p \leq P(N)$ where $P(N)$ is some sequence going to infinity with N. Such information imply that the sequence $\{m_N(p)\}_{p\geq 0}$ admits a limit point $\{m(p)\}_{p\geq 0}$. Equation (1.2.11) then implies that any such limit point $\{m(p)\}_{p\geq 0}$ must satisfy

$$m(p+1) = \sum_{l=0}^{p-1} m(l)\,m(p-1-l) - \sum_{j=1}^{2d} t_j\,m(p+j-1)\,. \tag{1.2.12}$$

Moreover, $m(p) \leq C^p$ for all p. It is then not hard to see that the limiting equation (1.2.12) has a unique solution such that $m(0) = 1$ provided the t_i's are small enough: indeed this is clear for $t_j = 0$ and the result is then obtained by a straightforward perturbation argument, see [38] for details. Finally, one can check that this unique solution is also given by

$$M(p) = \sum_{\ell_1,\dots,\ell_{2d}\geq 0}\left\{\prod_{j=1}^{2d}\frac{(-t_j)^{\ell_j}}{\ell_j!}\right\}\mathrm{Map}_0(p,\ell_1,\dots,\ell_{2d})\,,$$

where $\mathrm{Map}_0(p,\ell_1,\dots,\ell_{2d})$ is the number of connected planar maps with one marked, rooted face of degree p, and ℓ_j faces of degree j, for $1 \leq j \leq 2d$. Indeed, Tutte surgery—which consists in removing the root edge on the marked face, and describing all the possible maps ensuing from this removal—reveals that these numbers satisfy the recursive relation:

$$\begin{aligned}\mathrm{Map}_0(p+1,\ell_1,\dots,\ell_{2k}) = {} & \sum_{q=1}^{p-1}\sum_{\ell_j\geq l_j\geq 0}\left\{\prod_{j=1}^{2d}\binom{l_j}{\ell_j}\right\}\mathrm{Map}_0(q,l_1,\dots,l_{2d}) \\ & \times \mathrm{Map}_0(p-q-1,\ell_1-l_1,\dots,\ell_{2d}-l_{2d}) \\ & + \sum_{j=1}^{2d}\ell_j\,\mathrm{Map}_0(p+j-1,\ell_1,\dots,\ell_{j-1},\ell_j-1,\ell_{j+1},\dots,\ell_{2d})\,.\end{aligned}$$

which turns into the Schwinger–Dyson equation for the generating function $M(p)$.

This strategy to prove convergence to the generating function of planar maps is very similar to the so-called Stein's method [39] in classical probability, which is widely used to prove convergence to a Gaussian law. This method can roughly be summarised as follows. One considers a sequence of probability measures $\{\mu_N\}_{N\geq 0}$ on the real line and assumes that there exist differential operators $\{\mathcal{L}_N\}_{N\geq 0}$ such that for all N,

$$\mu_N\Big[\mathcal{L}_N[f]\Big] = 0$$

for a set of test functions. Assume moreover that $\{\mu_N\}_{N\geq 0}$ is tight and that $\mathcal{L}_N$ converges towards some operator $\mathcal{L}$. Then, if this convergence holds in a sufficiently strong sense, any limit point μ of the sequence $\{\mu_N\}_{N\geq 0}$ should satisfy

$$\mu\Big[\mathcal{L}[f]\Big] = 0\,.$$

If moreover there exists a unique probability measure μ satisfying these equations, then this entails the convergence of the sequence $\{\mu_N\}_N$ towards μ. For instance, convergence to the Gaussian law is proved when $\mathcal{L}[f](x) = f'(x) - xf(x)$. Higher order of the expansion can be obtained similarly in the case where one knows that $\mathcal{L}_N$ admits an asymptotic expansion:

$$\mathcal{L}_N = \mathcal{L} + \frac{\mathcal{L}^{(1)}}{N} + \frac{\mathcal{L}^{(2)}}{N^2} + \cdots$$

so that if $\mathcal{L}$ is invertible one could hope to prove an asymptotic expansion of the form:

$$\mu_N = \mu + \frac{\mu^{(1)}}{N} + \frac{\mu^{(2)}}{N^2} + \cdots$$

at least when integrated against a suitable class of test functions, and with:

$$\mu^{(1)}[f] = -\mu\Big[\mathcal{L}^{(1)} \circ \mathcal{L}^{-1}[f]\Big], \quad \mu^{(2)}[f] = -\mu^{(1)}\Big[(\mathcal{L}^{(1)}) \circ \mathcal{L}^{-1}[f]\Big] - \mu\big[\mathcal{L}^{(2)} \circ \mathcal{L}^{-1}[f]\big]. \tag{1.2.13}$$

The main difference in β ensembles is that the Schwinger–Dyson equation (1.2.11) is not closed on the $\{m_N(p)\}_{p\geq 0}$, namely that involves auxiliary quantities (covariances) which cannot be determined by the equation itself. In order to study the large-N behaviour of the moments by means of the Schwinger–Dyson equation (1.2.11) it is convenient to re-centre the quenched moment around their mean, leading to

$$\mathbb{E}_N^{V_{\text{pol}}}\left[\sum_{l=0}^{p-1}\left(\overline{m}_N(l)-m_N(l)\right)\left(\overline{m}_N(p-1-l)-m_N(p-1-l)\right)\right]-m_N(p+1)$$
$$+\frac{1}{N}\left(\frac{2}{\beta}-1\right)p\,m_N(p-1)-\sum_{j=1}^{2d}t_j\,m_N(p+j-1)$$
$$+\sum_{l=0}^{p-1}m_N(l)\,m_N(p-1-l)=0\,. \qquad (1.2.14)$$

If one then assumes that the covariance produces $o(1/N)$ contributions and that the moments admit the expansion:

$$m_N(p)=m(p)+\frac{\Delta m_N(p)}{N}+o(1/N)\,,$$

one would find

$$\Xi[\Delta m_N](p+1)\sim\left(\frac{2}{\beta}-1\right)p\,m_N(p-1)\,,$$

where Ξ is the endomorphism of $\mathbb{R}^{\mathbb{N}}$, which associates to a sequence $\{v(p)\}_{p\geq 0}$, the new sequence:

$$\Xi[v](p)=v(p)-2\sum_{l=0}^{p-2}m(l)v(p-l-2)+\sum_{j=1}^{2d}t_jv(p+j-2)\,. \qquad (1.2.15)$$

When all t_j's are equal to zero, Ξ is represented by a triangular (semi-infinite) matrix with diagonal elements equal to one, and therefore it is invertible. A perturbation argument shows that Ξ is still invertible when the t_j's are small enough. Hence, we deduce that

$$\lim_{N\to\infty}\Delta m_N=\left(\frac{2}{\beta}-1\right)p\,\Xi^{-1}[m](p-1)\,.$$

To get the next order of the corrections, one needs to be able to characterise the leading large-N behaviour of the covariance. To this end, following [40–42], one derives a "rank 2" Schwinger–Dyson equation, which will give access to the limit of the appropriately rescaled N covariance in the spirit of Stein's method. This equation is obtained by considering the effect, to the first order in ε, of an infinitesimal perturbation of the potential $V(x)\to V(x)+\varepsilon x^k$ in the first Schwinger–Dyson equation (1.2.11), *i.e.* $t_j\to t_j+2\delta_{j,k}k\epsilon/\beta$. It results in the insertion of a factor of $\overline{m}_N(k)$ in the expectation value in (1.2.11). If we introduced the centred random variable $\widetilde{m}_N=N(\overline{m}_N-m_N)$, this "rank 2" equation can be put in the form:

$$\mathbb{E}_N^{V_{\text{pol}}}\Bigg[\sum_{l=0}^{p-1}\widetilde{m}_N(k)\,\overline{m}_N(l)\,\overline{m}_N(p-1-l)-\widetilde{m}_N(k)\,\Xi[\widetilde{m}_N](p+1)$$
$$+\frac{1}{N}\left(\frac{2}{\beta}-1\right)p\,\widetilde{m}_N(k)\,\widetilde{m}_N(p-1)\Bigg]=k\,m_N(k-1+p)\,. \qquad (1.2.16)$$

In this equation, we simplified some terms exploiting the fact that $\widetilde{m}_N$ is centred and that the first Schwinger–Dyson equation (1.2.11) is satisfied. Again, assuming that the first term in (1.2.16) is negligible, we would deduce that:

$$\lim_{N\to\infty}\mathbb{E}_N^{V_{\text{pol}}}\big[\widetilde{m}_N(k)\,\widetilde{m}_N(p)\big]=-k\,\Xi^{-1}[S^{k-2}m](p):=w(k,p)\,.$$

where $S^{k-2}m$ is the sequence whose pth term is $m(k-2+p)$.

Plugging back this limit into the first Schwinger–Dyson equation yields the second order correction:

$$m_N(p)=m(p)+\frac{m^{(1)}(p)}{N}+\frac{m^{(2)}(p)}{N^2}+o\left(\frac{1}{N^2}\right),$$

with

$$\Xi[m^{(2)}](p+1)=\sum_{l=0}^{p-1}\Big\{w(l,p-1-l)+m^{(1)}(l)m^{(1)}(p-1-l)\Big\}+\left(\frac{2}{\beta}-1\right)pm^{(1)}(p-1)\,.$$

Again, one can check that when $\beta=2$, $m^{(2)}$ is the generating function for maps with genus 1 as its derivatives at the origin satisfy the same recursion relations, which in the case of maps are derived similarly by Tutte surgery. The same type of arguments can be carried on to all orders in $1/N$.

As a summary, to obtain the asymptotic expansion of moments, and hence of the partial derivatives of the partition function *c.f.* (1.2.9), we see that one needs uniqueness of the solution to the limiting equation (to obtain convergence of the observables), invertibility of the linearised operator Ξ (to solve recursively the linearised equations), a priori estimates on covariances, or more generally of the correlators (in order to be able to get approximately closed linearised equations for the observables). The expansion can then be established and computed recursively, and this recursion is the topological recursion of [43].

1.2.3.2 The Schwinger–Dyson Equations for a General Potential

We now expand the previous discussion of the Schwinger–Dyson equations, by considering moments of arbitrary test functions instead of polynomials. For the β-ensembles in presence of a general potential V the Schwinger–Dyson equation also arise from an integration by parts. The first Schwinger–Dyson equation takes the form:

$$\mathbb{E}_N^V\Big[\frac{\beta}{2}\int\frac{f(x)-f(y)}{x-y}\cdot \mathrm{d}L_N^{(\lambda)}(x)\cdot \mathrm{d}L_N^{(\lambda)}(y)+\frac{1}{N}\Big(1-\frac{\beta}{2}\Big)\int f'(x)\cdot \mathrm{d}L_N^{(\lambda)}(x)$$
$$-\int V'(x)f(x)\cdot \mathrm{d}L_N^{(\lambda)}(x)\Big]=0$$

Similar equations can be derived for test functions depending on n-variables, although we shall not present them explicitly here. We see that the first Schwinger–Dyson equation relates 1 and 2-linear statistics. More generally, the nth Schwinger–Dyson equations relates n-linear statistics to k-linear statistics with $k \leq (n+1)$. Therefore, as such, the Schwinger–Dyson equations do not allow one for the computation of the n-linear statistics. However, it turns out that these equations are still very useful in extracting the large-N asymptotic expansion of the n-linear statistics. For instance, the leading order as $N \to \infty$ of the first Schwinger–Dyson provides one with an equation satisfied by the equilibrium measure:

$$\frac{\beta}{2}\int\frac{f(x)-f(y)}{x-y}\cdot \mathrm{d}\mu_{\mathrm{eq}}(x)\cdot \mathrm{d}\mu_{\mathrm{eq}}(y)\;-\int V'(x)f(x)\cdot \mathrm{d}\mu_{\mathrm{eq}}(x)=0\,. \tag{1.2.17}$$

This equation is actually implied by differentiating the equality case in the Euler–Lagrange equation (1.2.3) for μ_{eq}, and then integrating the result against $f(x)\mathrm{d}\mu_{\mathrm{eq}}(x)$. The most important point, though, is that one can build on the Schwinger–Dyson equations so as to go beyond the leading order asymptotics. Doing so is achieved by carrying out a bootstrap analysis of the system of Schwinger–Dyson equations. The latter allows one to turn a rough estimate on the $(k+1)$-linear statistics into an improved estimate of the kth statistics. One repeats such a scheme until reaching the optimal order of magnitude estimates. On the technical level, the essential step of the bootstrap method consists in the inversion of a master operator $\mathcal{K}$, which appears in the "centring" of the Schwinger–Dyson equation around μ_{eq}, namely by substituting $L_N^{(\lambda)} = \mu_{\mathrm{eq}} + \mathcal{L}_N^{(\lambda)}$ in the first Schwinger–Dyson equation what, owing to the identity (1.2.17), leads to:

$$\mathbb{E}_N^V\Big[\int\mathcal{K}[f](x)\cdot \mathrm{d}\mathcal{L}_N^{(\lambda)}(x)+\frac{\beta}{2}\int\frac{f(x)-f(y)}{x-y}\cdot \mathrm{d}\mathcal{L}_N^{(\lambda)}(x)\cdot \mathrm{d}\mathcal{L}_N^{(\lambda)}(y)$$
$$+\frac{1}{N}\Big(1-\frac{\beta}{2}\Big)\int f'(x)\cdot \mathrm{d}\mathcal{L}_N^{(\lambda)}(x)\Big]$$
$$=-\frac{1}{N}\Big(1-\frac{\beta}{2}\Big)\int f'(x)\cdot \mathrm{d}\mu_{\mathrm{eq}}(x) \tag{1.2.18}$$

where:

$$\mathcal{K}[f](x) = \beta \int \frac{f(x) - f(y)}{x - y} \cdot \mathrm{d}\mu_{\mathrm{eq}}(y) - V'(x)f(x) = -V'_{\mathrm{eff}}(x)f(x) - \fint \frac{f(y)}{x-y} \cdot \mathrm{d}\mu_{\mathrm{eq}}(y) \,. \tag{1.2.19}$$

The centred around μ_{eq} nth Schwinger–Dyson equations, $n \geq 2$, all solely involve the operator $\mathcal{K}$.

Now, assuming that the operator $\mathcal{K}$ is invertible on some appropriate functional space, one can recast (1.2.18) in the form

$$\begin{aligned}\mathbb{E}_N^V\Big[\int f(x)\cdot \mathrm{d}\mathcal{L}_N^{(\lambda)}(x)\Big] = & -\frac{1}{N}\Big(1-\frac{\beta}{2}\Big)\int \partial_x\mathcal{K}^{-1}[f](x)\cdot \mathrm{d}\mu_{\mathrm{eq}}(x) \\ & -\frac{1}{N}\Big(1-\frac{\beta}{2}\Big)\,\mathbb{E}_N^V\Big[\int \partial_x\mathcal{K}^{-1}[\rho](x)\cdot \mathrm{d}\mathcal{L}_N^{(\lambda)}(x)\Big] \\ & -\frac{\beta}{2}\mathbb{E}_N^V\Big[\int \frac{\mathcal{K}^{-1}[f](x)-\mathcal{K}^{-1}[f](y)}{x-y}\cdot \mathrm{d}\mathcal{L}_N^{(\lambda)}(x)\cdot \mathrm{d}\mathcal{L}_N^{(\lambda)}(y)\Big] \,.\end{aligned}$$

The term arising in the first line is deterministic and produces an $\mathrm{O}(1/N)$ behaviour. The first term in the second line is given by a 1-linear centred statistic that is preceded by a factor of N^{-1}. It will thus be sub-dominant in respect to the deterministic term. In fact, its contribution to the asymptotic expansion of 1-linear statistics is the easiest to take into account. Indeed, assume that one knows the asymptotic expansion of 1-linear statistics up to $\mathrm{O}(N^{-k})$ and wants to push it one order in N further. Then, the term we are discussing will automatically admit an asymptotic expansion up to $\mathrm{O}(N^{-k-1})$ what readily allows one to identify its contribution to the next order in the expansion of 1-linear statistics. Taken this into account, it follows that the non-trivial part of the large-N expansion of 1-linear statistics will be driven by the one of 2-linear centred statistics. In all cases, if one assumes that the 2-linear statistics produce $\mathrm{o}(1/N)$ contributions, the first Schwinger–Dyson equation yields immediately the first term in the large-N expansion of 1-linear statistics. In order to push the expansion further, one should access to the first term in the large-N expansion of 2-linear centred statistics. The latter can be inferred from the second Schwinger–Dyson equation. We shall however, not go into more details.

The main point is that one can push the large-N expansion of k-linear statistics, this to the desired order of precision in N, by picking lower order corrections out of the higher order Schwinger–Dyson equations. The effectiveness of such a bootstrap analysis is due to the particular structure of the Schwinger–Dyson equations. It was indeed discovered in the early 90 s that the coefficients of topological expansions in $1/N$ of the n-point correlators are determined recursively by the Schwinger–Dyson equations. The calculation of the first sub-leading correction to (1.2.1) based on the use of Schwinger–Dyson equations for correlators was first carried out in the seminal papers of Ambjørn, Chekhov and Makeenko [41] and of these authors with Kristjansen [40]. The approach developed in these papers allowed, in principle, for a formal,[5] order-by-order computation of the large-N asymptotic behaviour of $\mathcal{Z}_N^{(2)}[V]$. However due to its combinatorial intricacy, the approach was quite complicated to set in practice. In [44], Eynard proposed a rewriting of the solutions of Schwinger–Dyson equations in a geometrically intrinsic form that strongly simplified the structure and intermediate calculations. Chekhov and Eynard then described the corresponding diagrammatics [45], and it led to the emergence of the so-called topological recursion fully developed by Eynard and Orantin in [43, 46]. It allows, in its present setting, for systematic order-by-order calculation of the coefficients arising in the large-N expansions of the β-ensemble partition functions, just as numerous other instances of multiple integrals, see *e.g.* the work of Borot, Eynard and Orantin [47]. Eynard, Chekhov, and subsequently these authors with Marchal have developed a similar[6] theory [42, 49, 50]. For $\beta = 2$, Kostov [51] has also developed an interpretation of the coefficient arising in the large-N expansions as conformal field theory amplitudes for a free boson living on a Riemann surface that is associated with the equilibrium measure. It was argued in [52] that Kostov's approach is indeed equivalent to the formalism of the topological recursion.

To summarise, the above works have elucidated the *a priori* structure of the large N expansions. We have not yet discussed the problem of actually proving the existence of an asymptotic expansion of $\ln \mathcal{Z}_N^{(\beta)}[V]$ to all algebraic orders in N, namely the fact that

$$\ln \mathcal{Z}_N^{(\beta)}[V] = c_1^{(\beta)} N \ln N + c_0^{(\beta)} \ln N + \sum_{k=0}^{K} N^{2-k} F_k^{(\beta)}[V] + O(N^{-K}) \quad (1.2.20)$$

for any $K \geq 0$ and with coefficients being some β-dependent functionals of the potential V. The existence and form of the expansion up to $o(1)$ when $\beta = 2$ was proven by Johansson [53] for polynomial V under the one-cut hypothesis, this by using the machinery described above and *a priori* bounds for the correlators that were first obtained by Boutet de Monvel, Pastur et Shcherbina [54]. Then, the existence of the all-order asymptotic expansion at $\beta = 2$ was proven by Albeverio, Pastur and Shcherbina [55] by combining Schwinger–Dyson equations and the bounds derived in [54]. In particular, this work proved that the coefficients of the asymptotic expansion coincide with the formal generating series enumerating ribbon graphs of [8]—also known under the name of "maps". Finally, Borot and Guionnet [31] systematised and extended to all $\beta > 0$ the approach of [55], hence establishing the existence of the all-order large-N asymptotic expansion of $\mathcal{Z}_N^{(\beta)}[V]$ at arbitrary β

[5] Namely based, among other things, on the assumption of the very existence of the asymptotic expansion.

[6] For potentials with logarithmic singularities and $\beta \neq 2$, some additional terms must be included [48].

and for analytic potentials under the one-cut hypothesis. This includes, in particular, the analytic convex potentials. The starting point always consists in establishing an *a priori* estimate for the fluctuations of linear statistics, on the basis of a statistical-mechanical analysis [54] or of concentration of measures [31], without assumptions on the potential beyond a sufficient regularity. This estimate takes the form:

$$\mathbb{P}_N^{(\beta)}\Big[\Big\{\boldsymbol{\lambda} \in \mathbb{R}^N \ : \ \Big|\int f(x)\cdot \mathrm{d}(L_N^{(\boldsymbol{\lambda})} - \mu_{\mathrm{eq}})(x)\Big| > t\Big\}\Big] \le \exp\Big\{ - C[f]\, N^2 t^2 + C' N \ln N\Big\}$$

for some constants $C[f] > 0$, C', and for t large enough independently of N. We remind that if $O_1, \ldots, O_n$ are random variables, their moment is:

$$\mathbb{E}_N^V\Big[\prod_{i=1}^n O_i\Big] = \partial_{t_1=0}\cdots\partial_{t_n=0}\mathbb{P}_N^{(\beta)}\Big[\exp\Big(\sum_{i=1}^n t_i O_i\Big)\Big]$$

while their cumulant is defined as:

$$\mathcal{C}_n\big[f_1,\ldots,f_n\big] = \partial_{t_1=0}\cdots\partial_{t_n=0}\ln\mathbb{P}_N^{(\beta)}\Big[\exp\Big\{\sum_{i=1}^n t_i O_i\Big\}\Big]\,.$$

In fact, the cumulants are enough for computing all the n-linear statistics. Indeed, one has the reconstruction

$$\mathbb{E}_N^V\Big[\int\prod_{a=1}^n f_a(x_a)\cdot\prod_{i=1}^n \mathrm{d}L_N^{(\boldsymbol{\lambda})}(x_i)\Big] = \sum_{s=1}^n \sum_{\substack{[\![1\,;\,n]\!] = \\ J_1\sqcup\cdots\sqcup J_s}} \prod_{a=1}^s \mathcal{C}_{|J_a|}\big[\{f_k\}_{k\in J_a}\big]\,.$$

The expression for n-linear statistics involving genuine test functions in n variables belonging to the test space $\mathcal{T}(\mathbb{R}^n)$ is then obtained by density of, say, $\mathcal{T}(\mathbb{R})\otimes\cdots\otimes\mathcal{T}(\mathbb{R})$ in $\mathcal{T}(\mathbb{R}^n)$.

It is advantageous to work with $\mathcal{C}_n\big[f_1,\ldots,f_n\big]$, since it is a homogeneous polynomial of degree n of the re-centred measure $L_N^{(\boldsymbol{\lambda})} - \mu_{\mathrm{eq}}$, and concentration occurs in this re-centred measure:

$$\Big|\mathcal{C}_n\big[f_1,\ldots,f_n\big]\Big| \le \Big\{\prod_{i=1}^n \mathcal{N}[f_i]\Big\}\cdot\Big(\frac{\ln N}{N}\Big)^{\frac{n}{2}}$$

where $\mathcal{N}$ is some norm. Then, under the one-cut assumptions, one shows that the master operator is invertible with a continuous inverse for a suitable norm $\mathcal{N}$. These two pieces of information allow one to neglect the contribution of some of the higher order cumulants in the system of Schwinger–Dyson equations and lead to a successive improvement of the *a priori* bounds up to the optimal scale

$$C_n[f_1,\ldots,f_n] = \frac{1}{N^{n-2}}\Big\{\mathcal{W}_n^{(0)}[f_1,\ldots,f_n] + \mathrm{o}(1)\Big\}. \tag{1.2.21}$$

There, $\mathcal{W}_n^{(0)}$ is some n-linear functional on the space of test functions that are pertinent for the analysis. The same method can be pushed further and establishes recursively an all-order asymptotic expansion for the cumulants:

$$C_n[f_1,\ldots,f_n] = \begin{cases} \sum_{g=0}^{[K/2]} N^{2-2g-n}\,\mathcal{W}_n^{(g)}[f_1,\ldots,f_n] + o(N^{2-n-2\lfloor K/2\rfloor}) & \text{if } \beta = 2 \\ \sum_{k=0}^{K} N^{2-k-n}\,\mathcal{W}_n^{[k]}[f_1,\ldots,f_n] + o(N^{2-K-n}) & \text{if } \beta \neq 2 \end{cases} \tag{1.2.22}$$

for any $K \geq 0$. The asymptotic expansion (1.2.20) for the free energy can then be obtained by the interpolation method that was outlined in Section 1.2.2.

Although this phenomenon will not occur in the present book, we would still like to mention for completeness that, when the support of μ_{eq} has several cuts, the form (1.2.20) of the asymptotic expansion is not valid any more: new bounded oscillatory in N contributions have to be included in $F_k^{(\beta)}[V]$ for $k \geq 0$. Heuristically speaking, this effect takes its roots in the possibility the particles have to tunnel from one cut to another [56, 57]. On the technical level, this takes its origin in the fact that the master operator $\mathcal{K}$ has a kernel whose dimension is given by the number of cuts minus one. For real-analytic off-critical potentials and general $\beta > 0$, the form of the all-order asymptotic expansion in the multi-cut case was conjectured in [57], and established in [58]. It was also established up to $o(1)$ by Shcherbina in [59] by a different technique, namely *via* a coupling to Brownian motion to replace the two-body interaction between different cuts with a linear but random one. We refer the reader to [58] for a longer discussion relative to the history of this particular problem. We also stress that, so far, the analysis of the double scaling limit around a critical potential where the support of the equilibrium measure changes its topology have not been addressed mainly due to difficulties such a transition induces on the level of constructing the pseudo-inverses of the master operator.

Above, we have focused our discussion solely on the approach based on the analysis of loop equations. However, when $\beta = 2$, the orthogonal polynomial based determinantal structure allows one to build on the Riemann–Hilbert problem characterisation of orthogonal polynomials [60] along with the non-linear steepest descent method [61, 62] based approach to characterising the large degree asymptotic behaviour of polynomials orthogonal with respect to varying weights [12, 63] so as to obtain the large-N asymptotic expansion of $\mathcal{Z}_N^{(2)}[V]$. Within the Riemann–Hilbert problem approach, Ercolani and McLaughlin [64] established the existence of the all order asymptotic expansion at $\beta = 2$ in the case of potentials that are a perturbation of the Gaussian interaction. Bleher and Its [65] obtained, up to a o(1) remainder the large-N expansion of $\mathcal{Z}_N^{(2)}[V]$ for polynomial V that give rise to a one-cut potential. Also, recently, Claeys, Grava and McLaughlin [66] developed the Riemann–Hilbert approach so as to obtain the large-N expansion of $\mathcal{Z}_N^{(2)}[V]$ in the case of a two-cut polynomial potential V.

1.2.4 The Asymptotic Expansion of the Free Energy up to $o(1)$

We shall now make some general remarks about the nature of the terms arising in the asymptotic expansion of $\ln \mathcal{Z}_N^{(\beta)}[V]$.

The terms diverging in N when $N \to \infty$ are not affected by the topology of the support of the equilibrium measure. In the case of regular potentials, the pre-factors of the $N \ln N$ and $\ln N$ corrections in (1.2.20) take the form:

$$c_1^{(\beta)} = \frac{\beta}{2}, \qquad c_0^{(\beta)} = \frac{1}{12}\left(3 + \frac{\beta}{2} + \frac{2}{\beta}\right).$$

They can be identified from the large-N asymptotics of the Gaussian partition function, *cf.* (1.2.6). Their presence in (1.2.20) is only the sign that there is a more natural normalisation of the N-fold integral (1.2.20) which would kill the logarithmic corrections in the large N limit. The terms of order N^2 and N are functionals of the equilibrium measure μ_{eq}, which depend in a non-local way on the density of this measure. As we have seen in Section 1.2.1, the prefactor of N^2 is given by a double integral involving μ_{eq}. The term of order N is also known to be given by a single integral involving μ_{eq}. More precisely, up to a universal—*viz.* μ_{eq}-independent—function of β, it is proportional to the von Neumann entropy of the equilibrium measure [42, 59, 67, 68]:

$$F_1^{(\beta)} = \frac{\beta}{2} \ln\left(\frac{\pi\beta}{e}\right) - \ln \Gamma\left(\frac{1+\beta}{2}\right) + \left(\frac{\beta}{2} - 1\right) \int \ln\left(\frac{\mathrm{d}\mu_{\mathrm{eq}(x)}}{\mathrm{d}x}\right) \cdot \mathrm{d}\mu_{\mathrm{eq}}(x)\,. \tag{1.2.23}$$

The fact that the entropy only appears as a sub-leading term is a typical feature of models having varying weights. The bounded term is affected by the topology of the support: it is a constant in the one-cut case and it contains an additional, oscillatory contribution in the multi-cut case. Unlike the non-decaying terms, the coefficient in front of N^{-k} with $k > 0$ only depends on the local behaviour (at an order increasing with k) of the equilibrium measure's density near the endpoints of its support.

Once the asymptotic expansion of the free energy is established up to $o(1)$, and in the case it does not contain oscillatory terms, one can deduce a central limit theorem for the fluctuations of linear statistics. The starting point is the formula for the Fourier transform of their distribution:

$$\mathbb{E}_N^V\left[\exp\left(\mathrm{i}sN \int f(x) \cdot \mathrm{d}L_N^{(\lambda)}\right)\right] = \frac{\mathcal{Z}_N^{(\beta)}\left[V - \frac{\mathrm{i}sf}{N}\right]}{\mathcal{Z}_N^{(\beta)}[V]}\,. \tag{1.2.24}$$

Resorting to arguments of complex analysis, one can fairly easily extend the validity of the asymptotic expansion (1.2.20) to potentials of the form $V_0 + \Delta V/N$, with V real-valued satisfying the previous assumptions, and ΔV complex-valued and differentiable enough. Besides, since the logarithmic correction does not depend on perturbations of the initial potential V_0 and $F_k^{(\beta)}[V]$ are smooth functionals of V away from critical points of the model, one deduces that:

$$\mathbb{E}_N^V\Big[\exp\Big(\mathrm{i}sN\int f(x)\cdot \mathrm{d}(L_N^{(\lambda)}-\mu_{\mathrm{eq}})(x)\Big)\Big] = \exp\Big\{\mathrm{i}s\,\delta_V F_1^{(\beta)}[f] - \frac{s^2}{2}\cdot\delta_V^2 F_0^{(\beta)}[f,f] + o(1)\Big\}$$

where the error $o(1)$ is uniform for s belonging to compact subsets of $\mathbb{C}$ and $\delta_V G[f]$ refers to the Gâteaux derivative of G at the point V and in the direction f. This implies that

$$\sum_{i=1}^{N} f(\lambda_i) - N\int f(x)\cdot \mathrm{d}\mu_{\mathrm{eq}}(x) = N\int f(x)\cdot \mathrm{d}(L_N^{(\lambda)} - \mu_{\mathrm{eq}})(x) \qquad (1.2.25)$$

converges in law to a Gaussian random variable, with covariance given by the Hessian (defined by the second-order functional derivative) of the energy functional introduced in Section 1.2.1, evaluated at V along the direction f:

$$F_0^{(\beta)}[V] = -\mathcal{E}^{(\beta)}[\mu_{\mathrm{eq}}]\,.$$

This central limit theorem for V polynomial, f differentiable enough, any $\beta = 2$ in the one-cut regime, was first obtained by Johansson [53]. The characteristics of the Gaussian variable can be computed solely from the knowledge of the functional $F_0^{(\beta)}[V]$ related to the energy functional, and of $F_1^{(\beta)}[V]$ related to the entropy. For $\beta = 2$, $F_1^{(\beta)}[V]$ approaches $\ln\left(\frac{2\pi}{e}\right)$ owing to its prefactor in (1.2.23), and (1.2.25) converges to a centred Gaussian variable.

We observe that the fluctuations of (1.2.25) are of order 1. In the multi-cut regime, the tunnelling of particles between different, far-apart segments of the support lead as well to order 1 fluctuations, which in general destroy the gaussianity of (1.2.25) and the central limit theorem. This phenomenon was predicted in [69], and given a precise form in [58, 59]: the Gaussian behaviour is in first approximation convoluted with the law of a discrete Gaussian variable, *i.e.* supported on a lattice of an arithmetic progression on $\mathbb{R}$ with step of order 1 and depending on f, and whose initial term drifts with N at a speed of order 1. As a result, the fluctuations display a Gaussian behaviour with interference fringes which are displaced when $N \to N+1$.

1.3 Generalisations

It is fair to say that there exists presently a pretty good understanding of large-N asymptotic expansions of β ensembles. The main remaining open questions concern the description of the asymptotic expansion uniformly around critical points (*viz.*

when the number of cuts changes) and the possibility to relax the regularity of the potential, for instance by allowing the existence of singularities, *e.g.* of the Fisher–Hartwig type.[7] What we would like to stress is that the techniques of asymptotic analysis described so far are effective in the sense that they allow, upon certain more or less obvious generalisations of technical details, treating various instances of other multiple integrals.

The framework of small enough perturbation of the Gaussian potential is, in general, the easiest to deal with. Asymptotic expansions for hermitian multi-matrix models have been obtained in such a setting. For instance, the expansion including the first sub-leading order was derived for a two-matrix model by Guionnet and Maurel–Segala [38], the one to all orders for multi-matrix models by Maurel–Segala [73] and the one to all orders for unitary random matrices in external fields was then obtained by an appropriate adaptation of the analysis of Schwinger–Dyson equations in [74]. Multi-matrix models are interesting in operator algebras, since their ring of observables in the large N limit give planar algebras. Without claiming exhaustiveness, they also appear in the theory of random tilings of arbitrary two-dimensional domains [75, 76], in gauge theories [77] and topological strings via the topological vertex [78, 79]. Most of the time, however, the technology for their asymptotic analysis is not sufficiently developed at present for the purposes of theoretical physics and algebraic geometry, even in the (rare) case where the integrand is real-valued. In fact, even the mere task of establishing the leading order asymptotics under fairly general assumptions is an open problem.

Another natural generalisation of β-ensembles consists in replacing the one-particle varying potential $N \cdot V$ by a regular and varying multi-particle potential

$$N \sum_{a=1}^{N} V(\lambda_a) \hookrightarrow \sum_{p=1}^{r} \frac{N^{2-p}}{p!} \sum_{1 \leq i_1, \ldots, i_p \leq N} V_p\big(\lambda_{i_1}, \ldots, \lambda_{i_p}\big) . \qquad (1.3.1)$$

For $r = 2$, such interactions were studied by Götze, Venker [80] and Venker [81] who showed that the bulk behaviour falls in the universality class of β-ensembles. For general r, Borot [82] has shown that the formal asymptotic expansion of the partition function subordinate to multi-particle potentials is captured by a generalisation of the topological recursion. The existence of the all-order asymptotic expansion was established by the authors in [83] under certain regularity assumptions on the multi-particle interactions. Note that for perturbations of the Gaussian potential of the form (1.3.1) the hypothesis of [83] are indeed satisfied. We give below a non-exhaustive list of physically interesting models with $r = 2$. The ones encountered in integrable models of statistical physics will be pointed out in Section 1.5.3.

[7] Although, even in these two cases, some partial progress has been achieved at $\beta = 2$ where one can build on the Riemann–Hilbert approach [70–72].

Biorthogonal Ensembles

For $r = 2$ and when $\beta = 2$, the structure of such models becomes determinantal in the special cases where the two-body interaction takes the form:

$$V_2(\lambda_1, \lambda_2) = \ln\left(\frac{\big(f(\lambda_2) - f(\lambda_1)\big)\big(g(\lambda_2) - g(\lambda_1)\big)}{(\lambda_2 - \lambda_1)^2}\right). \tag{1.3.2}$$

It is well known that, then, the associated multiple integrals can be fully characterised in terms of appropriate systems of bi-orthogonal polynomials in the sense of [84]. By bi-orthogonal polynomials, we mean two families of monic polynomials $\{P_n\}_{n\in\mathbb{N}}$ and $\{Q_n\}_{n\in\mathbb{N}}$ with $\deg[P_n] = \deg[Q_n] = n$ and which satisfy

$$\int_{\mathbb{R}} P_n\big(f(\lambda)\big) \cdot \big[g(\lambda)\big]^j \cdot \mathrm{e}^{-NV(\lambda)} \cdot \mathrm{d}\lambda = 0 \tag{1.3.3}$$

and

$$\int_{\mathbb{R}} Q_n\big(g(\lambda)\big) \cdot \big[f(\lambda)\big]^j \cdot \mathrm{e}^{-NV(\lambda)} \cdot \mathrm{d}\lambda = 0 \quad \text{for} \quad j \in \{0, \dots, n-1\} .$$

The system of bi-orthogonal polynomials subordinate to f and g exists and is unique for instance when f and g are real-valued and monotone functions. In that case, the multiple integral of interest can be recast as a determinant which, in turn, can be evaluated in terms of the overlaps involving the polynomials P_n and Q_n by carrying out linear combinations of lines and columns of the determinant:

$$\operatorname*{det}_{j,k\in[\![1\,;\,N]\!]}{}_N\left[\int_{\mathbb{R}} \big[f(\lambda)\big]^{j-1}\cdot\big[g(\lambda)\big]^{k-1}\cdot\mathrm{e}^{-NV(\lambda)}\cdot\mathrm{d}\lambda\right] = \prod_{n=0}^{N-1}\left\{\int_{\mathbb{R}} P_n\big(f(\lambda)\big)\cdot Q_n\big(g(\lambda)\big)\cdot\mathrm{e}^{-NV(\lambda)}\cdot\mathrm{d}\lambda\right\} .$$

It is due to their connections to bi-orthogonal polynomials that such multiple integrals are referred to as bi-orthogonal ensembles. The case $f(\lambda) = \lambda^{\theta}$ and $g(\lambda) = \lambda$ is of special interest, since the bi-orthogonal polynomials can be effectively described. In [85] Borodin was able to establish certain universality results for specific examples of confining potentials V. Furthermore, it was observed, first on a specific example by Claeys and Wang [86] and then in full generality by Claeys and Romano [87] that the bi-orthogonal polynomials can be characterised by means of a Riemann–Hilbert problem. However, for the moment, the Riemann–Hilbert problem-based machinery still did not lead to the asymptotic evaluation of the associated partition functions.

Statistical Physics in Two-Dimensional Random Lattices

For $\beta = 2$, with the same precise meaning that was discussed in Section 1.1.2, the N-fold integral:

$$\int_{\mathbb{R}^N} \prod_{1 \le a < b \le N} |\lambda_a - \lambda_b|^2$$

$$\times \exp\left\{\frac{1}{u}\left(-\sum_{a=1}^{N} \frac{N\lambda_a^2}{2} + \sum_{p=1}^{r}\sum_{h=0}^{H} \frac{N^{2-2h-p}}{p!} \sum_{\substack{m_1,\ldots,m_p \ge 1 \\ 1 \le a_1,\ldots,a_p \le N}} t^{(h)}_{m_1,\ldots,m_p} \prod_{i=1}^{p} \frac{\lambda_{a_i}^{m_i}}{m_i}\right)\right\} \cdot \mathrm{d}^N\boldsymbol{\lambda}$$

enumerates maps whose faces, instead of being restricted to be homeomorphic to disks, can have any topology. Each face homeomorphic to a surface of genus h, with p boundaries of respective perimeters $m_1, \ldots, m_p$, is counted with a local weight $t^{(h)}_{m_1,\ldots,m_p}$. This model was introduced in [82], and encompasses a large class of statistical physics models on two-dimensional random lattices. The simplest ones occur for $r = 2$, and the only other interesting cases we are aware of have $r = \infty$. The general $r = 2$ case is equivalent to the enumeration of maps carrying a configuration of self-avoiding loops crossing certain faces, each loop being counted with a Boltzmann weight n. The special case where the faces crossed by the loops are all triangles corresponds to:

$$V_2(x, y) = -n \ln\left|1 - z \cdot (\lambda_a + \lambda_b)\right| \tag{1.3.4}$$

where z is the local weight per triangle crossed by a loop. This is the $O(n)$ loop model which was first introduced by Kostov [88]. The 6-vertex model on a random lattice [89] is realised by:

$$V_2(x, y) = -4 \ln\left|\mathrm{e}^{\frac{\mathrm{i}\gamma}{2}}\lambda_a - \mathrm{e}^{-\frac{\mathrm{i}\gamma}{2}}\lambda_b\right| . \tag{1.3.5}$$

In (1.3.4) for $|n| < 2$ and in (1.3.5) for any real γ, the existence and uniqueness of the equilibrium measure is known, and the results of [83] for an all-order asymptotic analysis apply within the one-cut hypothesis.[8]

Chern–Simons Theory in Seifert Spaces

The perturbative expansions of SU(N) gauge theories lead to a weighted enumeration of Feynman graphs, which are dual to embedded graphs in surfaces, and t'Hooft observed that the N-dependence of the weights only comes from the topology of the surface. For this reason, N-fold integrals related to matrix models are very common in gauge theories, although the form of the probability measure might be complicated. For Chern–Simons theory on a simple class of three-dimensional manifolds $\mathscr{M}$ called Seifert spaces (which include the three-sphere), the measure can be explicitly computed [91, 92], and leads to a partition function for $\beta = 2$ and $r = 2$ with one-point and two-point interactions

[8] It was shown in [90], under fairly general hypothesis, that the equilibrium measures relevant for combinatorics are supported on a single segment.

$$V_1(\lambda) = c_2 \cdot \frac{(\ln \lambda)^2}{2u} + c_1 \cdot \ln \lambda, \qquad V_2(\lambda_1, \lambda_2) = \sum_{i=1}^{k} \ln\left(\frac{\lambda_1^{1/a_i} - \lambda_2^{1/a_i}}{\lambda_1 - \lambda_2}\right)$$

where $(a_1, \ldots, a_k)$ are integers, c_1, c_2 rational numbers related to the geometry of $\mathscr{M}$, and u is related to the coupling constant of the Chern–Simons theory. In this case of interest, the integration runs through $(\mathbb{R}^+)^N$. For the cases $\chi = 2 - k + \sum_{i=1}^{k} a_i^{-1} \geq 0$, the suitably defined energy functional is convex, and this implies existence and uniqueness of the equilibrium measure. The cases $2 - k + \sum_{i=1}^{k} a_i^{-1} < 0$ are rather interesting: although one does not know currently how to prove uniqueness of the equilibrium measure via potential theory by lack of convexity, Monte-Carlo simulations of the distribution of eigenvalues by Weiße seems to indicate that it should be unique [93, Appendix]. The all-order asymptotic expansion of the correlators in these models receive an interpretation in terms of perturbative knot invariants in $\mathscr{M}$, and by large N-dualities, they can be related to topological strings amplitudes in suitable target spaces [93, 94]. As an example of application of the existence of the all-order large-N asymptotic expansion established in [83], Borot and Eynard derived some arithmetic properties of these perturbative knot invariants in [93].

Multispecies β-Ensembles

The β-ensemble with the two-point interactions can be generalised to several types of particles, and appear in the study of coupling between conformal field theories with internal degrees of freedom (describing matter) and two-dimensional quantum gravity. From the probabilistic point of view, these models are potentially the source of new universality classes that can be more usually found in multi-matrix models. But, as they are already written as N-fold integrals, their study is much simpler than the multi-matrix models in which it is necessary to integrate over spaces of (non-commuting) matrices, without the possibility of simultaneous diagonalisation.

Let $\mathcal{D}$ be a graph having M vertices and possibly multiple edges, and let $\mathbf{A}$ stand for its adjacency matrix—*viz.* $A_{v,w}$ corresponds to the number of edges that link the vertices v and w–. Kharchev et al. [95] introduced the N-fold integral:

$$\int_{\mathbb{R}^N} \prod_{v \in \mathcal{D}} \prod_{a=1}^{N_v} \left\{ \mathrm{d}\lambda_a^{(v)}\, e^{-N V_v(\lambda_a^{(v)})} \right\} \cdot \prod_{a<b} |\lambda_a^{(v)} - \lambda_b^{(v)}|^2 \prod_{\substack{\{v,w\}\text{ edge} \\ \text{in } \mathcal{D}}} \prod_{\substack{1 \leq a \leq N_v \\ 1 \leq b \leq N_w}} |\lambda_a^{(v)} - \lambda_b^{(w)}|^{-\frac{A_{v,w}}{2}}. \tag{1.3.6}$$

The $\{\lambda_a^{(v)}\}_{a=1}^{N_v}$ are thought as the positions of N_v particles of type v, and $N = \sum_{v \in \mathcal{D}} N_v$ denote the total number of particles. These models appear in the $\mathcal{N} = 2$ supersymmetric gauge theories associated to ADE quivers [96]. Their study has been revived recently [77, 95] in view of the conjectures of Alday–Gaiotto–Tachikawa [97] which propose a precise relation between four-dimensional quiver gauge theories and the conformal blocks of Liouville theory of two-dimensional quantum gravity.

Kostov introduced the slightly different model:

$$\int_{(\mathbb{R}^+)^N} \prod_{v\in\mathcal{D}} \prod_{a=1}^{N_v} \left\{ d\lambda_a^{(v)}\, e^{-NV_v(\lambda_a^{(v)})} \right\} \prod_{1\leq a<b\leq N_v} |\lambda_a^{(v)} - \lambda_b^{(v)}|^2 \cdot \prod_{\substack{\{v,w\}\text{ edge}\\ \in\mathcal{D}}} \prod_{\substack{1\leq a\leq N_v\\ 1\leq b\leq N_w}} |\lambda_a^{(v)} + \lambda_b^{(w)}|^{-\frac{A_{v,w}}{2}}. \tag{1.3.7}$$

Its partition function enumerates maps in which each face has a colour chosen among the set of vertices of $\mathcal{D}$, restricted in such a way that faces of colour v and v' can be adjacent if and only if there is an edge between v and v' in $\mathcal{D}$. Each interface between a cluster of colour v and a cluster of colour v' is weighted by the number of edges between v and v' in $\mathcal{D}$. Up to an affine change of variables, we retrieve the $O(n)$-model (1.3.4) in the case in which $\mathcal{D}$ has a single vertex from which issue n loops. The q-Potts model corresponds to the case where $\mathcal{D}$ is the complete graph on q vertices, and the model corresponding to $\mathcal{D} =$ "the Dynkin diagram of type A_n" is called the (restricted) height model. One can show [47, Lemma 5.5] that the suitably defined energy functional for these models is strictly convex if and only if $(2 - \mathbf{A})$ is the Cartan matrix of a Dynkin diagram of type A, D, or E, or of the extended Dynkin diagrams $\widehat{A}$, $\widehat{D}$, or $\widehat{E}$, or of the cyclic graph. The N-fold integral (1.3.7) is simpler to analyse than (1.3.6) due to the absence [9] of logarithmic singularities when $\lambda_a^{(v)} = \lambda_b^{(w)}$. New universality classes remembering the ADE symmetries occur precisely when the confining potentials V_v are tuned so that the support S_v of the equilibrium measure for the particles of type v approaches 0. Indeed, at the vicinity of 0 the attractive interaction with $-\lambda_a^{(w)}$ of all other particles will change the local distribution of the particles. To our knowledge, the universal distributions governing these universality classes have not been derived (although the original work of [95] exhibits some integrable structure in these models), maybe because this model is not so well-known in the community working in random matrix theory from the point of view of probabilities.

Conduction in Disordered Wires

Experiments showed that the properties of quantum transport of electrons in chaotic cavities feature some universality, and therefore, one can expect to capture these properties as typical in an ensemble of random cavities. The simplest model consists of two cavities related by two wires, in which N modes can propagate. Landauer theory describes the conduction in such a system by a $2N \times 2N$ scattering matrix:

$$\mathbf{S} = \begin{pmatrix} \mathbf{r} & \mathbf{t}' \\ \mathbf{t} & \mathbf{r} \end{pmatrix}$$

such that the amplitudes of the N modes in the first cavity is related to the amplitudes of the N modes in the second cavity by multiplication by $\mathbf{S}$. Conservation of the current implies that S is unitary, and in turn this implies that the matrices $\mathbf{t}\mathbf{t}^{\dagger}$,

[9] The singularities at $\lambda_a^{(v)} = -\lambda_b^{(w)}$ are absent since the integration runs through $\mathbb{R}^+$.

$\mathbf{t}'(\mathbf{t}')^\dagger$, $\mathbf{1}-\mathbf{r}\mathbf{r}^\dagger$ and $\mathbf{1}-\mathbf{r}'(\mathbf{r}')^\dagger$ have identical spectrum consisting of N eigenvalues $\lambda_1,\ldots,\lambda_N \in [0\,;\,1]$. To understand the transport properties in this setting, it is necessary to investigate the behaviour of the linear statistics $\sum_{a=1}^N f(\lambda_a)$.

In the model, the distribution of the λ_a is drawn from a β-ensemble with $V(\lambda)$ proportional to $\ln \lambda$. In a model of non-ideal leads and for $\beta = 2$, the distribution depending on the mean free path ℓ and the length L of the wire is proportional to [98]:

$$\prod_{1\leq a<b\leq N} |\lambda_a-\lambda_b| \cdot \prod_{a=1}^N \mathrm{e}^{-NV(\lambda_a)} \cdot \det_{1\leq a,b\leq N} K_a\big[L/\ell N\,;\,\lambda_b\big]\cdot \mathrm{d}^N\boldsymbol{\lambda} \tag{1.3.8}$$

for kernels K_a which involve Gauss hypergeometric functions, whereas for $\beta \neq 2$ it is unknown. In the metallic regime, we have $1 \ll L/\ell \ll N$, and the distribution (1.3.8) simplifies drastically to a β-ensemble (here, for all $\beta = 1, 2, 4$) with a two-body interaction:

$$\prod_{a=1}^N \mathrm{e}^{-NV(\lambda_a)} \cdot \prod_{1\leq a<b\leq N} \big|\lambda_a-\lambda_b\big| \cdot \big|\mathrm{argsech}^2(\lambda_a^{1/2}) - \mathrm{argsech}^2(\lambda_b^{1/2})\big| \cdot \mathrm{d}^N\boldsymbol{\lambda} \tag{1.3.9}$$

where $\mathrm{sech}(x) = \frac{1}{\cosh(x)}$ and argsech is its reciprocal function, and V is some explicit one-body interaction. Whereas (1.3.9) falls into the class of models which can be treated with the existing methods of [83], the structure of (1.3.8) is much more involved and goes beyond the present technology based on Schwinger–Dyson equations.

We refer to [99] and references therein for a justification of these distributions, as well as for a deeper overview of the relations between random matrix theory (notably in the form of N-fold integrals) and quantum transport.

1.4 β-Ensembles with Non-varying Weights

In all the examples of the multiple integrals discussed so far, the interaction potential V is preceded by a power of N. This scaling ensures that, for typical configurations of the λ_a's, the logarithmic repulsion is of the same order of magnitude in N than the confining potential. As a consequence, with overwhelming probability when $N \to \infty$, the integration variables remain in a bounded region and exhibit a typical spacing $1/N$. The scheme developed in [12, 31, 55, 83] for the asymptotic analysis was adapted to this particular tuning of the interactions with N and, in general, breaks down if the nature of the balance between the interactions changes.

Serious problems relative to extracting the large-N asymptotic behaviour already start to arise in the case of *non-varying* weights, *i.e.* for multiple integrals:

$$\int_{\mathbb{R}^N} \prod_{a<b}^{N} |y_a - y_b|^\beta \prod_{a=1}^{N} \mathrm{e}^{-W(y_a)} \cdot \mathrm{d}^N \boldsymbol{y} \,. \tag{1.4.1}$$

Indeed, consider the integral (1.4.1) for N-large and focus on the contribution of a bounded domain of $\mathbb{R}^N$. In this case, the logarithmic interactions are dominant with respect to the confinement (and this by one order in N): the dominant contribution of such a region is obtained by spacing the y_a's as far apart as possible. Increasing the size of such a bounded region will increase the value of the dominant contribution, at least until the confining nature of the potential kicks in. Hence, to identify the configuration maximising the value of the integral, one should rescale the integration variables as $y_a = T_N \lambda_a$ with $T_N \to \infty$. The sequence T_N would then be chosen in such a way that the 2-body interaction and the confinement ensured by the potential have the same order of magnitude in N, the ideal situation being:

$$W(T_N \lambda) \;=\; N \cdot V_N(\lambda) \qquad \text{with} \qquad V_N(\lambda) \;=\; V_\infty(\lambda) \cdot \big(1 + \mathrm{o}(1)\big) \tag{1.4.2}$$

for some potential V_∞ and pointwise almost-everywhere in λ. These new variables λ are typically distributed in a bounded region and have a typical spacing $1/N$.

The simplest illustration of such a mechanism issues from the case of a polynomial potential $V(\lambda) = \sum_{a=1}^{2\ell} c_a \lambda^a$, $c_{2\ell} > 0$. In this case, the sequence T_N takes the form $T_N = N^{1/(2\ell)}$. Note that, up to a trivial prefactor, the two-body interaction $\lambda \mapsto |\lambda|^\beta$ is invariant under dilatations. As a consequence, for polynomial potentials, the asymptotic analysis can still be carried out by means of the previously described methods [63], with minor technical complications due to the handling of a N-dependent potential. Although illustrative, the polynomial case is by far not representative of the complexity represented by working with non-varying weights. Indeed, the genuinely hard part of the analysis stems form the fact that, in principle, in the expansion (1.4.2):

- the remainder may not be "sufficiently" uniform;
- the non-varying potential W may have singularities in the complex plane. This last scenario means that the singularities of the rescaled potential V_N given in (1.4.2) will collapse, with a N-dependent rate, on the integration domain.

In this situation, the usual scheme for obtaining sub-leading corrections breaks down. So far, the large-N asymptotic analysis of a "non-trivial" multiple integral of the type (1.4.1) were carried out only when $\beta = 2$ and this for only a handful of examples. Zinn–Justin [100] proposed an N-fold multiple integral representation of the type (1.4.2) for the partition function of the six-vertex model in its massless phase and subject to domain wall boundary conditions. By using a proper rescaling of the variables suggested in [100], Bleher and Fokin [101] carried out the large-N asymptotic analysis of the associated multiple integral within the Riemann–Hilbert problem approach to orthogonal polynomials. The most delicate point of their analysis was to absorb the contribution of the sequence of poles ζ_n/N, $n = 1, 2, \ldots,$ of the rescaled potential that were collapsing on $\mathbb{R}$. *In fine*, they obtained the asymptotic expansion of the logarithm of the integral up to $o(1)$ corrections.

The situation may, in fact, very easily be much worse than the scheme described above, simply because (1.4.2) might not even hold to the leading order with an N-independent V_∞. A simple example can be provided by $W(y) = y^2(3 + \cos(y))$ whose rescaled large-N leading behaviour has N-dependent oscillatory terms.

To conclude, it seems fair to state that despite the considerable developments that took place over the last 20 years in the field of large-N asymptotic expansion of N-dimensional integrals, the techniques of asymptotic analysis are still far from enabling one to grasp the large-N asymptotic behaviour of multiple integrals lacking the presence of a scaling of interactions. Such integrals arise quite naturally in concrete applications. For instance, it is well known that correlation functions in quantum integrable models are described by N-fold multiple integrals [102–105] or series thereof [106]. Usually, for reasons stemming from the physics of the underlying model, one is interested in the large-N behaviour of these integrals and, in particular, in the constant term arising in their asymptotics. However, for most cases of interest, the given N-fold integrals have a much too complicated integrand in order to apply any of the existing methods of analysis.

1.5 The Integrals Issued from the Method of Quantum Separation of Variables

1.5.1 The Quantum Separation of Variables for the Toda chain

The quantum separation of variables method refers to a technique allowing one the determination of the spectrum, eigenvectors and correlation functions of quantum integrable models. The method takes its roots in the 1985 work of Sklyanin [107] and applies to a wide range of lattice quantum integrable models such as spin chains [108–111], lattice discretisations of quantum field theories in $1+1$ dimensions [112–114] or multi-particle quantum Hamiltonians [107, 115, 116]. We will outline the main ideas of the method on the example of the open quantum Toda chain Hamiltonian with $(N+1)$-particles [117]:

$$\mathbf{H}_{\mathrm{Td}} = \sum_{a=1}^{N+1} \frac{\mathbf{p}_a^2}{2} + \mathrm{e}^{\mathbf{x}_{N+1}-\mathbf{x}_1} + \sum_{a=1}^{N} \mathrm{e}^{\mathbf{x}_a-\mathbf{x}_{a+1}} . \tag{1.5.1}$$

Above, $\mathbf{x}_a$ is to be understood as the operator of multiplication by the ath coordinate $\mathbf{x}_a \cdot \Phi(\boldsymbol{x}) = x_a\Phi(\boldsymbol{x})$ while $\mathbf{p}_a$ is the canonically conjugated operator, $\mathbf{p}_a \cdot \Phi(\boldsymbol{x}) = -\mathrm{i}\hbar\partial\Phi(\boldsymbol{x})/\partial x_a$, so that $[\mathbf{x}_a, \mathbf{p}_b] = \delta_{a,b}\mathrm{i}\hbar$. Here, $\boldsymbol{x}$ denotes a $N+1$ dimensional vector $\boldsymbol{x} = (x_1, \ldots, x_{N+1})$. Within such a realisation of the operators

$\mathbf{x}_n$ and $\mathbf{p}_n$, the Toda chain Hamiltonian is a multi-dimensional partial differential operator acting on the Hilbert space $\mathcal{H}_{\text{Toda}} = L^2\big(\mathbb{R}^{N+1}, \mathrm{d}^{N+1}\boldsymbol{x}\big)$.

The quantum Toda chain Hamiltonian is a quantum integrable model. This means, among other things, that $\mathbf{H}_{\text{Td}}$ can be embedded into a family $\{\mathbf{t}_k\}_{k=0}^{N}$ of operators in involution, conveniently collected as coefficients of the polynomial:

$$\mathbf{t}(\lambda) = \lambda^{N+1} + \sum_{k=0}^{N+1}(-1)^{N+1-k}\lambda^k\,\mathbf{t}_k \tag{1.5.2}$$

such that $\mathbf{H}_{\text{Td}} = \mathbf{t}_2 - \mathbf{t}_1^2$. Thus, solving the spectral problem associated with $\mathbf{H}_{\text{Td}}$ means, in fact, solving a multi-dimensional (due to the dimensionality of the ambient space) and multi-parameter (due to the necessity to keep track of eigenvalues t_k of the operators $\mathbf{t}_k$) spectral problem

$$\mathbf{t}_k \cdot \Phi(\boldsymbol{x}) = t_k \cdot \Phi(\boldsymbol{x})\,. \tag{1.5.3}$$

Since the model is translation invariant, its spectrum will contain a Lebesgue continuous part corresponding to the spectrum of the total momentum operator $\mathbf{P}_{\text{tot}} = \sum_{a=1}^{N+1}\mathbf{p}_a$. However, if one puts oneself in the centre of mass frame, or if one fixes the origin of coordinates at x_{N+1}, *viz.* sets $x_{N+1} = 0$, then the restricted operator has already a purely point-wise spectrum, see *e.g.* [118].

The idea of quantum separation of variables is to build—using the various symmetries stemming from the quantum integrability of the model—a unitary transform:

$$\mathscr{U} : \mathcal{H}_{\text{sep}} = L^2\big(\mathbb{R}^{N+1}, \mathrm{d}\nu\big) \longrightarrow \mathcal{H}_{\text{Toda}} = L^2\big(\mathbb{R}^{N+1}, \mathrm{d}^{N+1}\boldsymbol{x}\big) \tag{1.5.4}$$

such that the transformed operator $\mathscr{U}^{-1}\mathbf{t}(\lambda)\mathscr{U}$ becomes "separated". The L^2 space $\mathcal{H}_{\text{sep}}$ is endowed with a measure $\mathrm{d}\nu$ which is part of the unknowns in the problem of constructing $\mathscr{U}$. By "separated", we mean that one would like $\mathscr{U}$ to intertwine the $\mathbf{t}(\lambda)$ operator with a direct sum of finite difference operators in *one*-variable. It turns out that this problem can be solved. The measure $\mathrm{d}\nu$ on $\mathcal{H}_{\text{sep}}$ factorizes $\mathrm{d}\nu = \mathrm{d}\mu \otimes \mathrm{d}\varepsilon$ into a product of a "trivial" one-dimensional Lebesgue measure $\mathrm{d}\varepsilon$ that takes into account the spectrum ε of the total momentum operator, and a non-trivial measure $\mathrm{d}\mu$ which is absolutely continuous in respect to the N-dimensional Lebesgue measure $\mathrm{d}^N\boldsymbol{y}$ and takes the form

$$\mathrm{d}\mu(\boldsymbol{y}) = \mu(\boldsymbol{y})\cdot \mathrm{d}^N\boldsymbol{y} \quad \text{with} \quad \mu(\boldsymbol{y}) = \frac{1}{(2\pi\hbar)^N}\prod_{a<b}^{N}\left\{\frac{y_a - y_b}{\pi\hbar}\cdot\sinh\left[\frac{\pi(y_a - y_b)}{\hbar}\right]\right\}. \tag{1.5.5}$$

When applied to sufficiently well behaved functions $\widehat{\Phi} \in \mathcal{H}_{\text{sep}}$, the action of the unitary operator $\mathscr{U}$ takes the form of an integral transform

$$\mathscr{U}\big[\widehat{\Phi}\big](\boldsymbol{x}, x_{N+1}) \;=\; \int\limits_{\mathbb{R}^{N+1}} \varphi_{\boldsymbol{y}}(\boldsymbol{x}) \cdot \mathrm{e}^{\frac{\mathrm{i}}{\hbar}(\varepsilon-\overline{\boldsymbol{y}})x_{N+1}} \cdot \widehat{\Phi}(\boldsymbol{y};\varepsilon) \cdot \frac{\mathrm{d}\mu(\boldsymbol{y})}{\sqrt{N!}} \otimes \mathrm{d}\varepsilon \quad \text{with } \overline{\boldsymbol{y}}_N \;=\; \sum_{a=1}^{N} y_a \tag{1.5.6}$$

where $\boldsymbol{x}$ and $\boldsymbol{y}$ are N-dimensional vectors and the non-trivial ingredients $\varphi_{\boldsymbol{y}}(\boldsymbol{x})$ are the $\mathrm{GL}(N,\mathbb{R})$ Whittaker functions. The unitarity of the transform $\mathscr{U}$ as given by (1.5.6) has been established in [119] by means of harmonic analysis on groups, and in [120] by means of the quantum inverse scattering method.

One can show [118] that for the Toda chain, the original spectral problem attached to the family $\big\{\mathtt{t}(\lambda)\big\}_{\lambda\in\mathbb{C}}$ of operators in involution, is in one-to-one correspondence with the problem of finding all solutions $\big(t(\lambda)\,,\,q_t(\lambda)\big)$ to the below Baxter T-Q equation

$$t(\lambda)\cdot q_t(\lambda) \;=\; (\mathrm{i})^{N+1} q_t(\lambda+\mathrm{i}\hbar) \;+\; (-\mathrm{i})^{N+1} q_t(\lambda-\mathrm{i}\hbar) \tag{1.5.7}$$

that, furthermore, satisfy the conditions:

(i) $t(\lambda)$ is a polynomial of the form $t(\lambda) \;=\; \prod_{k=1}^{N+1}(\lambda-\tau_k)$ with $\big\{\tau_k\big\}_{k=1}^{N+1} \;=\; \big\{\tau_k^*\big\}_{k=1}^{N+1}$;

(ii) $\lambda \mapsto q_t(\lambda)$ is entire and satisfies, for some N-dependent $C>0$, to the bound

$$|q_t(\lambda)| \leq C\cdot\exp\bigg\{-\frac{(N+1)\pi}{2\hbar}|\mathrm{Re}\,\lambda|\bigg\}\,|\lambda|^{\frac{N+1}{2\hbar}(2|\mathrm{Im}\,\lambda|-\hbar)} \tag{1.5.8}$$

uniformly in $\lambda\,\in\{z\in\mathbb{C}:|\mathrm{Im}\,z|\leq\hbar/2\}$;

(iii) the roots $\{\tau_k\}_1^{N+1}$ satisfy to $\sum_{k=1}^{N+1}\tau_k=\varepsilon$.

The condition (iii) relates the τ_k's to the eigenvalue ε of the total momentum operator $\mathbf{P}_{\mathrm{tot}}$. This constraint issues from the fact that the Toda chain Hamiltonian is invariant under translation, hence making it more convenient to describe the spectrum of the chain directly in a sector corresponding to a fixed eigenvalue ε of $\mathbf{P}_{\mathrm{tot}}$. After such a reduction, q_t represents the "normalisable" part of the Toda chain eigenfunction, associated with the purely point-wise spectrum. More precisely, if $(t(\lambda), q_t(\lambda))$ is a solution to the $T-Q$ Eq. (1.5.7) then

$$\Phi_{\varepsilon;t}(\boldsymbol{x}, x_{N+1}) \;=\; \int\limits_{\mathbb{R}^{N+1}} \varphi_{\boldsymbol{y}}(\boldsymbol{x})\cdot\mathrm{e}^{\frac{\mathrm{i}}{\hbar}(\varepsilon-\overline{\boldsymbol{y}})x_{N+1}}\cdot\prod_{a=1}^{N}\big\{q_t(y_a)\big\}\cdot f(\varepsilon)\cdot\frac{\mathrm{d}\mu(\boldsymbol{y})}{\sqrt{N!}}\otimes\mathrm{d}\varepsilon \tag{1.5.9}$$

represents a wave packet having a dispersion in ε momentum space given by $f\in L^1(\mathbb{R})$. Further the function

$$\Phi_{\varepsilon;t}^{(\mathrm{norm})}(\boldsymbol{x}) \;=\; \int\limits_{\mathbb{R}^N} \varphi_{\boldsymbol{y}}(\boldsymbol{x})\cdot\prod_{a=1}^{N}\big\{q_t(y_a)\big\}\cdot\frac{\mathrm{d}\mu(\boldsymbol{y}_N)}{\sqrt{N!}} \tag{1.5.10}$$

represents the "normalisable" part of the generalised eigenfunction of the operators $\mathtt{t}(\lambda)$ associated with the eigenvalues $t(\lambda)$ and a total momentum ε. One speaks of a separation of variables since the normalisable part of the generalised eigenfunction is given by a product of functions in one variable $q_t(\lambda_a)$, $a = 1, \ldots, N$.

The spectral problem associated with the Baxter equation might seem underdetermined, in the sense that it contains too many unknowns. To convince oneself of the contrary, at least heuristically, it is helpful to make the parallel with the Sturm–Liouville spectral problem

$$\text{find all } \big(E, f\big) \in \mathbb{R} \times H_2(\mathbb{R}) \quad \text{such that} \quad -f''(x) + V(x)f(x) = E \cdot f(x) \tag{1.5.11}$$

with V sufficiently regular and growing fast enough at infinity and $H_2(\mathbb{R})$ is the second Sobolev space. Although the above ordinary differential equation admits two linearly independent solutions for any value of E, only for very specific values of E does one finds solutions belonging to $H_2(\mathbb{R})$. Regarding to (1.5.7), the regularity and growth requirements on q_t play the same role as the $H_2(\mathbb{R})$ space in the Sturm–Liouville problem: the T-Q equation admits solutions (t, q_t) belonging to the desired class only for well-tuned monic polynomials of degree $N + 1$. It is precisely this effect that gives rise to so-called quantisation conditions for the Toda chain.

In light of the above discussion, the quantum separation of variables may be thought of as a way to map a multi-parameter and multi-dimensional spectral problem onto a multi-parameter (so as to keep track of the different eigenvalues of the $\mathtt{t}_k$'s) but *one*-dimensional spectral problem. This results in a tremendous simplification of the problem.

Nekrasov and Shatashvilii conjectured in [121] and Kozlowski and Teschner later proved in [122] that it is possible to construct all solutions to the $T - Q$ equation for the Toda chain through solutions to non-linear integral equations. Namely, let $\boldsymbol{\sigma} = \{\sigma_k\}_{k=1}^{N+1}$ be complex numbers satisfying $|\mathrm{Im}\,\sigma_k| < \hbar/2$ and let $\ln Y_{\boldsymbol{\sigma}}$ denote the continuous and bounded on $\mathbb{R}$ solution (if it exists) to the non-linear integral equation:

$$\ln Y_{\boldsymbol{\sigma}}(\lambda) = \int_{\mathbb{R}} \mathrm{d}\mu \; K(\lambda - \mu) \ln\left(1 + \frac{Y_{\boldsymbol{\sigma}}(\mu)}{\vartheta(\mu - \mathrm{i}\hbar/2)\vartheta(\mu + \mathrm{i}\hbar/2)}\right), \tag{1.5.12}$$

where

$$K(\lambda) = \frac{\hbar}{\pi(\lambda^2 + \hbar^2)} \quad \text{and} \quad \vartheta(\lambda) = \prod_{k=1}^{N+1} (\lambda - \sigma_k) . \tag{1.5.13}$$

Starting from $Y_{\boldsymbol{\sigma}}$ one constructs the functions:

$$\ln \upsilon_{\uparrow}(\lambda) = -\int_{\mathbb{R}} \frac{\mathrm{d}\mu}{2\mathrm{i}\pi} \frac{1}{\lambda - \mu + \mathrm{i}\hbar/2} \cdot \ln\left(1 + \frac{Y_{\boldsymbol{\sigma}}(\mu)}{\vartheta(\mu - \mathrm{i}\hbar/2)\,\vartheta(\mu + \mathrm{i}\hbar/2)}\right), \tag{1.5.14}$$

and

$$\ln v_{\downarrow}(\lambda - \mathrm{i}\hbar) = \int_{\mathbb{R}} \frac{\mathrm{d}\mu}{2\mathrm{i}\pi} \frac{1}{\lambda - \mu - \mathrm{i}\hbar/2} \ln\left(1 + \frac{Y_{\sigma}(\mu)}{\vartheta(\mu - \mathrm{i}\hbar/2)\,\vartheta(\mu + \mathrm{i}\hbar/2)}\right) . \tag{1.5.15}$$

These auxiliary functions $v_{\uparrow/\downarrow}$ then give rise to the functions

$$\mathfrak{q}^{+}_{\sigma}(\lambda) = \frac{\hbar^{\mathrm{i}\frac{(N+1)\lambda}{\hbar}} \mathrm{e}^{-\frac{(N+1)\pi}{\hbar}\lambda} \cdot v_{\uparrow}(\lambda)}{\prod_{k=1}^{N+1}\left\{\Gamma\left(1 - \mathrm{i}\frac{\lambda - \sigma_k}{\hbar}\right)\right\}} , \qquad \mathfrak{q}^{-}_{\sigma}(\lambda) = \frac{\hbar^{-\mathrm{i}\frac{(N+1)\lambda}{\hbar}} \mathrm{e}^{-\frac{(N+1)\pi}{\hbar}\lambda} \cdot v_{\downarrow}(\lambda - \mathrm{i}\hbar)}{\prod_{k=1}^{N+1}\left\{\Gamma\left(1 + \mathrm{i}\frac{\lambda - \sigma_k}{\hbar}\right)\right\}} . \tag{1.5.16}$$

One can show that the ratio

$$t_{\sigma}(\lambda) = \frac{\mathfrak{q}^{+}_{\sigma}(\lambda - \mathrm{i}\hbar)\mathfrak{q}^{-}_{\sigma}(\lambda + \mathrm{i}\hbar) - \mathfrak{q}^{+}_{\sigma}(\lambda + \mathrm{i}\hbar)\mathfrak{q}^{-}_{\sigma}(\lambda - \mathrm{i}\hbar)}{\mathfrak{q}^{+}_{\sigma}(\lambda)\mathfrak{q}^{-}_{\sigma}(\lambda + \mathrm{i}\hbar) - \mathfrak{q}^{+}_{\sigma}(\lambda + \mathrm{i}\hbar)\mathfrak{q}^{-}_{\sigma}(\lambda)} \tag{1.5.17}$$

is, in fact, a monic polynomial in λ of degree $(N + 1)$ that has, furthermore, a self-conjugated set of roots. All these quantities being given one constructs the function, depending on $\boldsymbol{\sigma}$ and a parameter ζ:

$$q_{\sigma}(\lambda) = \frac{\mathfrak{q}^{+}_{\sigma}(\lambda) - \zeta\,\mathfrak{q}^{-}_{\sigma}(\lambda)}{\prod_{k=1}^{N+1}\left\{\mathrm{e}^{-\frac{\pi\lambda}{\hbar}} \sinh\frac{\pi}{\hbar}(\lambda - \sigma_k)\right\}} \tag{1.5.18}$$

which is a meromorphic solution to the T-Q equation associated with the polynomial $t_{\sigma}(\lambda)$ that, furthermore, satisfies to the growth estimates (1.5.8). The pair (t_{σ}, q_{σ}) provides ones with a solution to the Baxter T-Q equation if and only if the parameters $\{\sigma_l\}_{l=1}^{N+1}$ and ζ satisfy to the quantisation conditions, for any $k \in \{1, \ldots, N+1\}$:

$$\begin{aligned} 2\pi n_k = {} & \frac{(N+1)\sigma_k}{\hbar} \ln\hbar + \mathrm{i}\ln\zeta - \mathrm{i}\sum_{m=1}^{N+1} \ln\frac{\Gamma\big(1 + \mathrm{i}(\sigma_k - \sigma_m)/\hbar\big)}{\Gamma\big(1 - \mathrm{i}(\sigma_k - \sigma_m)/\hbar\big)} \\ & + \int_{\mathbb{R}} \frac{\mathrm{d}\tau}{2\pi} \left\{\frac{1}{\sigma_k - \tau + \mathrm{i}\hbar/2} + \frac{1}{\sigma_k - \tau - \mathrm{i}\hbar/2}\right\} \ln\left(1 + \frac{Y_{\sigma}(\tau)}{\vartheta(\tau - \mathrm{i}\hbar/2)\,\vartheta(\tau + \mathrm{i}\hbar/2)}\right) \end{aligned} \tag{1.5.19}$$

and $\sum_{k=1}^{N+1} \sigma_k = \varepsilon$.

It was shown in [122] that any solution to the T-Q equation gives rise to a solution Y_{σ} to (1.5.12) with a set of parameters $\boldsymbol{\sigma} = \{\sigma_k\}_{k=1}^{N+1}$ satisfying to the quantisation conditions (1.5.19) and, reciprocally, that any solution to (1.5.12) with parameters $\boldsymbol{\sigma}$ satisfying to (1.5.19) gives rise to the solution $(t_{\sigma}(\lambda), q_{\sigma}(\lambda))$ to the T-Q equations.

1.5.2 Multiple Integral Representations

The objects of main interest to the physics of a quantum integrable model are its correlation functions, namely expectation values of products of certain physically relevant operators taken between two eigenstates of the Hamiltonian of the model. The simplest such objects are the *form factors*, namely expectation values of so-called quasi-local operators. In the language of the Toda chain, such operators only act on a fixed subset of variables $(x_1, \ldots, x_r)$. The knowledge of form factors allows one, in principle, to access to all correlation functions involving products of quasi-local operators acting on different sets of variables: it is enough to insert the closure relation in between each of the operators. In order to compute the form factors of local operators within the quantum separation of variables method, one has to solve the inverse problem, that is to say find how the given local operator of interest is intertwined by the $\mathscr{U}$-transform. In other words, one should find how the given operator on $\mathcal{H}_{\text{Toda}}$ acts on the space $\mathcal{H}_{\text{sep}}$ where the separation of variables is realised. This inverse problem has been solved for different examples of quasi-local operators and for various models [110, 113, 123–128].

• The Toda Chain

The resolution of the inverse problem for the Toda chain has been pioneered by Babelon [123, 124] in 2002 and further developed in the works [125, 128]. These results, along with the unitarity of the separation of variables transform $\mathscr{U}$ lead to multiple integral representations for the form factors.

Let $\Phi_{\varepsilon;t}$ and $\Phi_{\varepsilon;t'}$ be two eigenfunctions of the Toda chain in the sector characterised by the total momentum ε and built up from solutions to the Baxter T-Q equation associated with the polynomials $t(\lambda)$ and $t'(\lambda)$. The associated finite part of the form factor[10] of the operator $\prod_{a=1}^{r}\left\{\mathrm{e}^{\mathbf{x}_a-\mathbf{x}_{N+1}}\right\}$ takes the form

$$\Big(\Phi_{\varepsilon;t'}, \prod_{a=1}^{r}\left\{\mathrm{e}^{\mathbf{x}_a-\mathbf{x}_{N+1}}\right\}\cdot\Phi_{\varepsilon;t}\Big)_{|x_{N+1}=0} = \frac{N!}{r!(N-r)!}\int\limits_{\mathbb{R}^N}\prod_{a<b}^{N}\Big\{\Big(\frac{y_a-y_b}{\pi\hbar}\Big)\sinh\Big[\frac{\pi(y_a-y_b)}{\hbar}\Big]\Big\}$$
$$\times\prod_{a=1}^{N}\Big\{\big(q_{t'}(y_a)\big)^{*}q_t(y_a)\Big\}$$
$$\times\prod_{a=1}^{r}\Big\{\frac{q_t(y_a+\mathrm{i}\hbar)}{q_t(y_a)}\cdot\prod_{b=r+1}^{N}\Big[\frac{\mathrm{i}}{y_a-y_b}\Big]\Big\}\cdot\frac{\mathrm{d}^N\boldsymbol{y}}{(2\pi\hbar)^N}\,. \qquad (1.5.20)$$

The index $|_{x_{N+1}=0}$ refers to the fact that the coordinates are chosen so that $x_{N+1}=0$.

---

[10]Namely the one built up from the normalisable part of the wave function, in contrast with the non-normalisable part associated with the continuous part of the spectrum described by ε.

• Lukyanov's Conjecture for the Sinh–Gordon model

Lukyanov [129] argued that the vacuum expectation value of the exponential of the field operator Φ in the quantum Sinh–Gordon model should be obtained from the properly normalised large-N limit

$$\langle \mathrm{e}^{\alpha\Phi} \rangle_R = \lim_{N\to+\infty} \left\{ \left(\frac{N}{mR} \right)^{\theta} \frac{\mathfrak{z}_\alpha}{\mathfrak{z}_0} \right\} \qquad \text{with} \qquad \theta = \frac{\alpha^2}{2(1+b^2)(1+b^{-2})} . \tag{1.5.21}$$

In the above expression m is the Sinh–Gordon mass parameter, $2\pi R$ is the volume, b is the coupling constant, and $\mathfrak{z}_\alpha$ is given by the N-fold multiple-integral representation

$$\mathfrak{z}_\alpha = \int_{\mathbb{R}^N} \prod_{k<\ell}^{N} \Big\{ \sinh\big[(1+b^2)(y_k - y_\ell)\big] \cdot \sinh\big[(1+b^{-2})(y_k - y_\ell)\big] \Big\} \cdot \prod_{a=1}^{N} \Big\{ \mathrm{e}^{-W_\alpha(y_a)} \Big\} \cdot \mathrm{d}^N \boldsymbol{y} . \tag{1.5.22}$$

The one-body interaction is given by the confining potential

$$W_\alpha(\lambda) = -\alpha\lambda + \frac{mR\cosh(\lambda)}{2\sin\left(\frac{\pi}{1+b^2}\right)} - \int_{\mathbb{R}} \frac{\ln\left[1+\mathrm{e}^{-E(\mu)}\right]}{\cosh(\lambda-\mu)} \cdot \frac{\mathrm{d}\mu}{\pi} . \tag{1.5.23}$$

Its expression involves the solution E to the non-linear integral equation

$$E(\lambda) = 2\pi mR\cosh(\lambda) - \int_{\mathbb{R}} \Phi(\lambda-\mu) \ln\left[1+\mathrm{e}^{-E(\mu)}\right] \cdot \mathrm{d}\mu \tag{1.5.24}$$

whose integral kernel takes the form

$$\Phi(\mu) = \frac{\cosh(\mu)\cos(\tau)}{\cosh(\lambda+\mathrm{i}\tau)\cosh(\lambda-\mathrm{i}\tau)} \qquad \text{where} \qquad \tau = \frac{\pi(b^2-b^{-2})}{2(2+b^2+b^{-2})} . \tag{1.5.25}$$

It is easy to see by using Banach's fixed point theorem that, at least for R large enough, the non-linear integral equations admits a unique solution $E \in L^\infty(\mathbb{R})$.

Per se Lukyanov's conjecture [129] takes its roots in semi-classical reasonings applied to the classical Sinh–Gordon model in finite volume. Later, Bytsko and Teschner [112] proposed a lattice discretised version of the quantum Sinh–Gordon model in finite volume $2\pi R$ and implemented the quantum separation of variables for this model. Teschner [130] provided a characterisation of the solutions to the T-Q equation which describes the spectrum of that model. The results of these two papers suggest a representation for the expectation value of the exponential of the field in the quantum Sinh–Gordon model slightly more complex than (1.5.21): the confining potential (1.5.23) arising in $\mathfrak{z}_\alpha$ should be replaced by a more involved expression which, in particular, depends on N. We will however not provide this explicit form here. It is an open question whether the limit (1.5.21) exists and if it exists whether it takes the same value when one inserts in $\mathfrak{z}_\alpha$ the potential conjectured by Lukyanov and the one suggested by Teschner's analysis. Note that a thorough characterisation of $\langle e^{\alpha\Phi}\rangle_R$ has been recently conjectured by Negro and Smirnov in [131] on completely independent grounds; it is an open question whether the limit (1.5.21) does indeed gives rise to the same object.

• General Structure of Form Factors in the Quantum Separation of Variables Method

It is possible to obtain multiple integral representations for the form factors arising in other models solvable by the quantum separation of variables. Although we shall not discuss the precise form taken by these representation, the main feature is that the form factors are either directly expressed—as in the case of the Lukyanov integral (1.5.22)—or very closely related—as in the case of the position operator form factor of the Toda chain (1.5.20)—to multiple integrals of the type

$$\int_{\mathscr{C}^N} \prod_{a<b}^{N} \Big\{ \sinh[\pi\omega_1(y_a - y_b)] \sinh[\pi\omega_2(y_a - y_b)] \Big\}^{\beta} \cdot \prod_{a=1}^{N} e^{-W(y_a)} \cdot d^N \boldsymbol{y} \qquad (1.5.26)$$

or their degenerations when some of the ω_k's are send to 0. There $\mathscr{C}$ is some curve in the complex plane which, in the simplest cases discussed above, coincides with $\mathbb{R}$. The coefficients ω_1, ω_2 are related to the given model's coupling constants. The confining potential W, which can be N-dependent, contains all the information on the eigenfunctions and the operator involved in the form factor.

1.5.3 The Goal of the Book

This work aims at developing the main features of a theory that would enable one to extract the large-N asymptotic behaviour out of the class of multiple integrals that naturally arises in the context of the so-called quantum separation of variables method:

$$\mathfrak{z}_N[W] = \int_{\mathbb{R}^N} \prod_{a<b}^{N} \Big\{ \sinh[\pi\omega_1(y_a - y_b)] \sinh[\pi\omega_2(y_a - y_b)] \Big\}^{\beta} \cdot \prod_{a=1}^{N} e^{-W(y_a)} \cdot d^N \boldsymbol{y} \,. \qquad (1.5.27)$$

As discussed earlier on, the examples issuing from the quantum separation of variables correspond to taking[11] $\beta = 1$ and specific choices of the potential W. Independently of its numerous potential applications to physics, should one only have in mind characterising the large-N behaviour of N-fold multiple integrals, it is precisely the class of integrals described by (1.5.27) that constitutes naturally the next one to investigate and understand after the β-ensembles issued ones (1.1.1) and (1.3.1). Indeed, on the one hand the integrand in (1.5.27) bears certain structural similarities with the one arising in β-ensembles. On the other hand, it brings two new features into the game. Therefore, $\mathfrak{z}_N[W]$ provides one with a good playground for pushing forward the methods of asymptotic analysis of N-fold integrals and learning how to circumvent or deal with certain of the problematic features mentioned in Section 1.4. To be more precise, the main features of the integrand in $\mathfrak{z}_N[W]$ which constitute an obstruction to applying the already established methods stem from the presence of

- a non-varying confining one-body potential W;
- a two-body interaction that has the same local (*viz.* when $\lambda_a \to \lambda_b$) singularity structure as in the β-ensemble case, while breaking other properties of the interaction such as the invariance under a re-scaling of all the integration variables.

Although the tools of asymptotic analysis discussed previously break down or have to be altered in a significant way, a certain analogy with matrix models and β-ensembles persists. Indeed, upon a proper rescaling in the spirit of Section 1.4, one can show for certain examples of potentials that the integral localises at a configuration of the integration variables in such a way that these condense, in the large-N limit, with a density ρ_{eq}. In fact, we show in Appendix B that it is possible to repeat, with some modifications, the large-deviation approach to β-ensemble integrals so as to obtain the leading asymptotic behaviour of $\ln \mathfrak{z}_N[W]$ for certain instances of confining potentials W. However, in order to go beyond the leading asymptotic behaviour of the logarithm, one has to alter the picture and work directly at the level of the rescaled model

$$Z_N[V] = \int_{\mathbb{R}^N} \prod_{a<b}^{N} \Big\{ \sinh\big[\pi\omega_1 N^{\alpha}(\lambda_a - \lambda_b)\big] \sinh\big[\pi\omega_2 N^{\alpha}(\lambda_a - \lambda_b)\big]\Big\}^{\beta}$$

$$\times \prod_{a=1}^{N} \Big\{ \mathrm{e}^{-N^{1+\alpha} V(\lambda_a)} \Big\} \cdot \prod_{a=1}^{N} \mathrm{d}\lambda_a \,. \qquad (1.5.28)$$

This integral is related to $\mathfrak{z}_N[W]$ by a rescaling of the integration variables. The exponent α is fixed by the growth of the original potential W at infinity. Finally, the potential V should depend on N and correspond to some rescaling of the original potential W. In fact, the main result obtained in the present paper deals with the

[11] Because of the two factors admitting a simple zero at $y_a = y_b$ the parameter β in Equation (1.5.27) should be compared to the parameter $\frac{\beta}{2}$ in the β ensembles. So, the special case $\beta = 1$ is put in parallel with the special case of $\beta = 2$-ensembles.

large-N asymptotic expansion of the rescaled partition function $Z_N[V]$ and this in the case where

- the potential V is smooth, strictly convex, has sub-exponential growth and is N-independent;
- $0 < \alpha < 1/6$;

The first assumption is more than enough to carry the large deviation analysis, which gives the leading order of $\ln Z_N[V]$, while the second assumption appears in the course of the bootstrap analysis of the Schwinger–Dyson equations. *Per se*, the application of our technique and results to computing the asymptotics of the original integral $\mathfrak{z}_N[W]$ would demand to take a N dependent potential and study $Z_N[V_N]$, which is technically much more involved. However, this problem is *not* conceptually different from the one studied in this book. Therefore, the setting we shall discuss is more fit for developing the method of asymptotic analysis of this class of integrals. We shall address the question of N-dependent potentials V_N related to specific applications to quantum integrable models in a separate publication.

Within our setting, in order to grasp sub-leading corrections to $\ln Z_N[V]$, one faces several difficulties:

(i) owing to the scaling N^{α}, the nature of the repulsive interaction between the λ_a's changes drastically between $N = \infty$ and N finite. Therefore, one has to keep track of the transition of scales between the *per-se* leading contribution—which feels, effectively, only the brute $N = \infty$ behaviour of the properly normalised two-body interaction—and the sub-leading corrections which experience the two-body interactions at all scales.

(ii) The presence of two scales N and N^{α} weakens a naive approach to the concentration of measures.

(iii) The derivative of the two-body interaction possesses a tower of poles that collapse down to the integration line, hence making the use of correlators and complex variables methods to study Schwinger–Dyson equations completely ineffective.

(iv) The master operator arising in the Schwinger–Dyson equations is a N-dependent singular integral operator of truncated Wiener–Hopf type. One has to invert this operator effectively and derive the fine, N-dependent bounds on its continuity constant as an operator between spaces of sufficiently differentiable functions.

(v) The large-N behaviour of one-point functions, as fixed by a successful large-N analysis of the Schwinger–Dyson equations, is expressed in terms of one and two dimensional integrals involving the inverse of the master operator. One has to extract the large-N asymptotic behaviour of such integrals.

The setting of methods enabling one to overcome these problems constitutes the main contributions of this book.

First, in order to strengthen the concentration of measures and, in fact, effectively absorb part of the asymptotic expansion into a single expression, one should work with N-dependent equilibrium measures, that is to say equilibrium measures associated with a minimisation problem of a quadratic N-dependent functional on the space of probability measures on $\mathbb{R}$. The density of such an N-dependent measure can be expressed as an integral transform whose kernel is given by a double integral involving the solution to an auxiliary matrix 2×2 Riemann–Hilbert problem. This very fact constitutes a crucial difference with the matrix model case in that, in the latter case, the density of equilibrium measure can be expressed in terms of the solution to a scalar Riemann–Hilbert problem, hence admitting an explicit, one-dimensional integral representation. On top of improving numerous bounds, the use of such N-dependent equilibrium measures turns out to be crucial in order to push the asymptotic expansion of $\ln Z_N[V]$ up to $o(1)$.

Second, the *per se* machinery of topological recursion mentioned earlier breaks down for this class of multiple integrals. In order to circumvent dealing with the collapsing of poles, we develop a distributional approach to the asymptotic analysis of Schwinger–Dyson equations. The latter demands, in particular, to have a much more precise control on its constituents.

Third, the inversion of the master operator is based on handlings of the inverse of the operator driving the singular integral equation for the density of equilibrium measure. Obtaining fine, N dependent bounds for this operator demands to go deep into the details of the solution of the 2×2 Riemann–Hilbert problem which arises as the building block of this inverse kernel. We develop techniques enabling one to do so.

Finally, the precise control on the objects issuing from Schwinger–Dyson equations yield, through usual interpolation by means of t-varying potentials, an N-dependent functional of the density of equilibrium measure—itself also depending on N—as an answer for the large-N asymptotics of $\ln Z_N[V]$. Setting forth methods for the asymptotic analysis of this functional demands, again, a very fine control of the inverse build through the Riemann–Hilbert problem approach. We develop such methods, in particular, by describing the new class of special functions related to our problem.

Putting in Perspective the Bi-Orthogonal Ensembles

At this point, it appears useful to make several comments with respect to the existing literature on bi-orthogonal ensembles. Indeed, the applications to the quantum separation of variables correspond to setting β to 1 in $\mathfrak{z}_N[W]$ and hence in $Z_N[V]$. In this case, these multiple integral corresponds to a bi-orthogonal ensemble. As such, they can be explicitly computed, at least in principle, by means of the system of bi-orthogonal polynomials associated with the $\mathrm{i}\omega_1^{-1}$ or $\mathrm{i}\omega_2^{-1}$ periodic functions $\mathrm{e}^{\pi\omega_1 y}$, $\mathrm{e}^{\pi\omega_2 y}$ and with respect to the weight $\mathrm{e}^{-W(y)}$ supported on $\mathbb{R}$. As shown by Claeys and Wang [86] for a specific degeneration (which corresponds basically to sending one of the ω's in (1.5.27) to zero) and then in full extent by Claeys and Romano [87], such a system of bi-orthogonal polynomials solves a vector Riemann–Hilbert problem. Furthermore, the non-linear steepest descent approach [12, 63] to the uniform in the variable large degree-N asymptotics of orthogonal polynomials can be generalised to such a bi-orthogonal setting, leading to Plancherel-Rotach like asymptotics for these bi-orthogonal polynomials [86]. In principle, by adapting the steps of [64], one should be able to derive the large-N asymptotic expansion of the integral $\mathfrak{z}_N[W]$

in presence of *varying* weights, *viz*. provided the replacement $W \hookrightarrow NV$ is made. However, such a result would by no means allow one for any easy generalisation to non-varying weights. Indeed, as we have argued, in the non-varying case, one rather needs to carry out the large-N analysis of the rescaled model $Z_N[V]$. However, starting from such a multiple integral would imply that one should study the system of bi-orthogonal polynomials associated with the functions $e^{\pi N^{\alpha}\omega_1 y}$, $e^{\pi N^{\alpha}\omega_2 y}$. The presence of N^{α} introduces a new scale in N to the Riemann–Hilbert analysis. Taking the latter into account would probably demand a quite non-trivial modification of the non-linear steepest descent method.

On top of all this, one needs to construct the equilibrium measure. For similar reasons of absorbing part of the asymptotic expansion, this measure will have to issue from the same N-dependent minimisation problem and hence correspond to the N-dependent equilibrium measure that we construct in the present paper. However, if one goes into the details of the work [87], one observes that these authors provide a one-fold integral representation for the density of the one-cut equilibrium measure arising in bi-orthogonal ensembles. The kernel of this transform involves the inverse of an explicit and basic transcendental function. Although extremely effective in the varying case, such an integral representation appears ineffective in the analysis of $Z_N[V]$. Indeed, then, one would have to manipulate N-dependent versions of this inverse and, in particular, obtain uniform in N local behaviours thereof. *A priori*, since this inverse does not seem to admit an explicit series expansion or a manageable integral representation, such a characterisation seems to be quite complicated. Furthermore, the transform constructed in [87] does not exhibit explicitly the factorisation of square root singularities at the edges—in contrast to the case of the one-fold integral representation arising in β ensembles, *c.f.* (1.2.4). This means that, just as in our setting, one would have to extract the square root behaviour by hand. Therefore, although one dimensional, we believe that this transform, in the present state of the art, is less effective then ours, at least from the point of view of our perspective of asymptotic analysis. In fact, when specialised to the construction of the equilibrium measure, the 2×2 Riemann–Hilbert analysis we use enables us, among other things, to provide the leading, up to exponentially small corrections in N, behaviour of the inverse of the N-rescaled map built in [87]. Thus, indirectly, our approach solves such a problem.

1.6 Notations and Basic Definitions

In this section, we introduce basic notations that we shall use throughout the book.

General Symbols

- o and O refer to standard domination relations between functions. In the case of matrix function $M(z)$ and $N(z)$, the relation $M(z) = \mathrm{O}(N(z))$ is to be understood entry-wise, *viz*. $M_{jk}(z) = \mathrm{O}(N_{jk}(z))$.
- $\mathrm{O}(N^{-\infty})$ means $\mathrm{O}(N^{-K})$ for arbitrarily large K's.

- Given a set $A \subseteq X$, $\mathbf{1}_A$ stands for the indicator function of A, and A^{c} denotes its complement in X.
- A Greek letter appearing in bold, *e.g.* $\boldsymbol{\lambda}$, will always denote an N-dimensional vector:

$$\boldsymbol{\lambda} = \big(\lambda_1, \dots, \lambda_N\big) \in \mathbb{R}^N .$$

  and $\mathrm{d}^N\boldsymbol{\lambda}$ denotes the product of Lebesgue measures $\prod_{a=1}^{N} \mathrm{d}\lambda_a$.
- given $x \in \mathbb{R}$, $\lfloor x \rfloor$ denotes the integer satisfying $\lfloor x \rfloor \le x < \lfloor x \rfloor + 1$
- Throughout the file, the curve $\mathscr{C}^{+}_{\mathrm{reg}}$ will denote the curve depicted in Fig. 5.1 appearing in Section 5.1.1. This curve is such that $2\varsigma = \mathrm{dist}\big(\mathbb{R}, \mathscr{C}^{+}_{\mathrm{reg}}\big) > 0$. Throughout the text, this distance will always be denoted by 2ς.
- I_2 is the 2×2 identity matrix while $\sigma^{\pm}$ and σ_3 stand for the Pauli matrices:

$$\sigma^{+} = \begin{pmatrix} 0\,1 \\ 0\,0 \end{pmatrix} \;,\quad \sigma^{-} = \begin{pmatrix} 0\,0 \\ 1\,0 \end{pmatrix} \quad \text{and} \quad \sigma_3 = \begin{pmatrix} 1 & 0 \\ 0 & -1 \end{pmatrix} .$$

Functional Spaces

- $\mathcal{M}^1(\mathbb{R})$ denotes the space of probability measures on $\mathbb{R}$. The weak topology on $\mathcal{M}^1(\mathbb{R})$ is metrized by the Vasershtein distance, defined for any two probability measure μ_1 and μ_2 by:

$$D_V[\mu_1, \mu_2] = \sup_{f \in \mathrm{Lip}_{1,1}(\mathbb{R})} \bigg| \int_{\mathbb{R}} f(\xi)\, \mathrm{d}(\mu_1 - \mu_2)(\xi) \bigg| \tag{1.6.1}$$

  where $\mathrm{Lip}_{1,1}(\mathbb{R})$ is the set of Lipschitz functions bounded by 1 and with Lipschitz constant bounded by 1. If f is a bounded, Lipschitz function, its bounded Lipschitz norm is:

$$||f||_{\mathrm{BL}} = ||f||_{L^{\infty}(\mathbb{R})} + \sup_{\xi \neq \eta \in \mathbb{R}} \bigg| \frac{f(\xi) - f(\eta)}{\xi - \eta} \bigg| .$$

- Given an open subset U of $\mathbb{C}^n$, $\mathcal{O}(U)$ refers to the ring of holomorphic functions on U. If f is a matrix of vector valued function, the notation $f \in \mathcal{O}(U)$ is to be understood entrywise, *viz.* $\forall\, a, b$ one has $f_{ab} \in \mathcal{O}(U)$.
- $C^k(A)$ refers to the space of function of k times continuously differentiable on the manifold A. $C^k_{\mathrm{c}}(A)$ refers to the spaces built out of functions in $C^k(A)$ that have a compact support.
- $L^p(A, \mathrm{d}\mu)$ refers to the space of pth-power integrable functions on a set A with respect to the measure μ. $L^p(A, \mathrm{d}\mu)$ is endowed with the norm

$$||f||_{L^p(A,\mathrm{d}\mu)} = \bigg\{ \int_A |f(x)|^p \, \mathrm{d}\mu(x) \bigg\}^{\frac{1}{p}} .$$

- More generally, given an n-dimensional manifold A, $W_k^p(A, \mathrm{d}\mu)$ refers to the pth Sobolev space of order k defined as

$$W_k^p(A, \mathrm{d}\mu) = \Big\{ f \in L^p(A, \mathrm{d}\mu) \;:\; \partial_{x_1}^{a_1} \dots \partial_{x_n}^{a_n} f \in L^p(A, \mathrm{d}\mu) \;, \quad \text{with} \;\; \sum_{\ell=1}^{n} a_\ell \le k \;\; \text{and} \;\; a_\ell \in \mathbb{N} \Big\} \,. \tag{1.6.2}$$

This space is endowed with the norm

$$||f||_{W_k^p(A,\mathrm{d}\mu)} = \max \Big\{ ||\partial_{x_1}^{a_1} \dots \partial_{x_n}^{a_n} f||_{L^p(A,\mathrm{d}\mu)} \;:\; a_\ell \in \mathbb{N}, \; \ell = 1, \dots, n, \quad \text{and satisfying} \; \sum_{\ell=1}^{n} a_\ell \le k \Big\} \,. \tag{1.6.3}$$

In the following, we shall simply write $L^p(A)$, $W_k^p(A)$ unless there will arise some ambiguity on the measure chosen on A.
- We shall also need the N-weighted norms of order ℓ for a function $f \in W_\ell^\infty(\mathbb{R}^n)$, which are defined as

$$\mathcal{N}_N^{(\ell)}[f] \;=\; \sum_{p=0}^{\ell} \frac{||f||_{W_p^\infty(\mathbb{R}^n)}}{N^{p\alpha}} \,.$$

In particular, we have the trivial bound $\mathcal{N}_N^{(\ell)}[f] \le (\ell + 1)||f||_{W_\ell^\infty(\mathbb{R}^n)}$. Also, the number of variables of f is implicit in this notation.
- The symbol $\mathcal{F}$ denotes the Fourier transform on $L^2(\mathbb{R})$ whose expression, versus $L^1 \cap L^2(\mathbb{R})$ functions, takes the form

$$\mathcal{F}[\varphi](\lambda) \;=\; \int_{\mathbb{R}} \varphi(\xi)\, \mathrm{e}^{\mathrm{i}\xi\lambda} \mathrm{d}\xi \,.$$

Given $\mu \in \mathcal{M}^1(\mathbb{R})$, we shall use the same symbol for denoting its Fourier transform, *viz.* $\mathcal{F}[\mu]$. The Fourier transform on $L^2(\mathbb{R}^n)$ is defined with the same normalisation.
- The sth Sobolev space on $\mathbb{R}^n$ is defined as

$$H_s(\mathbb{R}^n) = \Big\{ u \in \mathcal{S}'(\mathbb{R}^n) \;: \quad ||u||^2_{H_s(\mathbb{R}^n)} = \int_{\mathbb{R}^n} \Big(1 + \Big| \sum_{a=1}^{n} t_a^2 \Big|^{\frac{1}{2}} \Big)^{2s} \big| \mathcal{F}[u](t_1, \dots, t_n) \big|^2 \cdot \mathrm{d}^n \boldsymbol{t} \; < \; +\infty \Big\} \,, \tag{1.6.4}$$

in which S' refers to the space of tempered distributions. We remind that given a closed subset $F \subseteq \mathbb{R}^n$, $H_s(F)$ corresponds to the subspace of $H_s(\mathbb{R}^n)$ of functions whose support is contained in F.

- The subspace

$$\mathfrak{X}_s(A) = \left\{ H \in H_s(A) \ : \ \int_{\mathbb{R}+\mathrm{i}\epsilon} \chi_{11}(\mu)\mathcal{F}[H](N^\alpha\mu)\mathrm{e}^{-\mathrm{i}N^\alpha\mu b_N}\frac{\mathrm{d}\mu}{2\mathrm{i}\pi} = 0 \right\}$$

in which $A \subseteq \mathbb{R}$ is closed and $\epsilon > 0$ is arbitrary but small enough, will play an important role in the analysis. It is defined in terms of χ_{11}, the $(1, 1)$ entry of the unique solution χ to the 2×2 matrix valued Riemann–Hilbert problem given in Section 4.2.1.

- Given a smooth curve Σ in $\mathbb{C}$, the space $\mathcal{M}_\ell\big(L^2(\Sigma)\big)$ refers to $\ell \times \ell$ matrices with coefficients belonging to $L^2(\Sigma)$. It is endowed with the norm

$$||M||_{\mathcal{M}_\ell(L^2(\Sigma))} = \left\{ \int_\Sigma \sum_{a,b} \big[M_{ab}(s)\big]^* M_{ab}(s)\, \mathrm{d}\mu(s) \right\}^{\frac{1}{2}} .$$

and * denotes the complex conjugation.

Certain Standard Operators

- Given an oriented curve $\Sigma \subseteq \mathbb{C}$, $-\Sigma$ refers to the same curve but endowed with the opposite orientation.
- Given a function f defined on $\mathbb{C}\setminus\Sigma$, with Σ an oriented curve in $\mathbb{C}$, we denote—if it exists—by $f_\pm(s)$ the boundary values of $f(z)$ on Σ when the argument z approaches the point $s \in \Sigma$ non-tangentially and from the left $(+)$ or the right $(-)$ side of the curve. Furthermore, if one deals with vector or matrix-valued function, then this notation is to be understood entry-wise.
- $\mathbb{H}^\pm = \{z \in \mathbb{C} \ : \ \mathrm{Im}\,(\pm z) > 0\}$ is the upper/lower half-plane, and

$$\mathbb{R}^\pm = \{z \in \mathbb{R} \ : \ \pm z \geq 0\}$$

is the closed positive/negative real axis.

- The symbol C refers to the Cauchy transform on $\mathbb{R}$:

$$C[f](\lambda) = \int_{\mathbb{R}} \frac{f(s)}{s-\lambda}\cdot\frac{\mathrm{d}s}{2\mathrm{i}\pi} .$$

The $\pm$ boundary values $C_\pm$ define continuous operators on $H_s(\mathbb{R})$ and admit the expression

$$C_\pm[f](\lambda) = \frac{f(\lambda)}{2} + \frac{1}{2\mathrm{i}} \fint_{\mathbb{R}} \frac{f(s)\,\mathrm{d}s}{\pi(s-\lambda)} . \tag{1.6.5}$$

where $\fint$ denotes the principal value integral.

- Given a function f supported on a compact set A of $\mathbb{R}^n$, we denote by $f_\mathfrak{e}$ an extension of f onto some compact set K such that $A \subseteq \text{Int}(K)$. We do stress that the compact support is part of the data of the extension. As such, it can vary from one extension to another. However, the extension $f_\mathfrak{e}$ is always assumed to be of the same class as f. For instance, if f is $L^p(A)$, $W_k^p(A)$ or $C^k(A)$, then $f_\mathfrak{e}$ is $L^p(K)$, $W_k^p(K)$ or $C^k(K)$.

References

1. Mehta, M.L.: Random Matrices, 3^{éme} edn. Pure and Applied Mathematics, vol. 142. Elsevier/Academic, Amsterdam (2004)
2. Dumitriu, I., Edelman, A.: Matrix models for beta ensembles. J. Math. Phys. **43**(11), 5830–5847 (2002). arXiv:math-ph/0206043
3. Krishnapur, M., Rider, B., Virág, B.: Universality of the stochastic Airy operator. Commun. Pure Appl. Math. (2013). arXiv:math.PR/1306.4832
4. Anderson, G.W., Guionnet, A., Zeitouni, O.: An Introduction to Random Matrices. Cambridge Studies in Advances Mathematics, vol. 118. Cambridge University Press (2010)
5. Deift, P.A., Gioev, D.: Random Matrix Theory: Invariant Ensembles and Universality. Courant Lecture Notes in Mathematics, vol. 18 (2009)
6. Pastur, L., Shcherbina, M.: Eigenvalue Distribution of Large Random Matrices. Mathematical Surveys and Monographs, vol. 171. AMS, Providence, Rhode Island (2011)
7. Forrester, P.J.: Log-Gases and Random Matrices. Princeton University Press (2010)
8. Brézin, E., Itzykson, C., Parisi, G., Zuber, J.-B.: Planar diagrams. Commun. Math. Phys. **59**, 35–51 (1978)
9. Valkó, B., Virág, B.: Continuum limits of random matrices and the Brownian carousel. Invent. Math. **177**(3), 463–508 (2009). arXiv:math.PR/0712.2000
10. Ramírez, J.A., Rider, B., Virág, B.: Beta ensembles, stochastic Airy spectrum, and a diffusion. J. Am. Math. Soc. **24**(4), 919–944 (2011). arXiv:math.PR/0607331
11. Pastur, L., Shcherbina, M.: Universality of the local eigenvalue statistics for a class of unitary invariant random matrix ensembles. J. Stat. Phys. **86**, 109–147 (1997)
12. Deift, P.A., Kriecherbauer, T., McLaughlin, K.T.-R., Venakides, S., Zhou, X.: Uniform asymptotics for polynomials orthogonal with respect to varying exponential weights and application to universality questions in random matrix theory. Commun. Pure Appl. Math. **52**, 1335–1425 (1999)
13. Deift, P.A., Gioev, D.: Universality in Random Matrix Theory for orthogonal and symplectic ensembles. Int. Math. Res. Pap. (2007). arXiv:math-ph/0411075
14. Deift, P.A., Gioev, D.: Universality at the edge of the spectrum for unitary, orthogonal and symplectic ensembles of random matrices. Commun. Pure Appl. Math. **60**, 867–910 (2007). arXiv:math-ph/0507023
15. Bourgade, P., Erdös, L., Yau, H.-T.: Bulk universality of general β-ensembles with non-convex potential. J. Math. Phys. **53**, 095–221 (2012). arXiv:math-ph/1201.2283
16. Bourgade, P., Erdös, L., Yau, H.-T.: Universality of general β-ensembles. Duke Math. J. **163**, 1127–1190 (2014). arXiv:math.PR/1104.2272
17. Bourgade, P., Erdös, L., Yau, H.-T.: Edge universality of beta ensembles. Commun. Math. Phys. **332**, 261–353 (2014). arXiv:math.PR/1306.5728
18. Shcherbina, M.: Change of variables as a method to study general β-models: bulk universality. J. Math. Phys. **55**, 043–504 (2014). arXiv:math-ph/1310.7835
19. Bekerman, F., Figalli, A., Guionnet, A.: Transport maps for beta-matrix models and universality (2013). arXiv:math.PR/1311.2315
20. Mulase, M., Yu, J.: Non commutative matrix integrals and representations varieties of surface groups in a finite group. Ann. Inst. Fourier **55**, 1001–1036 (2005). arXiv:math.QA/0211127

21. Eynard, B., Marchal, O.: Topological expansion of the Bethe Ansatz, and non-commutative algebraic geometry. JHEP **0903**, 094 (2009). arXiv:math-ph/0809.3367
22. Eynard, B.: Formal matrix integrals and combinatorics of maps. In: Harnad, J. (ed.) CRM Series in Mathematical Physics, pp. 415–442. Springer, New York (2011)
23. Di Francesco, P., Ginsparg, P., Zinn-Justin, J.: 2D gravity and random matrices. Phys. Rep. **254**(1), 1–133 (1994). arXiv:hep-th/9306153
24. Mariño, M.: Lectures on non-perturbative effects in large N gauge theories, matrix models and strings (2012). arXiv:hep-th/1206.6272
25. Dorlas, T.C., Lewis, J.T., Pulé, J.V.: The Yang-Yang thermodynamic formalism and large deviations. Commun. Math. Phys. **124**(3), 365–402 (1989)
26. Landkof, N.S.: Foundations of Modern Potential Theory. Translated from Osnovy sovremennoi teorii potenciala, Nauka, Moscow. Springer, Berlin (1972)
27. Ben Arous, G., Guionnet, A.: Large deviations for Wigner's law and Voiculescu's non-commutative entropy. Probab. Theory Relat. Fields **108**, 517–542 (1997)
28. Deift, P.A., Kriecherbauer, T., McLaughlin, K.T.-R.: New results on the equilibrium measure for logarithmic potentials in the presence of an external field. J. Approx. Theory **95**, 388–475 (1998)
29. Saff, E.B., Totik, V.: Logarithmic Potentials with External Fields. Grundlehren der mathematischen Wissenschaften, vol. 316. Springer, Berlin, Heidelberg (1997)
30. Ben Arous, G., Dembo, A., Guionnet, A.: Ageing of spherical spin glasses. Probab. Theory Relat. Fields **120**(1), 1–67 (2001)
31. Borot, G., Guionnet, A.: Asymptotic expansion of beta matrix models in the one-cut regime. Commun. Math. Phys. **317**, 447–483 (2013). arXiv:math.PR/1107.1167
32. Wigner, E.P.: Characteristic vectors of bordered matrices with infinite dimensions. Ann. Math. **62**, 548–564 (1955)
33. Mhaskar, H.N., Saff, E.B.: Where does the sup norm of a weighted polynomial live? (A generalization of incomplete polynomials). Constr. Approx. **1**, 71–91 (1985)
34. Carleman, T.: Sur la résolution de certains equations intègrales. Arkiv för matematik, astronomi och fysik **16** (1922)
35. Tricomi, F.G.: Integral Equations. Interscience, London (1957). New edition, Dover Publications (1985)
36. Selberg, A.: Bemerkninger om et multiplet integral. Norsk. Mat. Tid. **26**, 71–78 (1944)
37. Leblé, T., Serfaty, S.: Large deviation principle for empirical fields of log and Riesz gases. arXiv:math.PR/1502.02970
38. Guionnet, A., Maurel-Segala, E.: Second order asymptotics for matrix models. Ann. Probab. **35**, 2160–2212 (2007)
39. Stein, C.: A bound for the error in the normal approximation to the distribution of a sum of dependent random variables. In: Proceedings of the Sixth Berkeley Symposium on Mathematical Statistics and Probability, vol. 2, pp. 583–602. University of California Press, Berkeley (1972)
40. Ambjørn, J., Chekhov, L., Kristjansen, C.F., Makeenko, Y.: Matrix model calculations beyond the spherical limit. Nucl. Phys. B **404**, 127–172 (1993); Erratum-ibid. B **449**, 681 (1995). arXiv:hep-th/9302014
41. Ambjørn, J., Chekhov, L., Makeenko, Y.: Higher genus correlatirs from the Hermitian one-matrix model. Phys. Lett. B **282**, 341–348 (1992). arXiv:hep-th/9203009
42. Chekhov, L., Eynard, B.: Matrix eigenvalue model: Feynman graph technique for all genera. JHEP **12**, 026 (2006). arXiv:math-ph/0604014
43. Eynard, B., Orantin, N.: Invariants of algebraic curves and topological expansion. Commun. Num. Theory Phys. **1**, 347–452 (2007). arXiv:math-ph/0702045
44. Eynard, B.: Topological expansion for the 1-Hermitian matrix model correlation functions. JHEP **11**, 031 (2004)
45. Chekhov, L., Eynard, B.: Hermitean matrix model free energy: Feynman graph technique for all genera. JHEP **0603**, 014 (2006). arXiv:hep-th/0504116

46. Eynard, B., Orantin, N.: Topological expansion of the 2-matrix model correlation functions: diagrammatic rules for a residue formula. JHEP **12**, 034 (2005). arXiv:math-ph/0504058
47. Borot, G., Eynard, B., Orantin, N.: Abstract loop equations, topological recursion and applications. Commun. Num. Theory Phys. **09**, 51–187 (2015). arXiv:math-ph/1303.5808
48. Chekhov, L.: Logarithmic potential beta-ensembles and Feynman graphs. Proc. Steklov Inst. Math. **272**, 58–74 (2011). arXiv:math-ph/1009.5940
49. Chekhov, L., Eynard, B., Marchal, O.: Topological expansion of the Bethe ansatz, and quantum algebraic geometry (2009). arXiv:math-ph/0911.1664
50. Chekhov, L., Eynard, B., Marchal, O.: Topological expansion of beta-ensemble model and quantum algebraic geometry in the sectorwise approach. Theor. Math. Phys. **166**, 141–185 (2011). arXiv:math-ph/1009.6007
51. Kostov, I.K.: Matrix models as conformal field theories. In: Brézin, E., Kazakov, V., Serban, D., Wiegmann, P., Zabrodin, A. (eds.) Matrix Models as Conformal Field Theories, Nato Science Series II, vol. 221. Springer, Netherland (2006)
52. Kostov, I.K., Orantin, N.: CFT and topological recursion. JHEP **11**, 056 (2010). arXiv:hep-th/1006.2028
53. Johansson, K.: On fluctuations of eigenvalues of random Hermitian matrices. Duke Math. J. **91**, 151–204 (1998)
54. de Monvel, A.B., Pastur, L., Shcherbina, M.: On the statistical mechanics approach in the random matrix theory. Integrated density of states. J. Stat. Phys. **79**, (3–4), 585–611 (1995)
55. Albeverio, S., Pastur, L., Shcherbina, M.: On the 1/N expansion for some unitary invariant ensembles of random matrices. Commun. Math. Phys. **224**, 271–305 (2001)
56. Bonnet, G., David, F., Eynard, B.: Breakdown of universality in multi-cut matrix models. J. Phys. A **33**, 6739–6768 (2000). arXiv:cond-mat/0003324
57. Eynard, B.: Large N expansion of convergent matrix integrals, holomorphic anomalies, and background independence. JHEP **12**, 0903:003 (2009). arXiv:math-ph/0802.1788
58. Borot, G., Guionnet, A.: Asymptotic expansion of beta matrix models in the multi-cut regime. arXiv:math-ph/1303.1045
59. Shcherbina, M.: Fluctuations of linear eigenvalue statistics of β-matrix models in the multi-cut regime. J. Stat. Phys. **151**, 1004–1034 (2013). arXiv:math-ph/1205.7062
60. Fokas, A.S., Its, A.R., Kitaev, A.V.: The isomonodromy approach to matrix models in 2D quantum gravity. Commun. Math. Phys. **147**, 395–430 (1992)
61. Deift, P.A., Zhou, X.: A steepest descent method for oscillatory Riemann-Hilbert problems. Bull. Am. Math. Soc. **26**, (1), 119–123 (1992). arXiv:math.AP/9201261
62. Deift, P.A., Zhou, X.: A steepest descent method for oscillatory Riemann-Hilbert problems. Asymptotics of the mKdV equation. Ann. Math. **137**, 297–370 (1993)
63. Deift, P.A., Kriecherbauer, T., McLaughlin, K.T.-R., Venakides, S., Zhou, X.: Strong asymptotics of orthogonal polynomials with respect to exponential weights. Commun. Pure Appl. Math. **52**, 1491–1552 (1999)
64. Ercolani, N.M., McLaughlin, K.D.T.-R.: Asymptotics of the partition function for random matrices via Riemann-Hilbert techniques and applications to graphical enumeration. Int. Math. Res. Not. **14**, 755–820 (2003). arXiv:math-ph/0211022
65. Bleher, P., Its, A.R.: Asymptotics of the partition function of a random matrix model. Ann. Inst. Fourier **55**, 1943–2000 (2005). arXiv:math-ph/0409082
66. Claeys, T., Grava, T., McLaughlin, K.-T.: Asymptotics for the partition function in two-cut random matrix models. arXiv:math.PR/1410.7001
67. Dyson, F.: Statistical theory of the energy levels of complex systems II. J. Math. Phys. **3**, 157–165 (1962)
68. Wiegmann, P., Zabrodin, A.: Large N expansion for the $2D$ Dyson gas. J. Phys. A **39**, 8933–8964 (2006). arXiv:hep-th/0601009
69. Pastur, L.: Limiting laws of linear eigenvalue statistics for Hermitian matrix models. J. Math. Phys. **47**, 10 (2006). arXiv:math.PR/math/0608719
70. Bleher, P., Its, A.R: Double scaling limit in the random matrix model: the Riemann-Hilbert approach. Commun. Pure Appl. Math. **56**, 433–516 (2003). arXiv:math-ph/0201003

71. Claeys, T., Its, A.R., Krasovsky, I.: Emergence of a singularity for Toeplitz determinants and Painlevé V. Duke Math. J. **160**, 207–262 (2011). arXiv:math-ph/1004.3696
72. Deift, P.A., Its, A.R., Krasovsky, I.: Asymptotics of Toeplitz, Hankel, and Toeplitz+Hankel determinants with Fisher-Hartwig singularities. Ann. Math. **172**, 1243–1299, (2011). arXiv:math.FA/0905.0443
73. Maurel-Segala, E.: High order expansion of matrix models and enumeration of maps. arXiv:math/0608192
74. Guionnet, A., Novak, J.: Asymptotics of unitary multimatrix models: the Schwinger-Dyson lattice and topological recursion (2014). arXiv:math.PR/1401.2703
75. Borodin, A., Gorin, V., Guionnet, A.: Gaussian asymptotics of discrete β ensembles. arXiv:math.PR/1505.03760
76. Eynard, B.: A matrix model for plane partitions. J. Stat. Mech. P10011 (2009). arXiv:math-ph/0905.0535
77. Bonelli, G., Maruyoshi, K., Tanzini, A., Yagi, F.: Generalized matrix models and AGT correspondance at all genera (2010). arXiv:hep-th/1011.5417
78. Eynard, B., Kashani-Poor, A.-K., Marchal, O.: A matrix model for the topological string I: deriving the matrix model. Ann. Henri Poincaré **15**, 1867–1901 (2014). arXiv:hep-th/1003.1737
79. Sułkowski, P.: Matrix models for β-ensembles from Nekrasov partition functions. JHEP **1004**, 063 (2010). arXiv:hep-th/0912.5476
80. Götze, F., Venker, M.: Local universality of repulsive particle systems and random matrices. Ann. Probab. **42**(6), 2207–2242 (2014). arXiv:math.PR/1205.0671
81. Venker, M.: Particle systems with repulsion exponent β and random matrices. Electron. Commun. Probab. **18**(83), 1–12 (2013). arXiv:math.PR/1209.3178
82. Borot, G.: Formal multidimensional integrals, stuffed maps, and topological recursion. Ann. Inst. Henri-Poincaré Comb. Phys. Interact. **1**, 225–264 (2014). arXiv:math-ph/1307.4957
83. Borot, G., Guionnet, A., Kozlowski, K.K.: Large-N asymptotic expansion for mean field models with Coulomb gas interaction. Int. Math. Res. Not. (2015). arXiv:math-ph/1312.6664
84. Konhauser, D.E.: Some properties of biorthogonal polynomials. J. Math. Anal. Appl. **11**, 242–260 (1965)
85. Borodin, A.: Biorthogonal ensembles. Nucl. Phys. B **536**(3), 704–732 (1998). arXiv:math.CA/9804027
86. Claeys, T., Wang, D.: Random matrices with equispaced external source. Commun. Math. Phys. **328**(3), 1023–1077 (2014). arXiv:math-ph/1212.3768
87. Claeys, T., Romano, S.: Biorthogonal ensembles with two-particle interactions (2013). arXiv:math.CA:1312.2892
88. Kostov, I.K.: $O(n)$ vector model on a planar random lattice: spectrum of anomalous dimensions. Mod. Phys. Lett. A **4**, 217 (1989)
89. Kostov, I.K.: Exact solution of the six-vertex model on a random lattice. Nucl. Phys. B **575**, 513–534 (2000). arXiv:hep-th/9911023
90. Borot, G., Bouttier, J., Guitter, E.: More on the $O(n)$ model on random maps via nested loops: loops with bending energy. J. Phys. A: Math. Theor. **45**, 206275 (2012). arXiv:math-ph/1202.5521
91. Bar-Natan, D., Lawrence, R.: A rational surgery formula for the LMO invariant. Isr. J. Math. **140**, 29–60 (2004). arXiv:math.GT/0007045
92. Mariño, M.: Chern-Simons theory, matrix integrals, and perturbative three-manifold invariants. Commun. Math. Phys. **253**, 25–49 (2004). arXiv:hep-th/0207096
93. Borot, G., Eynard, B.: Spectral curves, root systems, and application to $\mathrm{SU}(N)$ Chern-Simons theory on Seifert spaces. Set. Math. New Ser. (2016). arXiv:math-ph/1407.4500
94. Borot, G., Brini, A.: Chern-Simons theory on spherical Seifert manifolds, topological strings and integrable systems. Adv. Th. Math. Phys. (2015). arXiv:hep-th/1506.06887
95. Kharchev, S., Marshakov, A., Mironov, A., Morozov, A., Pakuliak, S.: Conformal matrix models as an alternative to conventional multi-matrix model. Nucl. Phys. B **404**, 717–750 (1993). arXiv:hep-th/9208044

96. Dijkgraaf, R., Vafa, C.: On geometry and matrix models. Nucl. Phys. B **644**, 21–39 (2002). arXiv:hep-th/0207106
97. Gaiotto, D., Alday, L.F., Tachikawa, Y.: Liouville correlation functions from four-dimensional gauge theories. Lett. Math. Phys. **91**, 167–197 (2010). arXiv:hep-th/0906.3219
98. Beenakker, C.W.J., Rajaei, B.: Nonlogarithmic repulsion of transmission eigenvalues in a disordered wire. Phys. Rev. Lett. **71**, 3689–3692 (1993)
99. Beenakker, C.W.J.: Random matrix theory of quantum transport. Rev. Mod. Phys. **69**, 731–808 (1997). arXiv:cond-mat.mes-hall/9612179
100. Zinn-Justin, P.: Six-vertex model with domain wall boundary conditions and one-matrix model. Phys. Rev. E **62**, 3411–3418 (2000). arXiv:math-ph/0005008
101. Bleher, P., Fokin, V.: Exact solution of the six-vertex model with domain wall boundary conditions. Disordered phase. Commun. Math. Phys. **268**, 223–284 (2006)
102. Jimbo, M., Kedem, R., Kojima, T., Konno, H., Miwa, T.: XXZ chain with a boundary. Nucl. Phys. B **441**, 437–470 (1995). arXiv:hep-th/9411112
103. Jimbo, M., Miki, K., Miwa, T., Nakayashiki, A.: Correlation functions of the XXZ model for $\Delta < -1$. Phys. Lett. A **168**, 256–263 (1992). arXiv:hep-th/9205055
104. Jimbo, M., Miwa, T.: qKZ equation with $|q| = 1$ and correlation functions of the XXZ model in the gapless regime. J. Phys. A **29**, 2923–2958 (1996). arXiv:hep-th/9601135
105. Kitanine, N., Maillet, J.-M., Terras, V.: Correlation functions of the XXZ Heisenberg spin-1/2 chain in a magnetic field. Nucl. Phys. B **567**, 554–582 (2000). arXiv:math-ph/9907019
106. Kitanine, N., Kozlowski, K.K., Maillet, J.-M., Slavnov, N.A., Terras, V.: Algebraic Bethe Ansatz approach to the asymptotics behavior of correlation functions. J. Stat. Mech.: Theory Exp. **04**, P04003 (2009). arXiv:math-ph/0808.0227
107. Sklyanin, E.K.: The quantum Toda chain. Lect. Notes Phys. **226**, 196–233 (1985)
108. Faldella, S., Kitanine, N., Niccoli, G.: The complete spectrum and scalar products for the open spin-1/2 XXZ quantum chains with non-diagonal boundary terms. J. Stat. Mech. P01011 (2014). arXiv:math-ph/1307.3960
109. Kitanine, N., Maillet, J.M., Niccoli, G.: Open spin chains with generic integrable boundaries: Baxter equation and Bethe ansatz completeness from SOV. J. Stat. Mech. P05015 (2014). arXiv:math-ph/1401.4901
110. Niccoli, G.: Non-diagonal open spin-1/2 XXZ quantum chains by separation of variables: complete spectrum and matrix elements of some quasi-local operators. J. Stat. Mech. P10025 (2012). arXiv:math-ph/1206.0646
111. Niccoli, G.: Antiperiodic dynamical 6-vertex model I: complete spectrum by SOV, matrix elements of the identity on separate states and connection to the periodic 8-vertex model. J. Phys. A: Math. Theor. **46**, 075003 (2013). arXiv:math-ph/1207.1928
112. Bytsko, A.G., Teschner, J.: Quantization of models with non-compact quantum group symmetry. Modular XXZ magnet and lattice sinh-Gordon model. J. Phys. A **39**, 12927–12982 (2006). arXiv:hep-th/0602093
113. Grosjean, N., Maillet, J.-M., Niccoli, G.: On the form factors of local operators in the lattice sine-Gordon model. J. Stat. Mech.: Theory Exp. P10006 (2012). arXiv:math-ph/1204.6307
114. Niccoli, G., Teschner, J.: The Sine-Gordon model revisited I. J. Stat. Mech. P09014 **1009** (2010). arXiv:hep-th/0910.3173
115. Kharchev, S., Lebedev, D.: Integral representation for the eigenfunctions of quantum periodic Toda chain. Lett. Math. Phys. **50**, 53–77 (1999). arXiv:hep-th/9910265
116. Kharchev, S., Lebedev, D., Semenov-Tian-Shansky, M.: Unitary representations of $U_q(\mathfrak{sl}(2, \mathbb{R}))$, the modular double and the multiparticle q-deformed Toda chains. Commun. Math. Phys. **225**, 573–609 (2002). arXiv:hep-th/0102180
117. Babelon, O., Bernard, D., Talon, M.: Introduction to Classical Integrable Systems. Cambridge Monographs on Mathematical Physics. Cambridge University Press (2002)
118. An, D.: Complete set of eigenfunctions of the quantum Toda chain. Lett. Math. Phys. **87**, 209–223 (2009)
119. Wallach, N.R.: Real Reductive Groups II. Pure and Applied Mathematics, vol. 132-II. Academic Press Inc. (1992)

120. Kozlowski, K.K.: Unitarity of the SoV transform for the Toda chain. Commun. Math. Phys. **334**(1), 223–273 (2015). arXiv:math-ph/1306.4967
121. Nekrasov, N.A., Shatashvili, S.L.: Quantization of integrable systems and four dimensional gauge theories. In: Exner, P. (ed.) Proceedings of the 16th International Congress on Mathematical Physics, Prague, World Scientific 2010, pp. 265–289 (2009). arXiv:hep-th/0908.4052
122. Kozlowski, K.K., Teschner, J.: TBA for the Toda chain. Festschrift volume for Tetsuji Miwa. In: Infinite Analysis 09: New Trends in Quantum Integrable Systems. arXiv:math-ph/1006.2906
123. Babelon, O.: Equations in dual variables for Whittaker functions. Lett. Math. Phys. **65**, 229–240 (2003). arXiv:math-ph/0307037
124. Babelon, O.: On the quantum inverse problem for the closed Toda chain. J. Phys. A **37**, 303–316 (2004). arXiv:hep-th/0304052
125. Kozlowski, K.K.: Aspects of the inverse problem for the Toda chain. J. Math. Phys. **54**, 121902 (2013). arXiv:nlin.SI/1307.4052
126. Niccoli, G.: Antiperiodic spin-1/2 XXZ quantum chains by separation of variables: form factors and complete spectrum. Nucl. Phys. B **870**, 397–420 (2013)
127. Niccoli, G.: Form factors and complete spectrum of XXX antiperiodic higher spin chains by quantum separation of variables. J. Math. Phys. **54**, 053516 (2013). arXiv:math-ph/1205.4537
128. Sklyanin, E.K.: Bispectrality for the quantum open Toda chain. Phys. A: Math. Theor. **46**, 382001 (2013). arXiv:nlin.SI/1306.0454
129. Lukyanov, S.: Finite temperature expectation values of local fields in the sinh-Gordon model. Nucl. Phys. B **612**, 391–412 (2001). arXiv:hep-th/0005027
130. Teschner, J.: On the spectrum of the Sinh-Gordon model in finite volume. Nucl. Phys. B **779**, 403–429 (2008). arXiv:hep-th/0702214
131. Negro, S., Smirnov, F.: On one-point functions for sinh-Gordon model at finite temperature. Nucl. Phys. B **875**, 166–185 (2013). arXiv:hep-th/1306.1476

Chapter 2
Main Results and Strategy of Proof

Abstract In the first part of this chapter we gather the main results which follow from the analysis developed in this book. To start with, in Section 2.1, we discuss an example, in Theorem 2.1.1, of the leading large-N asymptotic expansion of $\ln \mathfrak{z}_N[W]$ where $\mathfrak{z}_N[W]$ is the unscaled partition function defined by (1.5.27). We shall also argue that the large-N asymptotic behaviour of (1.5.27)—whose integrand does not depend explicitly on N—can be deduced from the one of the rescaled model (2.5.1)—whose integrand depends explicitly on N—that we propose to study. Then, after presenting the *per se* model of interest and listing the assumptions on which our analysis builds in Section 2.2, we shall discuss the form of the large-N asymptotic expansion of the logarithm of the rescaled partition function $\ln Z_N[V]$ in Section 2.3. Then, in Section 2.4, we shall discuss the characterisation of the N-dependent equilibrium measure that is pertinent for our study as well as the form of the inverse $\mathcal{W}_N$ of a fundamental singular integral operator $\mathcal{S}_N$ that arises naturally in the study. Finally, Section 2.5, we outline the main steps of the proof.

2.1 A Baby Integral as a Motivation

Let $\mathcal{E}_{(\text{ply})}$ be the functional, defined in $\mathbb{R} \cup \{+\infty\}$ for any probability measure $\mu \in \mathcal{M}^1(\mathbb{R})$ by:

$$\mathcal{E}_{(\text{ply})}[\mu] = \int \Big\{ \frac{c_q}{2}\Big(|\xi|^q + |\eta|^q \Big) - \frac{\beta\pi(\omega_1+\omega_2)}{2} |\xi - \eta| \Big\} \mathrm{d}\mu(\xi)\mathrm{d}\mu(\eta) \, . \tag{2.1.1}$$

Theorem 2.1.1 *$\mathcal{E}_{(\text{ply})}$ is a lower semi-continuous good rate function. Furthermore, given a potential W such that*

$$\lim_{|\xi| \to +\infty} |\xi|^{-q}\, W(\xi) = c_q > 0 \quad \textit{for some } q > 1 \, , \tag{2.1.2}$$

G. Borot et al., *Asymptotic Expansion of a Partition Function Related to the Sinh-model*, Mathematical Physics Studies,
DOI 10.1007/978-3-319-33379-3_2

it holds

$$\lim_{N\to+\infty} \frac{\ln \mathfrak{z}_N[W]}{N^{2+\frac{1}{q-1}}} = - \inf_{\mu\in\mathcal{M}^1(\mathbb{R})} \mathcal{E}_{(\mathrm{ply})}\big[\mu\big] . \tag{2.1.3}$$

This infimum is attained at a unique probability measure $\mu_{\mathrm{eq}}^{(\mathrm{ply})}$. *This measure is continuous with respect to the Lebesgue measure and has density*

$$\rho_{\mathrm{eq}}^{(\mathrm{ply})}(\xi) = \frac{q(q-1)|\xi|^{q-2}}{2\pi\beta(\omega_1+\omega_2)} \cdot \mathbf{1}_{[a\,;b]}(\xi) . \tag{2.1.4}$$

$\mu_{\mathrm{eq}}^{(\mathrm{ply})}$ *is supported on the interval* $[a\,;b]$, *with* (a,b) *being the unique solution to the set of equations*

$$|b|^{q-1} = |a|^{q-1} = \frac{\pi\beta(\omega_1+\omega_2)}{q} . \tag{2.1.5}$$

We have, explicitly:

$$\lim_{N\to+\infty} \frac{\ln \mathfrak{z}_N[W]}{N^{2+\frac{1}{q-1}}} = \big(c_q\big)^{\frac{1}{q}} \cdot \bigg(\frac{\pi\beta}{q}(\omega_1+\omega_2)\bigg)^{\frac{q+1}{q}} \cdot \frac{2q^2-9q+6}{2(2q-1)} . \tag{2.1.6}$$

The proof of this proposition is postponed to Appendix B, and follows similar steps to, *e.g.* [1]. We now provide heuristic arguments to justify the occurrence of scaling in N in this problem. Just as discussed in the introduction, the repulsive effect of the sinh-2 body interactions will dominate over the confining effect of the potential as long as the integration variables will be located in some bounded set. Furthermore, in the same situation, the Lebesgue measure should contribute to the integral at most as an exponential in N. We thus look for a rescaling of the variables $y_a = T_N\lambda_a$ where the effects of the confining potential and the sinh-2 body interactions will be of the same order of magnitude in N. This recasts the partition function as

$$\mathfrak{z}_N[W] = \Big(T_N\Big)^N \int_{\mathbb{R}^N} \prod_{a<b}^{N} \Big\{ \sinh\big[\pi\omega_1 T_N(\lambda_a-\lambda_b)\big] \sinh\big[\pi\omega_2 T_N(\lambda_a-\lambda_b)\big]\Big\}^{\beta}$$
$$\times \prod_{a=1}^{N} \Big\{ \mathrm{e}^{-W(T_N\lambda_a)} \Big\} \, \mathrm{d}^N\boldsymbol{\lambda} , \tag{2.1.7}$$

Taking into account the large-variable asymptotics of the potential, we have:

$$\sum_{a=1}^{N} W\big(T_N\lambda_a\big) \sim T_N^q\, N , \tag{2.1.8}$$

where the symbol $\sim$ means that for a "typical" distribution of the variables $\{\lambda_a\}_1^N$, the leading in N asymptotic behaviour of the sum in the right-hand side should be of the order of the left-hand side. Similarly, assuming a typical distribution of the variables $\{\lambda_a\}_1^N$ such that most of the pairs $\{\lambda_a, \lambda_b\}$ satisfy $T_N|\lambda_a - \lambda_b| \gg 1$, one has

$$\sum_{a<b}^{N} \beta \ln \Big\{ \sinh\big[\pi\omega_1 T_N(\lambda_a - \lambda_b)\big] \sinh\big[\pi\omega_2 T_N(\lambda_a - \lambda_b)\big]\Big\} \sim C N^2 T_N . \tag{2.1.9}$$

Thus, the confining potential and the two-body interaction will generate a comparable order of magnitude in N as soon as $N^2 \cdot T_N = T_N^q \cdot N$, *i.e.*

$$T_N = N^{\frac{1}{q-1}} . \tag{2.1.10}$$

Theorem 2.1.1 indeed justifies that the empirical distribution $L_N^{(\lambda)}$ of $\lambda_a = N^{\frac{-1}{q-1}} y_a$ concentrates around the equilibrium measure, with a large deviation principle governed by the rate function (2.1.1).

This observation implies that, in fact, $Z_N[V_N]$ with $V_N(\lambda) = N^{-\frac{q}{q-1}} \cdot W(N^{\frac{1}{q-1}}\lambda)$ is the good object to study in that it involves interactions that are already tuned to the proper scale in N. Due to the relation $\mathfrak{z}_N[W] = N^{\frac{N}{q-1}} \cdot Z_N[V_N]$, one readily has access to the large-N asymptotic expansion of $\mathfrak{z}_N[W]$.

2.2 The Model of Interest and the Assumptions

It follows from the arguments given in the previous section that, effectively, the analysis of the unrescaled model boils down to the one subordinated to the partition function

$$Z_N[V] = \int_{\mathbb{R}^N} \prod_{a<b}^{N} \Big\{ \sinh\big[\pi\omega_1 N^{\alpha}(\lambda_a - \lambda_b)\big] \sinh\big[\pi\omega_2 N^{\alpha}(\lambda_a - \lambda_b)\big]\Big\}^{\beta} \prod_{a=1}^{N} \mathrm{e}^{-N^{1+\alpha} V(\lambda_a)} \cdot \mathrm{d}^N\boldsymbol{\lambda} , \tag{2.2.1}$$

with α some parameter—equal to $1/(q-1)$ in the previous paragraph—and V a potential that possibly depends on N. Due to such an effective reduction, in this book, we shall develop the general formalism to extract the large-N asymptotic behaviour. Therefore, we shall keep the complexity at minimum. In particular, we shall *not* consider the case of N-dependent potentials which would put the analysis of $Z_N[V]$ in complete correspondence with the one of $\mathfrak{z}_N[W]$. Indeed, this would lead to numerous technical complication in our arguments, without bringing more light on the underlying phenomena. By focusing on (2.2.1), we believe that the new features and ideas of our methods are better isolated and illustrated.

In the present paper we obtain the large-N asymptotic expansion of $\ln Z_N[V]$ up to $o(1)$ under four hypothesis

- the potential V is confining, *viz.* there exists $\epsilon > 0$ such that

$$\liminf_{|\xi|\to+\infty} |\xi|^{-(1+\epsilon)} V(\xi) = +\infty ; \tag{2.2.2}$$

- the potential V is smooth and strictly convex on $\mathbb{R}$;
- the potential is sub-exponential, namely there exists $\epsilon > 0$ and $C_V > 0$ such that

$$\forall \xi \in \mathbb{R}, \qquad \sup_{\eta\in[0\,;\epsilon]} \big|V'(\xi+\eta)\big| \le C_V\big(|V(\xi)|+1\big) , \tag{2.2.3}$$

  and given any $\kappa > 0$ and $p \in \mathbb{N}$, there exists $C_{\kappa,p}$ such that

$$\forall \xi \in \mathbb{R}, \qquad \big|V^{(p)}(\xi)\big|\,\mathrm{e}^{-\kappa V(\xi)} \le C_{\kappa,p} . \tag{2.2.4}$$

- the exponent α in N^{α} is neither too large nor too small:

$$0 < \alpha < 1/6 . \tag{2.2.5}$$

The first hypothesis guarantees that the integral (2.2.1) is well-defined, and that the λ's will typically remain in a compact region of $\mathbb{R}$ independent of N. It could be weakened to study weakly confining potentials, for the price of introducing more technicalities, similar to those already encountered for β ensembles—see *e.g.* [2].

In the second assumption, V could be assumed $\mathcal{C}^k$ for k large enough. The convexity assumption guarantees that the support of the equilibrium measure is a single segment.[1] In principle, the multi-cut regime that may arise when the potential is not strictly convex can be addressed by importing the ideas of [3] to the present framework. We expect that the analysis of the Riemann–Hilbert problem in the multi-cut regime is very similar to the present case, but with a larger range of degrees for the polynomial freedom appearing in the solution (4.3.14). Though it would certainly represent some amount of work, the ideas we develop here should also be applicable to derive the fine large N analysis of the solution of the Riemann–Hilbert problem in the bulk and in the vicinity of all the edges of the support of the equilibrium measure.

The third assumption is not essential, but allows some simplification of the intermediate proofs concerning the equilibrium measure and the large deviation estimates, *e.g.* Theorem 3.1.6 and Corollary 3.1.10. It is anyway satisfied in physically relevant problems.

[1] See *e.g.* the expression of the $N = \infty$ equilibrium measure (2.4.3). Its proof is given in Appendix C.

In the fourth assumption, $\alpha = 0$ can already be addressed with existing methods [4]. The analysis of the transition regime $\alpha \to 0$ appears to be difficult. The upper limit $\alpha < \alpha^* = 1/6$ has a purely technical origin. The value of α^* could be increased by entering deeper into the fine structure of the analysis of the Schwinger–Dyson equation, and by finding more precise local and global bounds for the large N behaviour of the inverse of the master operator $\mathcal{U}_N^{-1}$, in more cunning norms. Intuitively, the genuine upper limit should be $\alpha^* = 1$, since in the $\alpha > 1$ case, we reach a regime where the particles do not feel the local repulsion any more. However, obtaining microscopic estimates is usually a difficult question—for β ensembles, it has been addressed *e.g.* in [5, 6]. So, one can expect important technical difficulties to extend our result to values of α increasing up to 1.

This set of hypothesis offers a convenient framework for our purposes, enabling us to focus on the technical aspects (i)–(vi) listed in Section 1.5.3 without adding extra complications.

2.3 Asymptotic Expansion of $Z_N[V]$ at $\beta = 1$

We now state one of the main results of the paper, namely the large-N asymptotic expansion of the partition function $Z_N[V]$ which holds for any potential V satisfying the hypothesis stated above

Theorem 2.3.1 *The below asymptotic expansion holds*

$$\ln\left(\frac{Z_N[V]}{Z_N[W_{G;N}]}\right)_{|\beta=1}$$

$$= -N^{2+\alpha}\sum_{p=0}^{\lfloor 2/\alpha\rfloor+1}\frac{\mathfrak{F}_p[V]}{N^{\alpha p}} + N^{\alpha}\cdot\beth_0\cdot\Big(\mathfrak{Q}[V,W_{G;N}](b_N)-\mathfrak{Q}[V,W_{G;N}](a_N)\Big)$$

$$+\aleph_0\cdot\Big(\mathfrak{Q}[V,W_{G;N}]'(b_N)+\mathfrak{Q}[V,W_{G;N}]'(a_N)\Big)+\mathrm{o}(1). \tag{2.3.1}$$

The whole V-dependence of this expansion is encoded in the coefficients $\mathfrak{F}_p[V]$ and in the function $\mathfrak{Q}[V,W_{G;N}](\xi)$. $\beth_0$ and $\aleph_0$ are numerical coefficients given, respectively, in terms of a single and four-fold integral. Also, the answer involves the Gaussian potential

$$W_{G;N}(\xi) = \frac{\pi\beta(\omega_1+\omega_2)\cdot\Big[\xi^2-(a_N+b_N)\xi\Big]}{b_N-a_N+\frac{1}{N^{\alpha}}\sum\limits_{p=1}^{2}\frac{1}{\pi\omega_p}\ln\left(\frac{\omega_1\omega_2}{\omega_p(\omega_1+\omega_2)}\right)} \tag{2.3.2}$$

and sequences a_N and b_N that are given in Theorem 2.4.3. If we denote $V_N^{\pm} = V \pm W_{G;N}$, the coefficients $\mathcal{T}_p[V]$ take the form

$$\mathcal{T}_0[V] = \frac{-1}{4\pi(\omega_1+\omega_2)} \int_{a_N}^{b_N} V_N^-(\xi) \cdot (V_N^-)''(\xi)\,\mathrm{d}\xi \tag{2.3.3}$$

when $p = 0$ and, for any $p \geq 1$:

$$\begin{aligned}\mathcal{T}_p[V] = u_{p+1} \int_{a_N}^{b_N} V_N^-(\xi) \cdot V^{(p+2)}(\xi)\,\mathrm{d}\xi \\ + \sum_{\substack{s+\ell=p-1 \\ s,\ell\geq 0}} \frac{\daleth_{s,\ell}}{s!} \Big\{ (-1)^{\ell} (V_N^-)^{(\ell+1)}(a_N) \cdot (V_N^+)^{(s+1)}(a_N) \\ + (-1)^{s} (V_N^-)^{(\ell+1)}(b_N) \cdot (V_N^+)^{(s+1)}(b_N) \Big\}.\end{aligned} \tag{2.3.4}$$

The coefficients $\daleth_{s,\ell}$ are defined by:

$$\daleth_{s,\ell} = \frac{\mathrm{i}^{s+\ell+1}}{2\pi} \sum_{r=1}^{\ell+1} \frac{s!}{r!(s+\ell+1-r)!} \cdot \frac{\partial^r}{\partial\mu^r}\Big(\frac{\mu}{R_\downarrow(-\mu)}\Big)_{|\mu=0} \cdot \frac{\partial^{s+1+\ell-r}}{\partial\mu^{s+1+\ell-r}}\Big(\frac{1}{R_\downarrow(\mu)}\Big)_{|\mu=0}, \tag{2.3.5}$$

$R_\downarrow$ is the $\mathbb{H}^-$ Wiener–Hopf factor of $1/\mathcal{F}[S](\lambda)$, with S defined in (2.4.15), that reads

$$R_\downarrow(\lambda) = \frac{\lambda}{2\pi\sqrt{\omega_1+\omega_2}} \cdot \Big(\frac{\omega_2}{\omega_1+\omega_2}\Big)^{-\frac{\mathrm{i}\lambda}{2\pi\omega_1}} \cdot \Big(\frac{\omega_1}{\omega_1+\omega_2}\Big)^{-\frac{\mathrm{i}\lambda}{2\pi\omega_2}} \cdot \frac{\Gamma\Big(\frac{\mathrm{i}\lambda}{2\pi\omega_1}\Big)\cdot\Gamma\Big(\frac{\mathrm{i}\lambda}{2\pi\omega_2}\Big)}{\Gamma\Big(\frac{\mathrm{i}\lambda(\omega_1+\omega_2)}{2\pi\omega_1\omega_2}\Big)}. \tag{2.3.6}$$

The function $\mathfrak{L}$ describing the constant term is defined as

$$\mathfrak{L}[V, W_{G;N}](\xi) = \frac{V'(\xi) - W'_{G;N}(\xi)}{V''(\xi) - W''_{G;N}(\xi)} \ln\Big(\frac{V''(\xi)}{W''_{G;N}(\xi)}\Big). \tag{2.3.7}$$

The V-independent coefficient $\beth_0$ in front of the term N^{α} reads

$$\beth_0 = \int_0^{+\infty} \frac{\mathrm{d}u\, J(u)}{2\pi}\big(uS'(u) + S(u)\big)$$

with

$$J(u) = \int_{\mathscr{C}_{\mathrm{reg}}^{(+)}} \frac{2\sinh\big[\lambda/(2\omega_1)\big]\sinh\big[\lambda/(2\omega_2)\big]}{\sinh\big[\lambda(\omega_1+\omega_2)/(2\omega_1\omega_2)\big]} \cdot \frac{\mathrm{e}^{\mathrm{i}\lambda u}\,\mathrm{d}\lambda}{2\mathrm{i}\pi}. \tag{2.3.8}$$

Finally, the numerical prefactor $\aleph_0$ *is expressed in terms of the four-fold integral*

$$\aleph_0 = -\frac{\omega_1+\omega_2}{2}\int_{\mathbb{R}}\frac{\mathrm{d}u\, J(u)}{2\pi}\int_{|u|}^{+\infty}\mathrm{d}v\,\partial_u\Big\{S(u)\cdot\Big(\mathfrak{r}\big[\tfrac{v-u}{2}\big]-\mathfrak{r}\big[\tfrac{v+u}{2}\big]\Big)\Big\}$$
$$+\int_{\mathscr{C}_{\mathrm{reg}}^{(+)}}\frac{\mathrm{d}\lambda\,\mathrm{d}\mu}{(2\mathrm{i}\pi)^2}\frac{\mu(\omega_1+\omega_2)}{(\lambda+\mu)R_{\downarrow}(\lambda)R_{\downarrow}(\mu)}$$
$$\times\int_0^{+\infty}\mathrm{d}x\,\mathrm{d}y\,\mathrm{e}^{\mathrm{i}\lambda x+\mathrm{i}\mu y}\,\partial_x\Big\{S(x-y)\Big(\mathfrak{r}(x)-\mathfrak{r}(y)-\frac{x-y}{2\pi(\omega_1+\omega_2)}\Big)\Big\}. \tag{2.3.9}$$

The integrand of $\aleph_0$ *involves the function* $\mathfrak{r}$ *which is given by*

$$\mathfrak{r}(x) = \frac{\mathfrak{c}_1(x)+\mathfrak{c}_0(x)\Big[\sum_{p=1}^{2}\frac{1}{2\pi\omega_p}\ln\Big(\frac{\omega_1\omega_2}{\omega_p(\omega_1+\omega_2)}\Big)\Big]}{1+2\pi\beta(\omega_1+\omega_2)\mathfrak{c}_0(x)} \tag{2.3.10}$$

with

$$\mathfrak{c}_p(x) = \frac{\mathrm{i}^p}{2\mathrm{i}\pi\sqrt{\omega_1+\omega_2}}\int_{\mathscr{C}_{\mathrm{reg}}^{(+)}}\frac{\mathrm{e}^{\mathrm{i}\lambda x}}{\lambda}\frac{\partial^p}{\partial\lambda^p}\Big(\frac{1}{R_{\downarrow}(\lambda)}\Big)\cdot\frac{\mathrm{d}\lambda}{2\mathrm{i}\pi}. \tag{2.3.11}$$

The result for $\beta \neq 1$ contains two more terms, and is given in the body of the book, by Proposition 3.4.1, in terms of N-dependent simple and double integrals $\mathfrak{I}^{(2)}_{s;\beta}$. The final form for the asymptotics up to $o(1)$ of these extra terms can be worked out following the steps of Section 6.3, although we decided to leave it out of the scope of this book, since $\beta \neq 1$ does not seem to appear in quantum integrable systems.

2.4 The N-dependent Equilibrium Measure and the Master Operator

It is not hard to generalise the proof of Theorem 2.1.1 to the present setting so as to obtain the below characterisation of the leading in N asymptotic behaviour for $Z_N[V]$.

Theorem 2.4.1 *Let* $\mathcal{E}_\infty$ *be the lower semi-continuous good rate function*

$$\mathcal{E}_\infty[\mu] = \frac{1}{2}\int\big(V(\eta)+V(\xi)\; -\;\pi\beta(\omega_1+\omega_2)|\xi-\eta|\big)\,\mathrm{d}\mu(\xi)\mathrm{d}\mu(\eta)\,. \tag{2.4.1}$$

Then, one has that

$$\lim_{N\to+\infty} \frac{\ln Z_N[V]}{N^{2+\alpha}} = - \inf_{\mu\in\mathcal{M}^1(\mathbb{R})} \mathcal{E}_\infty\big[\mu\big] . \tag{2.4.2}$$

The infimum is attained at a unique probability measure μ_{eq}. *This measure is continuous with respect to the Lebesgue measure, and has density*

$$\rho_{\mathrm{eq}}(\xi) = \frac{V''(\xi)}{2\pi\beta(\omega_1+\omega_2)} \cdot \mathbf{1}_{[a\,;b]}(\xi) \tag{2.4.3}$$

supported on the interval $[a\,;b]$, *with* (a,b) *being the unique solution to the set of equations*

$$V'(b) = -V'(a) = \pi\beta(\omega_1+\omega_2) . \tag{2.4.4}$$

One has, explicitly,

$$\lim_{N\to+\infty} \frac{\ln Z_N[V]}{N^{2+\alpha}} = -\frac{V(a)+V(b)}{2} + \frac{\big(V'(b)\big)^2(b-a) + \int_a^b \big(V'(\xi)\big)^2 \mathrm{d}\xi}{4\pi\beta(\omega_1+\omega_2)} . \tag{2.4.5}$$

The strict convexity of V guarantees that the density (2.4.3) is positive and that it reaches a non-zero limit at the endpoints of the support. This behaviour differs from the situation usually studied in β ensembles with analytic potentials which leads to a generic square root (or inverse square root) vanishing (or divergence) of the equilibrium density at the edges.

Note that the function $\mathcal{E}_\infty$ defined in (2.4.1) arises as a good rate function in the large deviation estimates for the empirical measure $L_N^{(\lambda)}$, *c.f.* (1.2.2). In fact, a refinement of Theorem 2.4.1 would lead to the more precise estimates

$$\ln Z_N[V] = -N^{2+\alpha}\mathcal{E}_\infty\big[\mu_{\mathrm{eq}}\big] + \mathrm{O}(N^2) . \tag{2.4.6}$$

Thus, with respect to the usual varying weight β-ensemble case, there is a loss of precision by a $N^{1-\alpha}$ factor. This, in fact, takes its origin in that the purely asymptotic rate function $\mathcal{E}_\infty\big[\mu_{\mathrm{eq}}\big]$ does not absorb enough of the fine structure of the saddle-point. As a consequence, the remainder $\mathrm{O}(N^2)$ mixes both types of contributions: the deviation of the saddle-point with respect to its asymptotic position and the fluctuation of the integration variables around the saddle-point.

The fine, N-dependent, structure of the saddle-point is much better captured by the N-dependent deformation[2] of the rate functions $\mathcal{E}_\infty$:

[2]The property of lower semi-continuity along with the fact that $\mathcal{E}_N$ has compact level sets is verified exactly as in the case of β-ensembles, so we do not repeat the proof here.

$$\mathcal{E}_N[\mu] = \frac{1}{2} \int \bigg(V(\xi) + V(\eta) - \frac{\beta}{N^\alpha} \ln \bigg\{ \prod_{p=1}^{2} \sinh\big[\pi N^\alpha \omega_p(\xi - \eta)\big] \bigg\} \bigg) \mathrm{d}\mu(\xi)\mathrm{d}\mu(\eta) . \tag{2.4.7}$$

This N-dependent rate functions appear extremely effective for the purpose of our analysis. Namely, it allows us re-summing a whole tower of contributions into a single term. The use of $\mathcal{E}_N$ should not be considered as a mere technical simplification of the intermediate steps; it is, in fact, of prime importance. The use of the more classical object $\mathcal{E}_\infty$ would render the analysis of the Schwinger–Dyson equations impossible. This fact will become apparent in the core of the text.

As usual, this minimiser admits a characterisation in terms of a variational problem:

Proposition 2.4.2 *For any strictly convex potential V, the N-dependent rate function $\mathcal{E}_N$ admits its minimum on $\mathcal{M}^1(\mathbb{R})$ at a unique probability measure $\mu_{\mathrm{eq}}^{(N)}$. This equilibrium measure is supported on a segment $[a_N ; b_N]$ and corresponds to the unique solution to the integral equations*

$$V(\xi) - \frac{\beta}{N^\alpha} \int \ln \bigg\{ \prod_{p=1}^{2} \sinh\big[\pi N^\alpha \omega_p(\xi - \eta)\big] \bigg\} \mathrm{d}\mu_{\mathrm{eq}}^{(N)}(\eta) = C_{\mathrm{eq}}^{(N)} \quad \text{on } [a_N ; b_N] \tag{2.4.8}$$

$$V(\xi) - \frac{\beta}{N^\alpha} \int \ln \bigg\{ \prod_{p=1}^{2} \sinh\big[\pi N^\alpha \omega_p(\xi - \eta)\big] \bigg\} \mathrm{d}\mu_{\mathrm{eq}}^{(N)}(\eta) > C_{\mathrm{eq}}^{(N)} \quad \text{on } \mathbb{R} \setminus [a_N ; b_N] , \tag{2.4.9}$$

with $C_{\mathrm{eq}}^{(N)}$ a constant whose determination is part of the problem (2.4.8)–(2.4.9). *The equilibrium measure admits a density $\rho_{\mathrm{eq}}^{(N)}$, which is $\mathcal{C}^{k-2}$ in the interior $]a_N ; b_N[$ if V is $\mathcal{C}^k$. Finally, one has the behaviour at the edges:*

$$\rho_{\mathrm{eq}}^{(N)}(\xi) \underset{\xi \to a_N^+}{=} \mathrm{O}\big(\sqrt{\xi - a_N}\big) , \qquad \rho_{\mathrm{eq}}^{(N)}(\xi) \underset{\xi \to b_N^-}{=} \mathrm{O}\big(\sqrt{b_N - \xi}\big) . \tag{2.4.10}$$

The proof of this proposition is rather classical. It follows, for instance, from [4, Section 2.3] in what concerns the regularity, and from a convexity argument (see [7, Theorem 2.2]) in what concerns connectedness of the support and the strict inequality in (2.4.9). Elements of proof are nevertheless gathered in Appendix C. In fact, regarding to the equilibrium measure, we can be much more precise when N is large enough:

Theorem 2.4.3 *In the $N \to \infty$ regime, the equilibrium measure $\mu_{\mathrm{eq}}^{(N)}$:*

- *is supported on the single interval $[a_N ; b_N]$ whose endpoints admit the asymptotic expansion*

$$a_N \;=\; a + \sum_{\ell=1}^{k} \frac{a_{N;\ell}}{N^{\ell\alpha}} + \mathrm{O}\Big(\frac{1}{N^{(k+1)\alpha}}\Big) \quad \text{and} \quad b_N \;=\; b + \sum_{\ell=1}^{k} \frac{b_{N;\ell}}{N^{\ell\alpha}} + \mathrm{O}\Big(\frac{1}{N^{(k+1)\alpha}}\Big)\,, \tag{2.4.11}$$

where $k \in \mathbb{N}^*$ is arbitrary, (a, b) are as defined in (2.4.4) *while*

$$\begin{pmatrix} b_{N;1} \\ a_{N;1} \end{pmatrix} = \Big\{ \sum_{p=1}^{2} \frac{1}{2\pi\omega_p} \ln\Big(\frac{\omega_1\omega_2}{\omega_p(\omega_1+\omega_2)} \Big) \Big\} \cdot \begin{pmatrix} V''(a)\cdot\{V''(b)\}^{-1} \\ -V''(b)\cdot\{V''(a)\}^{-1} \end{pmatrix}; \tag{2.4.12}$$

- *is continuous with respect to Lebesgue. Its density is $\rho_{\mathrm{eq}}^{(N)}$ vanishes like a squareroot at the edges:*

$$\begin{aligned} \rho_{\mathrm{eq}}^{(N)}(\xi) &\underset{\xi\to a_N^+}{\sim} \bigg(\frac{V''(a_N) + \mathrm{O}(N^{-\alpha})}{\pi\beta\sqrt{\pi(\omega_1+\omega_2)}} \bigg) \sqrt{\xi - a_N}\,, \\ \rho_{\mathrm{eq}}^{(N)}(\xi) &\underset{\xi\to b_N^-}{\sim} \bigg(\frac{V''(b_N) + \mathrm{O}(N^{-\alpha})}{\pi\beta\sqrt{\pi(\omega_1+\omega_2)}} \bigg) \sqrt{b_N - \xi}\,, \end{aligned} \tag{2.4.13}$$

and there exists a constant $C > 0$ independent of N such that:

$$||\rho_{\mathrm{eq}}^{(N)}||_{L^\infty([a_N\,;b_N])} \;\leq\; C\,||V''||_{L^\infty([a_N\,;b_N])}\,. \tag{2.4.14}$$

This density takes the form $\rho_{\mathrm{eq}}^{(N)} = \mathcal{W}_N[V']$, with $\mathcal{W}_N$ as defined in (2.4.17).

If the potential V defining the equilibrium measures satisfies $V \in \mathcal{C}^k([a_N\,;b_N])$, then the density is of class $\mathcal{C}^{k-2}$ on $]a_N\,;b_N[$.

Note that the characterisation of $\rho_{\mathrm{eq}}^{(N)}$ in the theorem above comes from the fact that it is solution to the singular integral equation $\mathcal{S}_N\big[\rho_{\mathrm{eq}}^{(N)}\big](\xi) = V'(\xi)$ on $[a_N\,;b_N]$, where

$$\mathcal{S}_N\big[\phi\big](\xi) \;=\; \fint_{a_N}^{b_N} S\big[N^\alpha(\xi-\eta)\big]\phi(\eta)\,\mathrm{d}\eta \qquad \text{and} \qquad S(\xi) \;=\; \sum_{p=1}^{2} \beta\pi\omega_p \operatorname{cotanh}\big[\pi\omega_p\xi\big]. \tag{2.4.15}$$

The unknowns in this equation $(\rho_{\mathrm{eq}}^{(N)}, a_N, b_N)$ should be picked in such a way that $\rho_{\mathrm{eq}}^{(N)}$ has mass 1 on $[a_N\,;b_N]$ and is regular at the endpoints a_N, b_N. Thus, determining the equilibrium measure boils down to an inversion of the singular integral operator $\mathcal{S}_N$. In fact, the singular integral operator $\mathcal{S}_N$ also intervenes in the Schwinger–Dyson equations. The precise control on its inverse $\mathcal{W}_N$—defined between appropriate functional spaces—plays a crucial role in the whole asymptotic analysis.

These pieces of information can be obtained by exploiting the fact that the operator $\mathcal{S}_N$ is of truncated Wiener–Hopf type. As such, its inversion is equivalent to solving a 2×2 matrix valued Riemann–Hilbert problem. This Riemann–Hilbert problem admits a solution for N large enough that can be constructed by means of a variant of the non-linear steepest descent method. By doing so, we are able to describe, quite explicitly, the inverse $\mathcal{W}_N$ by means of the unique solution χ to the 2×2 matrix valued Riemann–Hilbert problem given in Section 4.2.1. We will not discuss the structure of this solution here and, rather, refer the reader to the relevant section. We will, however, provide the main consequence of this analysis, *viz.* an explicit representation for the operator $\mathcal{W}_N$. For this purpose, we need to announce that χ_{11}, the $(1, 1)$ matrix entry of χ, is such that $\mu \mapsto \mu^{1/2} \cdot \chi_{11}(\mu) \in L^{\infty}(\mathbb{R})$.

Theorem 2.4.4 *Let $0 < s < 1/2$. The operator $\mathcal{S}_N : H_s([a_N ; b_N]) \to \mathfrak{X}_s(\mathbb{R})$ is continuous and invertible where, for any closed $A \subseteq \mathbb{R}$,*

$$\mathfrak{X}_s(A) = \Big\{ H \in H_s(A) \; : \; \int\limits_{\mathbb{R}+\mathrm{i}\epsilon} \chi_{11}(\mu)\mathcal{F}[H](N^{\alpha}\mu)\mathrm{e}^{-\mathrm{i}N^{\alpha}\mu b_N} \frac{\mathrm{d}\mu}{2\mathrm{i}\pi} = 0 \Big\} \tag{2.4.16}$$

is a closed subspace of $H_s(A)$ such that $\mathcal{S}_N\big(H_s([a_N ; b_N])\big) = \mathfrak{X}_s(\mathbb{R})$. The inverse is given by the integral transform $\mathcal{W}_N$ which takes, for $H \in \mathcal{C}^1([a_N ; b_N]) \cap \mathfrak{X}_s(\mathbb{R})$, the form

$$\mathcal{W}_N[H](\xi) = \frac{N^{2\alpha}}{2\pi\beta} \int\limits_{\mathbb{R}+2\mathrm{i}\epsilon} \frac{\mathrm{d}\lambda}{2\mathrm{i}\pi} \int\limits_{\mathbb{R}+\mathrm{i}\epsilon} \frac{\mathrm{d}\mu}{2\mathrm{i}\pi} \frac{\mathrm{e}^{-\mathrm{i}N^{\alpha}(\xi-a_N)\lambda}}{\mu-\lambda} \Big\{ \chi_{11}(\lambda)\chi_{12}(\mu) - \frac{\mu}{\lambda} \cdot \chi_{11}(\mu)\chi_{12}(\lambda) \Big\}$$
$$\times \int\limits_{a_N}^{b_N} \mathrm{d}\eta \, \mathrm{e}^{\mathrm{i}N^{\alpha}\mu(\eta-b_N)} H(\eta) \, . \tag{2.4.17}$$

In the above integral representations the parameter $\epsilon > 0$ is small enough but arbitrary. Furthermore, for any $H \in \mathcal{C}^1([a_N ; b_N])$, the transform $\mathcal{W}_N$ exhibits the local behaviour

$$\mathcal{W}_N[H](\xi) \underset{\xi \to a_N^+}{\sim} C_L H'(a_N)\sqrt{\xi - a_N} \quad \text{and} \quad \mathcal{W}_N[H](\xi) \underset{\xi \to b_N^-}{\sim} C_R H'(b_N)\sqrt{b_N - \xi} \, . \tag{2.4.18}$$

where $C_{L/R}$ are some H-independent constants.

Note that, within such a framework, the density of the equilibrium measure $\mu_{\mathrm{eq}}^{(N)}$ is expressed in terms of the inverse as $\rho_{\mathrm{eq}}^{(N)} = \mathcal{W}_N[V']$. In this case, the pair of endpoints (a_N, b_N) of the support of $\mu_{\mathrm{eq}}^{(N)}$ corresponds to the unique solution to the system of equations

$$\int\limits_{a_N}^{b_N} \mathcal{W}_N[V'](\xi)\,\mathrm{d}\xi = 1 \quad \text{and} \quad \int\limits_{\mathbb{R}+\mathrm{i}\epsilon} \frac{\mathrm{d}\mu \, \chi_{11}(\mu)}{2\mathrm{i}\pi} \int\limits_{a_N}^{b_N} \mathrm{e}^{\mathrm{i}\mu N^{\alpha}(\eta-b_N)} V'(\eta)\,\mathrm{d}\eta = 0 \, . \tag{2.4.19}$$

The first condition guarantees that $\mu_{\mathrm{eq}}^{(N)}$ has indeed mass 1, while the second one ensures that its density vanishes as a square root at the edges a_N, b_N. Using fine properties of the inverse, these conditions can be estimated more precisely in the large-N limit, hence enabling one to fix the large-N asymptotic expansion of the endpoints a_N, b_N as announced in (2.4.11) and (2.4.12).

2.5 The Overall Strategy of the Proof

In the following, we shall denote by $p_N(\boldsymbol{\lambda})$ the probability density on $\mathbb{R}^N$ associated with the partition function $Z_N[V]$ defined in (2.2.1):

$$p_N(\boldsymbol{\lambda}) = \frac{1}{Z_N[V]} \prod_{a<b}^{N} \Big\{ \sinh\big[\pi\omega_1 N^{\alpha}(\lambda_a - \lambda_b)\big] \sinh\big[\pi\omega_2 N^{\alpha}(\lambda_a - \lambda_b)\big] \Big\}^{\beta} \prod_{a=1}^{N} \mathrm{e}^{-N^{1+\alpha} V(\lambda_a)} . \tag{2.5.1}$$

$p_N(\boldsymbol{\lambda})$ gives rise to a probability measure $\mathbb{P}_N$ on $\mathbb{R}^N$. We also agree that, throughout the book, $L_N^{(\boldsymbol{\lambda})}$ refers to the empirical measure

$$L_N^{(\boldsymbol{\lambda})} = \frac{1}{N} \sum_{a=1}^{N} \delta_{\lambda_a} \tag{2.5.2}$$

associated with the stochastic vector $\boldsymbol{\lambda}$.

Definition 2.5.1 Let $\nu_1, \ldots, \nu_\ell$ be any (possibly depending on the stochastic vector $\boldsymbol{\lambda}$) measures and ψ a function in ℓ variables. Then we agree upon

$$\big\langle \psi \big\rangle_{\nu_1 \otimes \cdots \otimes \nu_\ell} \equiv \big\langle \psi(\xi_1, \ldots, \xi_\ell) \big\rangle_{\nu_1 \otimes \cdots \otimes \nu_\ell} \equiv \mathbb{P}_N\Big[\int_{\mathbb{R}^\ell} \psi(\xi_1, \ldots, \xi_\ell)\, \mathrm{d}\nu_1 \otimes \cdots \otimes \mathrm{d}\nu_\ell \Big] \tag{2.5.3}$$

whenever it makes sense. We shall add the superscript V whenever the functional dependence of the probability measure on the potential V needs to be made clear.

Note that if none of the measures $\nu_1, \ldots, \nu_\ell$ is stochastic, then the expectation versus $\mathbb{P}_N$ in (2.5.3) can be omitted.

As explained in Section 1.2.3, the Schwinger–Dyson equations constitute a tower of equations which relate expectation values of functions in many, non necessarily fixed, variables that are integrated versus the empirical measure (2.5.2). More precisely, the Schwinger–Dyson equations at rank k ($k \geq 1$) yield exact relations between various expectation values of a function in k variables and its transforms, this versus the empirical measure. The knowledge of these expectation values, yields an access to the derivatives of the partition function with respect of external parameters. For instance, if $\{V_t\}_t$ is a smooth one parameter family of potentials, then

$$\partial_t \ln Z_N[V_t] = -N^{2+\alpha} \langle \partial_t V_t \rangle_{L_N^{(\lambda)}}^{V_t} . \tag{2.5.4}$$

The exponent V_t appearing in the right-hand side is there so as to emphasise that the expectation value is computed with respect to the probability measure subordinate to the t-dependent potential V_t.

Thus the problem boils down to obtaining a sufficiently precise control on the behaviour in N of the one-point expectation values. This can be achieved on the basis of a careful analysis of the system of Schwinger–Dyson equations associated with the present model. Since this machinery does not simplify much in the $\beta = 1$ case, we do this for general β. The result for some sufficiently regular function H and potentials V satisfying to the general hypothesis, is our Proposition 3.3.6.

In the $\beta = 1$ case, Proposition 3.3.6 reads:

$$-N^{2+\alpha} \langle H \rangle_{L_N^{(\lambda)}}^{V} = -N^{2+\alpha} \int_{a_N}^{b_N} H(\xi) \cdot \mathcal{W}_N\big[V'\big](\xi)\, \mathrm{d}\xi + \frac{1}{2} \mathfrak{I}_{\mathrm{d}}\big[H, V\big] + \mathrm{o}(1) . \tag{2.5.5}$$

and the proof shows that the remainder o(1) is uniform in H and V provided that H is regular enough and that V satisfies to the hypothesis given in (2.2.2)–(2.2.4).

Furthermore, the expansion (2.5.5) involves

$$\mathfrak{I}_{\mathrm{d}}[H, V] = \int_{a_N}^{b_N} \mathcal{W}_N\Big[\partial_\xi \big\{ S\big(N^\alpha(\xi - *)\big) \cdot \mathcal{G}_N\big[H, V\big](\xi, *)\big\}\Big](\xi)\, \mathrm{d}\xi , \tag{2.5.6}$$

with

$$\mathcal{G}_N\big[H, V\big](\xi, \eta) = \frac{\mathcal{W}_N[H](\xi)}{\mathcal{W}_N[V'](\xi)} - \frac{\mathcal{W}_N[H](\eta)}{\mathcal{W}_N[V'](\eta)} . \tag{2.5.7}$$

Note that, in (2.5.6), the $*$ indicates the variable of the function on which the operator $\mathcal{W}_N$ acts. Given sufficiently regular functions H, V, we obtain in Section 6.3 and more precisely in Proposition 6.3.10 the large-N asymptotic behaviour of $\mathfrak{I}_{\mathrm{d}}[H, V]$. We then have all the elements to calculate the large-N asymptotic behaviour of the partition function $Z_N[V]$. For this purpose, we observe that, when $\beta = 1$, the partition function associated to a quadratic potential can be explicitly evaluated as shown in Proposition D.0.19. One can also show (*cf.* Lemma D.0.18) that there exists a unique, up to a constant, quadratic potential $V_{G;N}$ such that its associated equilibrium measure has the same support $[a_N, b_N]$ as the one associated with V. Then $V_t = (1-t)V_{G;N} + tV$ is a one parameter t smooth family of strictly convex potentials, and $\mu^{(N)}_{\mathrm{eq};V_t} = (1-t)\mu^{(N)}_{\mathrm{eq};V_{G;N}} + t\mu^{(N)}_{\mathrm{eq};V}$. Furthermore, if follows from the details of the analysis that led to (2.5.5) that the remainder o(1) will be uniform in

$t \in [0\,;\,1]$. As a consequence, by combining all of the above results and integrating equation (2.5.4) over t, we get that, in the asymptotic regime,

$$\ln\left(\frac{Z_N[V]}{Z_N[V_{G;N}]}\right) = -N^{2+\alpha}\int_0^1 \mathrm{d}t \int \partial_t V_t(\xi)\,\mathrm{d}\mu^{(N)}_{\mathrm{eq};V_t}(\xi)$$
$$+ N^{\alpha}\cdot \beth_0 \cdot \Big(\mathfrak{Q}[V, V_{G;N}](b_N) - \mathfrak{Q}[V, V_{G;N}](a_N)\Big)$$
$$+ \aleph_0 \cdot \Big(\mathfrak{Q}'[V, V_{G;N}](b_N) + \mathfrak{Q}'[V, V_{G;N}](a_N)\Big) + \mathrm{o}(1). \tag{2.5.8}$$

The constants $\beth_0$ and $\aleph_0$ were defined respectively in (2.3.8) and (2.3.9), while $\mathfrak{Q}$ is as given by (2.3.7).

Note also that the first integral can be readily evaluated (integration of rational functions in t) on the asymptotic level by means of Proposition 6.1.6. It produces an expansion into inverse powers of N^{α} and, as such, does not contribute to the constant term unless α is of the form $2/n$ for some integer n. Note that it is this integral that gives rise to the functional $\mathfrak{S}_p[V]$ in (2.3.1). Finally, the answer for the large-N asymptotic behaviour of the partition function $Z_N[V_{G;N}]_{|\beta=1}$ can be found in Proposition D.0.19. As follows from Lemma D.0.18 the quadratic potential $V_{G;N}$ is such that $V_{G;N} - W_{G;N} = \mathrm{O}(N^{-\infty})$ with $W_{G;N}$ as defined in 2.3.2 and where the remainder is uniform on some N-independent relatively compact open neighbourhood of $[a_N\,;\,b_N]$. This allows one to replace $V_{G;N}$ by $W_{G;N}$ in the right hand side of (2.5.8). Also, it is clear form the large-N expansion of the partition function associated with quadratic potentials given in Proposition D.0.19 that $\ln Z_N[V_{G;N}]_{|\beta=1} - \ln Z_N[W_{G;N}]_{|\beta=1} = \mathrm{o}(1)$.

For $\beta \neq 1$, (2.5.5) is modified by the addition of two more terms $\mathfrak{I}^{(2)}_{\mathrm{s};\beta}$ and $\mathfrak{I}_{\mathrm{d};\beta}$. Their large N behaviour can be determined without difficulty—but with some algebra—along the lines of Section 6.1.6 and Section 6.3. Then, to arrive to a final answer for $Z_N[V]_{\beta\neq 1}$ similar to (2.5.8), we would need to compute exactly the partition function for the Gaussian potential $Z_N[V_{G;N}]_{\beta\neq 1}$. We do not know at present how to perform such a calculation. Thus, we would be able to derive the asymptotic behaviour of the partition function at $\beta \neq 1$ up to a universal, *i.e.* not depending on the potential V, function of β. However, since the values $\beta \neq 1$ do not seem to appear in quantum integrable systems, we shall limit ourselves in this book to the result of Proposition 3.4.1 for the case $\beta \neq 1$.

References

1. Anderson, G.W., Guionnet, A., Zeitouni, O.: An Introduction to Random Matrices. Cambridge Studies in Advances Mathematics, vol. 118. Cambridge University Press (2010)

2. Hardy, A., Kuijlaars, A.B.J.: Weakly admissible vector equilibrium problems. J. Approx. Theory **164** 854–868 (2012). arXiv:math.CA/1110.6800
3. Borot, G., Guionnet, A.: Asymptotic expansion of beta matrix models in the multi-cut regime. arXiv:math-ph/1303.1045
4. Borot, G., Guionnet, A., Kozlowski, K.K.: Large-N asymptotic expansion for mean field models with Coulomb gas interaction. Int. Math. Res. Not. (2015). arXiv:math-ph/1312.6664
5. Bourgade, P., Erdös, L., Yau, H.-T.: Edge universality of beta ensembles. Commun. Math. Phys. **332**, 261–353 (2014). arXiv:math.PR/1306.5728
6. Bourgade, P., Erdös, L., Yau, H.-T.: Universality of general β-ensembles. Duke Math. J. **163**, 1127–1190 (2014). arXiv:math.PR/1104.2272
7. Mhaskar, H.N., Saff, E.B.: Where does the sup norm of a weighted polynomial live? (A generalization of incomplete polynomials). Constr. Approx. **1**, 71–91 (1985)
8. Ambjørn, J., Chekhov, L., Kristjansen, C.F., Makeenko, Y.: Matrix model calculations beyond the spherical limit. Nucl. Phys. B **404**, 127–172 (1993); Erratum-ibid. B **449**, 681 (1995). arXiv:hep-th/9302014

Chapter 3
Asymptotic Expansion of $\ln Z_N[V]$—The Schwinger–Dyson Equation Approach

Abstract In the present chapter we develop all the necessary tools to prove the large-N asymptotic expansion for $\ln Z_N[V]$ up to $o(1)$ terms, in the form described in (2.5.5). This asymptotic expansion contains N-dependent functionals of the equilibrium measure whose large-N asymptotic analysis will be carried out in Sections 6.1–6.3. We shall first obtain some *a priori* bounds on the fluctuations of linear statistics around their means computed *versus* the N-dependent equilibrium measure $\mu_{\mathrm{eq}}^{(N)}$. In other words, we consider observables given by integration against products of the centred measure: $\mathcal{L}_N^{(\boldsymbol{\lambda})} = L_N^{(\boldsymbol{\lambda})} - \mu_{\mathrm{eq}}^{(N)}$. Then we shall build on a bootstrap approach to the Schwinger–Dyson equations so as to improve these *a priori* bounds. We shall use these improved bounds so as to identify the leading and sub-leading terms in the Schwinger–Dyson equations what, eventually, leads to an analogue, at $\beta \neq 1$, of the representation (2.5.5) which will be given in Proposition 3.3.6. Finally, upon integrating the relation (2.5.4) so as to to interpolate the partition function between a Gaussian and a general potential, we will get the N-dependent large-N asymptotic expansion of $\ln Z_N[V]$ in Proposition 3.4.1.

3.1 A Priori Estimates for the Fluctuations Around $\mu_{\mathrm{eq}}^{(N)}$

For simplification, we use the notation:

$$s_N(\xi) = \frac{\beta}{2N^{\alpha}} \ln\Big[\sinh\big(\pi\omega_1 N^{\alpha}\xi\big)\sinh\big(\pi\omega_2 N^{\alpha}\xi\big)\Big] \tag{3.1.1}$$

for the two-body interaction kernel. This kernel provides a natural way of comparing two probability measures:

Definition 3.1.1 If $\mu, \nu \in \mathcal{M}_1(\mathbb{R})$, we set:

$$\mathfrak{D}^2\big[\mu,\nu\big] \equiv -\int s_N(\xi-\eta)\,\mathrm{d}(\mu-\nu)(\xi)\,\mathrm{d}(\mu-\nu)(\eta)\,, \tag{3.1.2}$$

G. Borot et al., *Asymptotic Expansion of a Partition Function Related to the Sinh-model*, Mathematical Physics Studies,
DOI 10.1007/978-3-319-33379-3_3

with s_N as given in (3.1.1). $\mathfrak{D}^2\big[\mu,\nu\big]$ is a well-defined number in $\mathbb{R}\cup\{+\infty\}$.
The notation is justified by the property $\mathfrak{D}^2 \geq 0$ following from:

Lemma 3.1.2 *We have the representation:*

$$\mathfrak{D}^2\big[\mu,\nu\big] \;=\; \int \bigg\{ \frac{\pi\beta}{2N^{\alpha}\varphi} \sum_{p=1}^{2} \mathrm{cotanh}\Big[\frac{\varphi}{2\omega_p N^{\alpha}}\Big] \bigg\} \cdot \big|\mathcal{F}[\mu-\nu](\varphi)\big|^2 \cdot \frac{\mathrm{d}\varphi}{2\pi}\,, \qquad (3.1.3)$$

where $\mathcal{F}[\mu](\xi)$ is the Fourier transform of the measure μ.

Proof The claim follows in virtue of the formula

$$\mathcal{F}\big[f_t\big](\varphi) \;=\; -\frac{\pi}{\varphi}\bigg(\mathrm{cotanh}\Big(\frac{\pi\varphi}{2t}\Big) - \frac{2t}{\pi\varphi}\bigg)$$

with $f_t(x) \;=\; \ln|\sinh(tx)| - t|x| + \ln 2$. ■

Definition 3.1.3 The classical positions x_i^N for the measure $\mu_{\mathrm{eq}}^{(N)}$ are defined by

$$\frac{i}{N} \;=\; \int_{-\infty}^{x_i^N} \mathrm{d}\mu_{\mathrm{eq}}^{(N)}(y) \qquad \text{for} \qquad i \in [\![\, 1\,;\, N \,]\!] \qquad \text{and} \qquad x_0^N = a_N \quad , \quad x_N^N = b_N\,. \qquad (3.1.4)$$

Our first task is to derive a lower bound for the partition function (2.2.1), by restricting to configurations of points close to their classical positions:

Lemma 3.1.4 $Z_N[V] \;\geq\; \exp\Big\{ - N^{2+\alpha}\mathcal{E}_N\big[\mu_{\mathrm{eq}}^{(N)}\big] \;+\; \mathrm{O}\big(N^{1+\alpha}\big)\Big\}\,.$

We stress on this occasion that using the N-dependent rate function $\mathcal{E}_N$ allows the gain of a factor $1/N$ in the remainder with respect to the leading term, while using $\mathcal{E}_\infty$ would lead to a weaker estimate $\mathrm{O}(N^2)$ for the remainder in the lower bound. This is of particular importance to simplify the analysis of Schwinger–Dyson equations that will follow.

Proof It follows from the local expressions obtained in Chapter 5 that $\mu_{\mathrm{eq}}^{(N)}$ is continuous with respect to Lebesgue measure with density bounded by a constant M independent of N, as shown in (2.4.14). This ensures that

$$\big|x_{i+1}^N - x_i^N\big| \geq \frac{1}{MN}\,, \qquad i \in [\![\, 0\,;\, N-1 \,]\!]\,. \qquad (3.1.5)$$

We obtain our lower bound by keeping only configurations in

$$\Omega \;=\; \big\{\boldsymbol{\lambda} \in \mathbb{R}^N \;:\; \sup_a |\lambda_a - x_a^N| \;\leq\; \frac{1}{4MN}\big\}\,.$$

Let σ_ϵ be some N-independent ϵ-neighbourhood of $[a_N\,;\,b_N]$. Since $V \in \mathcal{C}^1(\mathbb{R})$, it follows that

$$\big|V(\lambda_a) - V(x_a^N)\big| \;\le\; \frac{||V'||_{L^\infty(\sigma_\epsilon)}}{4MN} \quad viz. \quad -\,V(x_a^N) - \frac{||V'||_{L^\infty(\sigma_\epsilon)}}{4MN} \;\le\; -V(\lambda_a) \tag{3.1.6}$$

for $a \in [\![\, 1\,;\, N \,]\!]$ and for any $\boldsymbol{\lambda} \in \Omega$. Thus, upon a re-centring at x_a^N of the integration with respect to λ_a, we get

$$\begin{aligned} Z_N[V] \ge & \prod_{a=1}^{N}\Big\{\mathrm{e}^{-N^{1+\alpha}V(x_a^N)}\Big\}\cdot \mathrm{e}^{-\frac{N^{1+\alpha}}{4M}||V'||_{L^\infty(\sigma_\epsilon)}} \\ & \times \int\limits_{[-1/(4MN),1/(4MN)]^N} \mathrm{d}^N\nu \cdot \prod_{a<b}^{N}\Big\{\prod_{p=1}^{2}\sinh\big[\pi\omega_p N^\alpha(\nu_a - \nu_b + x_a^N - x_b^N)\big]\Big\}^\beta \\ \ge & \prod_{a=1}^{N}\Big\{\mathrm{e}^{-N^{1+\alpha}V(x_a^N)}\Big\}\cdot \mathrm{e}^{-\frac{N^{1+\alpha}}{4M}||V'||_{L^\infty(\sigma_\epsilon)}} \times \prod_{a<b}^{N}\mathrm{e}^{2N^\alpha s_N(x_b^N - x_a^N)} \\ & \times \int\limits_{\nu_1<\cdots<\nu_N} \mathrm{d}^N\nu \prod_{a=1}^{N}\Big\{\mathbf{1}_{|\xi|<\frac{1}{4MN}}(\nu_a)\Big\}\,. \end{aligned} \tag{3.1.7}$$

We remind that s_N has been defined in (3.1.1). The second line is obtained by keeping only the configurations where $i \mapsto \nu_i$ is increasing, and then using that sinh is an increasing function. Finally:

$$Z_N[V] \;\ge\; \prod_{a=1}^{N}\Big\{\mathrm{e}^{-N^{1+\alpha}V(x_a^N)}\Big\}\cdot \mathrm{e}^{-\frac{N^{1+\alpha}}{4M}||V'||_{L^\infty(\sigma_\epsilon)}}\cdot \prod_{a<b}^{N}\mathrm{e}^{2N^\alpha s_N(x_b^N - x_a^N)}\cdot \frac{1}{N!}\Big(\frac{1}{4NM}\Big)^N\,. \tag{3.1.8}$$

We rewrite the first product involving the potential by comparison between the Riemann sum and the integral:

$$\frac{1}{N}\sum_{a=1}^{N} V(x_a^N) \;=\; \int\limits_{\mathbb{R}} V(\xi)\,\mathrm{d}\mu_{\text{eq}}^{(N)}(\xi) \;+\; \delta_N, \qquad |\delta_N| \le \frac{||V'||_{L^\infty(\sigma_\epsilon)}}{N}\cdot(b_N - a_N)\,. \tag{3.1.9}$$

it thus remains to bound from below the β-exponent part. Using that s_N is increasing on $\mathbb{R}^+$, we get:

$$\int\limits_{x<y} s_N(y-x)\,\mathrm{d}\mu_{\mathrm{eq}}^{(N)}(x)\,\mathrm{d}\mu_{\mathrm{eq}}^{(N)}(y)$$

$$= \sum_{a,b=0}^{N-1} \int\limits_{x_a^N}^{x_{a+1}^N} \int\limits_{x_b^N}^{x_{b+1}^N} \mathbf{1}_{x<y}(x,y) s_N(y-x)\,\mathrm{d}\mu_{\mathrm{eq}}(x)\,\mathrm{d}\mu_{\mathrm{eq}}(y)$$

$$\leq \frac{1}{N^2}\sum_{a=0}^{N-1}\sum_{b=a+1}^{N-1} s_N(x_{b+1}^N - x_a^N) + \sum_{a=0}^{N-1} s_N(x_{a+1}^N - x_a^N)\cdot\frac{1}{2N^2}\,. \tag{3.1.10}$$

The first sum can be recast as

$$\sum_{a=0}^{N-1}\sum_{b=a+2}^{N} s_N(x_b^N - x_a^N) = \sum_{a=1}^{N-1}\sum_{b=a+1}^{N} s_N(x_b^N - x_a^N) + \sum_{b=1}^{N} s_N(x_b^N - x_0^N)$$
$$- \sum_{a=0}^{N-1} s_N(x_{a+1}^N - x_a^N)\,. \tag{3.1.11}$$

It follows from (3.1.5) and from $|x_a^N - x_b^N| < |b_N - a_N| < C$ for some $C > 0$ independent of N, that:

$$\max_{0\leq a\leq N-1} |s_N(x_{a+1}^N - x_a^N)|$$
$$= N^{-\alpha}\mathrm{O}\big(\ln N + N^{\alpha}\big) \quad \text{and} \quad \max_{1\leq a\leq N} |s_N(x_a^N - x_0^N)| = \mathrm{O}\big(1\big)\,. \tag{3.1.12}$$

Hence, it follows that

$$N^2 \int\limits_{x<y} s_N(y-x)\,\mathrm{d}\mu_{\mathrm{eq}}^{(N)}(x)\,\mathrm{d}\mu_{\mathrm{eq}}^{(N)}(y) \leq \mathrm{O}\big(N\big) + \sum_{a<b}^{N} s_N(x_b^N - x_a^N)\,, \tag{3.1.13}$$

thus leading to the claim. ■

We now estimate the fluctuations of linear statistics by using an idea introduced in [1].

Definition 3.1.5 Given a configuration of points $\lambda_1 \leq \cdots \leq \lambda_N$, we build a sequence $\widetilde{\lambda}_1 < \cdots < \widetilde{\lambda}_N$ defined as

$$\widetilde{\lambda}_1 = \lambda_1 \quad \textit{and} \quad \widetilde{\lambda}_{k+1} = \widetilde{\lambda}_k + \max\big(\lambda_{k+1} - \lambda_k, \mathrm{e}^{-(\ln N)^2}\big)\,. \tag{3.1.14}$$

Further, for any $\boldsymbol{\lambda} \in \mathbb{R}^N$, we associate a vector $\widetilde{\boldsymbol{\lambda}} \in \mathbb{R}^N$ by ordering the λ's with a permutation σ, apply the previous construction to obtain a N-uple $\widetilde{\boldsymbol{\lambda}}$, and put them in original order with the permutation σ^{-1}. The corresponding empirical measure is:

$$L_N^{(\widetilde{\boldsymbol{\lambda}})} = \frac{1}{N} \sum_{a=1}^{N} \delta_{\widetilde{\lambda}_a}$$

and we denote $L_{N;u}^{(\widetilde{\boldsymbol{\lambda}})}$ the convolution of $L_N^{(\widetilde{\boldsymbol{\lambda}})}$ with the uniform probability measure on $[0\,;\mathrm{e}^{-(\ln N)^2}/N]$.

The new configuration has been constructed such that, for $\ell \neq k$,

$$\big|\widetilde{\lambda}_k - \widetilde{\lambda}_\ell\big| \geq \mathrm{e}^{-(\ln N)^2}, \quad \big|\lambda_k - \lambda_\ell\big| \leq \big|\widetilde{\lambda}_k - \widetilde{\lambda}_\ell\big| \;\; \text{and} \;\; \big|\lambda_k - \widetilde{\lambda}_k\big| \leq (k-1)\cdot \mathrm{e}^{-(\ln N)^2} . \tag{3.1.15}$$

The advantage of working with $L_{N;u}^{(\widetilde{\boldsymbol{\lambda}})}$ is that it is Lebesgue continuous; as such it can appear in the argument of $\mathcal{E}_N$ or $\mathfrak{D}^2$ and yield finite results. The scale of regularisation $e^{-(\ln N)^2}$ is somewhat arbitrary, but in any case negligible compared to $N^{-\alpha}$.

We introduce the effective potential associated to the N-dependent equilibrium measure:

$$V_{N;\text{eff}}(\xi) = V(\xi) - 2\int s_N(\xi-\eta)\,\mathrm{d}\mu_{\text{eq}}^{(N)}(\eta) - C_{\text{eq}}^{(N)} . \tag{3.1.16}$$

By the characterisation of the equilibrium measure (Theorem 2.4.2), $V_{N;\text{eff}} = 0$ in the support $[a_N\,; b_N]$, while $V_{N;\text{eff}} > 0$ outside $[a_N\,; b_N]$.

Proposition 3.1.6 *Assume that*

- *the partition function $Z_N[V]$ satisfies a lower-bound of the form*

$$Z_N[V] \geq \exp\Big\{ - N^{2+\alpha}\mathcal{E}_N\big[\mu_{\text{eq}}^{(N)}\big] + \delta_N \Big\}, \qquad \delta_N = \mathrm{o}\big(N^{2+\alpha}\big)\,; \tag{3.1.17}$$

- *the potential is sub-exponential, viz. there exists $\epsilon > 0$ and $C_V > 0$ such that*

$$\forall x \in \mathbb{R}, \qquad \sup_{t\in[0\,;\epsilon]} \big|V'(x+t)\big| \leq C_V\Big(\big|V(x)\big| + 1\Big) . \tag{3.1.18}$$

Then, given any $0 < \eta < 1$, we have for all $\boldsymbol{\lambda} \in \mathbb{R}^N$ that

$$p_N(\boldsymbol{\lambda}) \leq \exp\Big\{ - N^{2+\alpha}\mathfrak{D}^2\big[L_{N;u}^{(\widetilde{\boldsymbol{\lambda}})}, \mu_{\text{eq}}^{(N)}\big] - \delta_N - N^{2+\alpha}(1-\eta)\int_{\mathbb{R}} V_{N;\text{eff}}(\xi)\,\mathrm{d}L_{N;u}^{(\widetilde{\boldsymbol{\lambda}})}(\xi) + \mathrm{O}\big(N(\ln N)^2\big)\Big\} . \tag{3.1.19}$$

The effective potential $V_{N;\mathrm{eff}}$ *has been defined in* (3.1.16) *while* $\mathfrak{D}^2[\mu,\nu]$ *is as given in (3.1.1).*

Proof The partition function takes the form:

$$Z_N[V] = \int_{\mathbb{R}^N} \mathrm{d}^N\lambda \, \exp\Big\{ - N^{2+\alpha}\Big(\int V(x)\,\mathrm{d}L_N^{(\lambda)}(x) - \Sigma_{\mathrm{diag}}[L_N^{(\lambda)}]\Big)\Big\},$$

$$\Sigma_{\mathrm{diag}}[\mu] \; = \; \int_{x\neq y} s_N(x-y)\,\mathrm{d}\mu(x)\mathrm{d}\mu(y)\,.$$

where s_N defined in (3.1.1). We are going to estimate the cost of replacing $L_N^{(\lambda)}$ by $L_{N;u}^{(\tilde{\lambda})}$ in the above integration. We start with the term involving the potential. Since we assumed V sub-exponential, we have:

$$\begin{aligned}
&\Big| \int V(x)\,\mathrm{d}L_N^{(\lambda)}(x) - \int V(x)\,\mathrm{d}L_N^{(\tilde{\lambda})}(x) \Big| \\
&\leq \frac{1}{N}\sum_{a=1}^{N} \frac{(a-1)}{\mathrm{e}^{(\ln N)^2}} \cdot \sup\Big\{ |V'(\tilde{\lambda}_a + t)| \;:\; t \in \Big[0;\frac{(a-1)}{\mathrm{e}^{(\ln N)^2}}\Big]\Big\} \\
&\leq \frac{N C_V}{\mathrm{e}^{(\ln N)^2}} \Big(\int |V(x)|\,\mathrm{d}L_N^{(\tilde{\lambda})}(x) \,+\, 1\Big)\,.
\end{aligned} \tag{3.1.20}$$

Further, since $V(x) \to +\infty$ when $|x| \to \infty$, there exists $C'_{\mathrm{eff}} > 0$ such that

$$\forall x \in \mathbb{R}, \qquad C'_{\mathrm{eff}}\big(1 + V_{N;\mathrm{eff}}(x)\big) \;\geq\; C_V\big(|V(x)| + 1\big)\,. \tag{3.1.21}$$

As a consequence,

$$\begin{aligned}
\exp\Big\{ - N^{2+\alpha} \int V(x)\,\mathrm{d}L_N^{(\lambda)}(x)\Big\} \;\leq\; \exp\Big\{ & \frac{N^{3+\alpha}\,C'_{\mathrm{eff}}}{e^{(\ln N)^2}}\Big[1 + \int V_{N;\mathrm{eff}}(x)\,\mathrm{d}L_N^{(\tilde{\lambda})}(x)\Big] \\
& - N^{2+\alpha}\int V(x)\mathrm{d}L_N^{(\tilde{\lambda})}(x)\Big\}\,.
\end{aligned} \tag{3.1.22}$$

Now, let us consider the term involving the sinh interaction. Since s_N is increasing on $\mathbb{R}^+$ and the spacings between $\tilde{\lambda}_a$'s are larger than those between the λ_a's, it follows that $\Sigma_{\mathrm{diag}}\big[L_N^{(\lambda)}\big] \leq \Sigma_{\mathrm{diag}}\big[L_N^{(\tilde{\lambda})}\big]$. Furthermore, we have:

$$\Sigma_{\mathrm{diag}}\big[L_N^{(\widetilde{\lambda})}\big] - \Sigma_{\mathrm{diag}}\big[L_{N;u}^{(\widetilde{\lambda})}\big]$$
$$= \int\limits_{x\neq y} \mathrm{d}L_N^{(\widetilde{\lambda})}(x)\,\mathrm{d}L_N^{(\widetilde{\lambda})}(y) \int\limits_{[0;1]^2} \mathrm{d}^2u \Big\{ s_N(x-y)$$
$$- s_N\big(x - y + N^{-1}e^{-(\ln N)^2}(u_1-u_2)\big)\Big\}$$
$$- \frac{1}{N}\int\limits_{[0;1]^2} \mathrm{d}^2u\, s_N\big[N^{-1}e^{-(\ln N)^2}(u_1-u_2)\big]\,. \qquad (3.1.23)$$

When N is large enough, we can use the Lipschitz behaviour of s_N on $[e^{-(\ln N)^2}/2;+\infty[$ for the first term. Indeed:

$$|s_N'(x)| = \sum_{p=1}^{2} \frac{\beta\pi\omega_p}{2}\,\mathrm{cotanh}[\pi\omega_p N^\alpha|x|] \leq c' N^{-\alpha} e^{(\ln N)^2} \qquad (3.1.24)$$

for some $c' > 0$. Besides, we exploit that s_N is increasing to bound the second term. This leads to:

$$\big|\Sigma_{\mathrm{diag}}\big[L_N^{(\widetilde{\lambda})}\big] - \Sigma_{\mathrm{diag}}\big[L_{N;u}^{(\widetilde{\lambda})}\big]\big| \leq CN^{-\alpha-1} + C' N^{-(1+\alpha)}(\ln N)^2\,. \qquad (3.1.25)$$

Since the measure $L_{N;u}^{(\widetilde{\lambda})}$ is continuous with respect to Lebesgue, it is not any more necessary to take care of the diagonal singularity in s_N, and we obtain:

$$\exp\bigg\{ - N^{2+\alpha}\Big(\int\limits_{\mathbb{R}} V(x)\mathrm{d}L_N^{(\lambda)}(x) - \Sigma_{\mathrm{diag}}[L_N^{(\lambda)}]\Big)\bigg\}$$
$$\leq \exp\Big\{ - N^{2+\alpha}\mathcal{E}_N[L_{N;u}^{(\widetilde{\lambda})}] + \mathrm{O}\big(N(\ln N)^2\big)\Big\}$$
$$\times \exp\Big\{ e^{-(\ln N)^2} N^{3+\alpha} C'_{\mathrm{eff}} \int V_{N;\mathrm{eff}}(x)\mathrm{d}L_{N;u}^{(\widetilde{\lambda})}(x)\Big\}\,. \qquad (3.1.26)$$

Since $\mu_{\mathrm{eq}}^{(N)}$ is also continuous with respect to Lebesgue, $\mathcal{E}_N[\mu_{\mathrm{eq}}^{(N)}]$ is finite and we can expand the first term around $\mu_{\mathrm{eq}}^{(N)}$:

$$\mathcal{E}_N\big[L_{N;u}^{(\widetilde{\lambda})}\big] = \mathcal{E}_N\big[\mu_{\mathrm{eq}}^{(N)}\big] + \mathfrak{D}^2\big[L_{N;u}^{(\widetilde{\lambda})}, \mu_{\mathrm{eq}}^{(N)}\big]$$
$$+ \int\limits_{\mathbb{R}} \mathrm{d}(L_{N;u}^{(\widetilde{\lambda})} - \mu_{\mathrm{eq}}^{(N)})(x)\Big\{V(x) - 2\int\limits_{\mathbb{R}} \mathrm{d}\mu_{\mathrm{eq}}^{(N)}(y)\, s_N(x-y)\Big\}\,.$$

We recognize in the last integral $V_{N;\mathrm{eff}}(x) + C_{\mathrm{eq}}^{(N)}$ integrated against a measure of mass 0. So, we can omit the constant C_{eq}, and since $V_{N;\mathrm{eff}} = 0$ on the support of $\mu_{\mathrm{eq}}^{(N)}$, we actually find:

$$\mathcal{E}_N\big[L^{(\widetilde{\lambda})}_{N;u}\big] = \mathcal{E}_N\big[\mu^{(N)}_{\mathrm{eq}}\big] + \mathfrak{D}^2\big[L^{(\widetilde{\lambda})}_{N;u}, \mu^{(N)}_{\mathrm{eq}}\big] + \int_{\mathbb{R}} V_{N;\mathrm{eff}}(x)\,\mathrm{d}L^{(\widetilde{\lambda})}_{N;u}(x)$$

If we plug this relation in (3.1.26), we obtain a similar bound but now with $V_{N;\mathrm{eff}}$ having the prefactor $N^{2+\alpha} - e^{-(\ln N)^2} N^{3+\alpha} C'_{\mathrm{eff}} \leq (1-\eta)N^{2+\alpha}$, this for any $0 < \eta < 1$, provided that N is large enough. ■

In order to bound the one and multi-point expectation values and in particular the various terms arising in the Schwinger–Dyson equations, we introduce the exponential regularisation of a function. This regularisation allows one to deal with functions that are unbounded at infinity but whose expectation values are still well defined.

Definition 3.1.7 Given a function f in n variables, its exponential regularisation with growth κ is defined by

$$\mathcal{K}_\kappa[f](\xi_1, \ldots, \xi_n) = \Big\{\prod_{a=1}^{n} \mathrm{e}^{-\kappa V(\xi_a)}\Big\} \cdot f(\xi_1, \ldots, \xi_n)\,. \tag{3.1.27}$$

Definition 3.1.8 We define the centred empirical measure as:

$$\mathcal{L}^{(\lambda)}_N = L^{(\lambda)}_N - \mu^{(N)}_{\mathrm{eq}}\,. \tag{3.1.28}$$

Prior to establishing the simplest *a priori* bounds on the multi-point expectation values $\langle f\rangle_{\bigotimes_1^n \mathcal{L}^{(\lambda)}_N}$, we need to establish a convenient decomposition thereof. The latter is written in such a way that the leading in N behaviour comes from the part involving a restriction of f to a compactly supported function.

Lemma 3.1.9 *There exists $t > 0$ and a functional $\mathcal{A}^{(n)}_N$ on the space of functions f such that $\mathcal{K}_\kappa[f] \in W^\infty_0(\mathbb{R}^n)$ for some $\kappa > 0$ satisfying*

$$\sup_{N\in\mathbb{N}} \Big\{\mathrm{supp}\big[\mu^{(N)}_{\mathrm{eq}}\big]\Big\} \subset [-t/2\,;t/2] \quad \text{and} \quad \Big|\mathcal{A}^{(n)}_N[f]\Big| \leq c_n \mathrm{e}^{-cN^{1+\alpha}} \cdot ||\mathcal{K}_\kappa[f]||_{W^\infty_0(\mathbb{R}^n)} \tag{3.1.29}$$

and such that

$$\langle f\rangle_{\bigotimes_1^n \mathcal{L}^{(\lambda)}_N} = \langle f_{|\mathrm{c}}\rangle_{\bigotimes_1^n \mathcal{L}^{(\lambda)}_N} + \mathcal{A}^{(n)}_N[f]\,. \tag{3.1.30}$$

In the above decomposition,

$$f_{|\mathrm{c}}(\xi_1, \ldots, \xi_n) = f(\xi_1, \ldots, \xi_n) \cdot \prod_{a=1}^{n} \phi(\xi_a) \tag{3.1.31}$$

where $\phi \in \mathcal{C}^\infty_{\mathrm{c}}(\mathbb{R})$ is such that

$$0 \leq \phi \leq 1\,, \quad \phi_{|[-t\,;t]} = 1 \ \textit{and} \ \mathrm{supp}[\phi] \subset [-(t+1)\,;t+1]\,. \tag{3.1.32}$$

Proof We first claim that the constant $C_{\text{eq}}^{(N)}$ arising in the minimisation problem for the equilibrium measure (2.4.8) is bounded in N. Indeed, it follows from (2.4.8) that

$$C_{\text{eq}}^{(N)} = \int_{a_N}^{b_N} V(\xi)\,\mathrm{d}\mu_{\text{eq}}^{(N)}(\xi) - \frac{\beta}{N^{\alpha}} \int_{[a_N\,;b_N]^2} \ln\Big\{\prod_{p=1}^{2} \sinh[\pi\omega_p N^{\alpha}(\xi-\eta)]\Big\}\,\mathrm{d}\mu_{\text{eq}}^{(N)}(\xi)\mathrm{d}\mu_{\text{eq}}^{(N)}(\eta)\,. \tag{3.1.33}$$

Therefore, we have:

$$\big|C_{\text{eq}}^{(N)}\big| \le ||V||_{L^{\infty}([a_N\,;b_N])} + \widetilde{C}\,||V''||^2_{L^{\infty}([a_N\,;b_N])} \int_{[a_N\,;b_N]^2} \frac{1}{N^{\alpha}}\Big|\ln\Big\{\prod_{p=1}^{2}\sinh[\pi\omega_p N^{\alpha}(\xi-\eta)]\Big\}\Big|\,\mathrm{d}\xi\mathrm{d}\eta\,, \tag{3.1.34}$$

where we have used that $\mu_{\text{eq}}^{(N)}$ is a probability measure and that its density is bounded by (2.4.14). The double integral remaining in (3.1.34) can be bounded by an N-independent constant. Such bounds are obtained by using that the function

$$g_N(\xi) = \frac{1}{N^{\alpha}}\Big|\ln\Big\{\prod_{a=1}^{2}\sinh[\pi\omega_a N^{\alpha}\xi]\Big\}\Big| - \pi(\omega_1+\omega_2)|\xi| \tag{3.1.35}$$

approaches 0 point-wise in $\xi \in [a_N - b_N\,; b_N - a_N]\setminus\{0\}$ and is bounded as $|g_N(\xi)| \le C\big(1 + \big|\ln|\xi|\big|\big)$. Since the endpoints a_N and b_N are bounded in N in virtue of (2.4.11), we can apply the dominated convergence theorem to $(\xi,\eta)\mapsto g_N(\xi-\eta)$ on $[a_N\,; b_N]^2$.

The finiteness in N of $C_{\text{eq}}^{(N)}$ along with the confinement hypothesis (2.2.2) on the potential implies the existence of $t > 0$ independent of N such that:

$$\forall \xi \in \mathbb{R}\setminus[-t,t], \qquad V_{N;\text{eff}}(\xi) \ge \frac{V(\xi)}{2} \ge \frac{|\xi|}{2}\,. \tag{3.1.36}$$

where the effective potential is defined by (3.1.16). In virtue of (2.4.11), one can always choose t such that it also holds $\text{supp}[\mu_{\text{eq}}^{(N)}] \subset [-t/2\,; t/2]$.

Since

$$\big\langle f\big\rangle_{\otimes_1^n \mathcal{L}_N^{(\lambda)}} = \big\langle f_{\text{sym}}\big\rangle_{\otimes_1^n \mathcal{L}_N^{(\lambda)}} \quad \text{with} \quad f_{\text{sym}}(\xi_1,\dots,\xi_n) = \frac{1}{n!}\sum_{\sigma\in\mathfrak{S}_n} f\big(\xi_{\sigma(1)},\dots,\xi_{\sigma(n)}\big) \tag{3.1.37}$$

we may assume that f is a completely symmetric function. Then, one gets

$$\big\langle f \big\rangle_{\otimes_1^n \mathcal{L}_N^{(\lambda)}} = \big\langle f_{|c} \big\rangle_{\otimes_1^n \mathcal{L}_N^{(\lambda)}} + \mathcal{A}_N^{(n)}[f] \quad , \quad \mathcal{A}_N^{(n)}[f] = \sum_{p=1}^{n} \binom{n}{p} \mathcal{A}_{N;p}^{(n)}\big[f_{\mathrm{sym}}\big] \tag{3.1.38}$$

where

$$\mathcal{A}_{N;p}^{(n)}\big[f_{\mathrm{sym}}\big] = \mathbb{P}_N \Bigg[\int\limits_{[-t\,;t]^{\mathrm{c}}} \prod_{a=1}^{p} \mathrm{d}L_N^{(\lambda)}(\xi_a) \int\limits_{-(t+1)}^{(t+1)} \prod_{a=p+1}^{n} \mathrm{d}\mathcal{L}_N^{(\lambda)}(\xi_a)$$
$$\times \prod_{a=1}^{p} \Big(1 - \phi(\xi_a)\Big) \prod_{a=p+1}^{n} \phi(\xi_a) f_{\mathrm{sym}}(\xi_1, \dots, \xi_n) \Bigg] \,. \tag{3.1.39}$$

Note that, in the intermediate steps, we have used that $\mathrm{supp}\big[\mu_{\mathrm{eq}}^{(N)}\big] \cap [-t\,;t]^{\mathrm{c}} = \emptyset$. Hence, one gets the bound

$$\Big|\mathcal{A}_{N;p}^{(n)}\big[f_{\mathrm{sym}}\big]\Big| \leq ||\mathcal{K}_{\kappa}[f]||_{W_0^{\infty}(\mathbb{R}^n)} \cdot 2^p \cdot ||\mathrm{e}^{\kappa V}||^{n-p}_{L^{\infty}([-(t+1)\,;t+1])} \cdot A_p \tag{3.1.40}$$

with

$$A_p = \frac{1}{N^p} \int\limits_{\mathbb{R}^N} p_N(\boldsymbol{\lambda}) \sum_{\substack{a_1,\dots,a_p \\ =1}}^{N} \prod_{\ell=1}^{p} \Big\{ \mathrm{e}^{\kappa V(\lambda_{a_\ell})} \cdot \mathbf{1}_{[-t;t]^{\mathrm{c}}}\big(\lambda_{a_\ell}\big) \Big\} \cdot \mathrm{d}^N \boldsymbol{\lambda} \,. \tag{3.1.41}$$

Observe that given any symmetric function g in p-variables, one has the decomposition

$$\sum_{a_1,\dots,a_p=1}^{N} g\big(\lambda_{a_1}, \dots, \lambda_{a_p}\big) = \sum_{\ell=1}^{p} \sum_{\substack{r_1,\dots,r_\ell \geq 1 \\ r_1+\dots+r_\ell = p}} C_{r_1,\dots,r_\ell}^{(\ell,p)}$$
$$\times \sum_{\substack{1 \leq b_1,\dots,b_\ell \leq N \\ \text{pairwise disjoint}}} g\Big(\underbrace{\lambda_{b_1}, \dots, \lambda_{b_1}}_{r_1}, \dots, \underbrace{\lambda_{b_\ell}, \dots, \lambda_{b_\ell}}_{r_\ell} \Big) \tag{3.1.42}$$

where $C_{r_1,\dots,r_\ell}^{(\ell,p)} > 0$ are purely combinatorial coefficients. The latter implies that, for some p-dependent constant C_p:

$$\sum_{\ell=1}^{p} \sum_{\substack{r_1,\dots,r_\ell \geq 1 \\ r_1+\dots+r_\ell = p}} C_{r_1,\dots,r_\ell}^{(\ell,p)} \leq C_p \,. \tag{3.1.43}$$

As a consequence, we get

$$A_p = \frac{1}{N^p} \sum_{\ell=1}^{p} \sum_{\substack{r_1,\cdots,r_\ell \geq 1 \\ r_1+\cdots+r_\ell=p}} C^{(\ell,p)}_{r_1,\ldots,r_\ell} \int_{\mathbb{R}^N} p_N(\boldsymbol{\lambda}) \sum_{\substack{b_1,\ldots,b_\ell \geq 1 \\ \text{pairwise disjoint}}} \prod_{s=1}^{\ell} \Big\{ \mathrm{e}^{\kappa r_s V(\lambda_{b_s})} \cdot \mathbf{1}_{|x|>t}(\lambda_{b_s}) \Big\} \cdot \mathrm{d}^N \boldsymbol{\lambda}$$

$$\leq \sum_{\ell=1}^{p} \frac{N \cdots (N-\ell+1)}{N^p} \sum_{\substack{r_1+\cdots+r_\ell \\ =p}} C^{(\ell,p)}_{r_1,\ldots,r_\ell}$$

$$\times \int_{([-t\,;t]^{\mathrm{c}})^\ell} \prod_{a=1}^{\ell} \mathrm{d}\lambda_a \int_{\mathbb{R}^{N-\ell}} \prod_{a=\ell+1}^{N} \mathrm{d}\lambda_a \cdot p_N(\boldsymbol{\lambda}) \prod_{a=1}^{\ell} \Big\{ \mathrm{e}^{\kappa r_s V(\lambda_a)} \Big\} . \tag{3.1.44}$$

It follows from (3.1.19) with $\eta = 1/2$ given in Proposition 3.1.6 that:

$$\big| p_N(\boldsymbol{\lambda}) \, \mathbf{1}_{\Omega}(\boldsymbol{\lambda}) \big| \leq \prod_{a=1}^{N} \mathrm{e}^{-\frac{1}{2} N^{1+\alpha} V_{N;\mathrm{eff}}(\lambda_a)} \cdot \exp\Big\{ - N^{2+\alpha} \inf_{\boldsymbol{\lambda} \in \Omega} \mathfrak{D}^2\big[L^{(\widetilde{\boldsymbol{\lambda}})}_{N;u}, \mu^{(N)}_{\mathrm{eq}} \big] \Big\} . \tag{3.1.45}$$

This bound leads to:

$$A_p \leq C_p \cdot \max_{\ell=1,\ldots,p} \Big\{ \int_{[-t\,;t]^c} \mathrm{e}^{-\frac{1}{2} N^{1+\alpha} V_{N;\mathrm{eff}}(\xi) + \kappa V(\xi)} \mathrm{d}\xi \Big\}^{\ell} \cdot \max_{\ell=1,\ldots,p} \Big\{ \int_{\mathbb{R}} \mathrm{e}^{-\frac{1}{2} N^{1+\alpha} V_{N;\mathrm{eff}}(\xi)} \mathrm{d}\xi \Big\}^{N-\ell} . \tag{3.1.46}$$

Further, in virtue of (3.1.36) we have, for N large enough,

$$\Big| \int_{[-t\,;t]^c} \mathrm{e}^{-\frac{1}{2} N^{1+\alpha} V_{N;\mathrm{eff}}(\xi) + \kappa V(\xi)} \mathrm{d}\xi \Big| \leq \Big| \int_{[-t\,;t]^c} \mathrm{e}^{-\frac{1}{8} N^{1+\alpha} V(\xi)} \mathrm{d}\xi \Big|$$

$$\leq \Big| \int_{[-t\,;t]^c} \mathrm{e}^{-\frac{1}{8} N^{1+\alpha} |\xi|} \mathrm{d}\xi \Big| = \mathrm{O}\big(\mathrm{e}^{-c N^{1+\alpha}} \big) . \tag{3.1.47}$$

The integral over $\mathbb{R}$ in (3.1.46) is bounded uniformly by a constant A, since $V_{N;\mathrm{eff}} \geq 0$, and $V_{N;\mathrm{eff}}$ grows at least linearly at infinity. All in all, for any $p \in [\![1\,;\,n]\!]$, (3.1.46) is bounded by $C' A^N \mathrm{e}^{-cN^{1+\alpha}} = o(\mathrm{e}^{-c' N^{1+\alpha}})$, whence the result. ■

Corollary 3.1.10 *Let $\kappa \geq 0$. There exist constants $C_n > 0$ depending on n and κ such that the below bounds hold for any f satisfying $\mathcal{K}_\kappa[f] \in W_1^\infty(\mathbb{R}^n)$*

$$\Big| \big\langle f(\xi_1, \ldots, \xi_n) \big\rangle_{\bigotimes_1^n \mathcal{L}_N^{(\boldsymbol{\lambda})}} \Big| \leq C_n \Big\{ N^{-n} \, ||\mathcal{K}_\kappa[f]||_{W_1^\infty(\mathbb{R}^n)}$$

$$+ N^{(\alpha-1)n/2} \, ||\mathcal{K}_\kappa[f]||^{1/2}_{W_n^\infty(\mathbb{R}^n)} \cdot ||\mathcal{K}_\kappa[f]||^{1/2}_{W_0^\infty(\mathbb{R}^n)} \Big\} . \tag{3.1.48}$$

Proof Using the decomposition (3.1.30) it is enough to obtain the bounds (3.1.48) for the compact restriction $f_{|c}$ of f defined in (3.1.31). Upon decomposing $\mathcal{L}_N^{(\lambda)} = \mathcal{L}_{N;u}^{(\widetilde{\lambda})} + (L_N^{(\lambda)} - L_{N;u}^{(\widetilde{\lambda})})$, we can write:

$$\begin{aligned}\langle f_{|c}\rangle_{\otimes_1^n \mathcal{L}_N^{(\lambda)}} &= \sum_{\ell=1}^{n} \sum_{\substack{i_1<\cdots<i_\ell \\ =1}}^{n} \mathbb{P}_N\Big[\int_{\mathbb{R}^n} f_{|c}(\xi_1,\ldots,\xi_n)\prod_{a=1}^{\ell}\mathrm{d}\mathcal{L}_{N;u}^{(\widetilde{\lambda})}(\xi_{i_a}) \prod_{\substack{a=1\\ \neq i_1,\ldots,i_\ell}}^{n} \mathrm{d}\Big(L_{N;u}^{(\widetilde{\lambda})} - L_N^{(\lambda)}\Big)(\xi_a)\Big] \\ &\qquad + \langle f\rangle_{\otimes_1^n \mathcal{L}_{N;u}^{(\widetilde{\lambda})}} .\end{aligned} \tag{3.1.49}$$

Since $\widetilde{\lambda}_a$'s are not far from λ_a's according to (3.1.15), we can bound for any $\ell \leq n-1$,

$$\begin{aligned}&\Big|\mathbb{P}_N\Big[\int_{\mathbb{R}^n} f_{|c}(\xi_1,\ldots,\xi_n)\prod_{a=1}^{\ell}\mathrm{d}\mathcal{L}_{N;u}^{(\widetilde{\lambda})}(\xi_{i_a}) \prod_{\substack{a=1\\ \neq i_1,\ldots,i_\ell}}^{n} \mathrm{d}(L_{N;u}^{(\widetilde{\lambda})} - L_N^{(\lambda)})(\xi_a)\Big]\Big| \\ &\quad \leq 2^n \cdot ||\mathcal{K}_\kappa[f]||_{W_1^\infty(\mathbb{R}^n)} \frac{N(N-1)}{2} \cdot \frac{\mathrm{e}^{-(\ln N)^2}}{N} .\end{aligned} \tag{3.1.50}$$

To get the second factor, we used the chain of bounds

$$||f_{|c}||_{W_1^\infty(\mathbb{R}^n)} \;\leq\; C_1||\mathcal{K}_\kappa[f_{|c}]||_{W_1^\infty(\mathbb{R}^n)} \;\leq\; C_2||\mathcal{K}_\kappa[f]||_{W_1^\infty(\mathbb{R}^n)} . \tag{3.1.51}$$

As a consequence, the first sum in (3.1.49) will only give rise to $||\mathcal{K}_\kappa[f]||_{W_1^\infty(\mathbb{R}^n)} \cdot \mathrm{O}(N^{-\infty})$ corrections. This being settled, Proposition 3.1.6 ensures the existence of $M > 0$ and a constant $C > 0$ such that, for N large enough:

$$\mathbb{P}_N\Big[\Omega_{M;N}\Big] = \mathrm{O}\Big(\mathrm{e}^{-CM\,N^{1+\alpha}}\Big) \;\; \text{with} \;\; \Omega_{M;N} = \Big\{\boldsymbol{\lambda} \in \mathbb{R}^N : \mathfrak{D}^2\big[L_{N;u}^{(\widetilde{\lambda})}, \mu_{\mathrm{eq}}^{(N)}\big] > M/N\Big\} . \tag{3.1.52}$$

This ensures that

$$\Big|\langle f_{|c}\rangle_{\otimes_1^n \mathcal{L}_{N;u}^{(\widetilde{\lambda})}}\Big| \;\leq\; C'\,||\mathcal{K}_\kappa[f]||_{L^\infty(\mathbb{R}^n)}\,\mathrm{e}^{-C''M\,N^{1+\alpha}} \;+\; \mathfrak{R}_{N;u}[f_{|c}] \tag{3.1.53}$$

with

$$\mathfrak{R}_{N;u}[f_{|c}] \;=\; \Big|\mathbb{P}_N\Big[\mathbf{1}_{\Omega_{M;N}^c}\int_{\mathbb{R}^n} f_{|c}(\xi_1,\ldots,\xi_n)\prod_{a=1}^{n}\mathrm{d}\mathcal{L}_{N;u}^{(\widetilde{\lambda})}(\xi_a)\Big]\Big| . \tag{3.1.54}$$

Finally, using Cauchy–Schwarz inequality to make the distance $\mathfrak{D}$ appear:

$$\begin{aligned}\mathfrak{R}_{N:u}[f] &= \bigg|\mathbb{P}_N\bigg[\mathbf{1}_{\Omega^c_{M:N}}\int_{\mathbb{R}^n}\mathcal{F}[f_{|c}](\varphi_1,\dots,\varphi_n)\prod_{a=1}^{n}\mathcal{F}[\mathcal{L}^{(\widetilde{\lambda})}_{N;u}](-\varphi_a)\cdot\frac{\mathrm{d}^n\boldsymbol{\varphi}}{(2\pi)^n}\bigg]\bigg| \\ &\leq \bigg\{\int_{\mathbb{R}^n}\frac{\big|\mathcal{F}[f_{|c}](\varphi_1,\dots,\varphi_n)\big|^2}{\prod_{i=1}^n\Big\{\frac{\pi\beta}{2N^\alpha\varphi_i}\sum_{p=1}^{2}\mathrm{cotanh}\Big[\frac{\varphi_i}{2\omega_pN^\alpha}\Big]\Big\}}\cdot\frac{\mathrm{d}^n\boldsymbol{\varphi}}{(2\pi)^n}\bigg\}^{\frac{1}{2}} \\ &\quad\times\mathbb{P}_N\bigg[\mathbf{1}_{\Omega^c_{M:N}}\mathfrak{D}^n\big[L^{(\widetilde{\lambda})}_{N;u},\mu^{(N)}_{\mathrm{eq}}\big]\bigg]\,. \end{aligned} \tag{3.1.55}$$

The last factor, because it is evaluated on the complement on $\Omega_{M;N}$, is at most $O(N^{-n/2})$. The Fourier transform part of the bound can be estimated with the bound:

$$\begin{aligned}\prod_{i=1}^{n}\bigg|\frac{\pi\beta}{2N^\alpha\varphi_i}\sum_{p=1}^{2}\mathrm{cotanh}\Big[\frac{\varphi_i}{2\omega_pN^\alpha}\Big]\bigg|^{-1} &\leq \prod_{i=1}^{n}\big(C\,N^\alpha|\varphi_i|\big) \\ &\leq (CN^\alpha)^n\bigg(1+\Big\{\sum_{i=1}^{n}\varphi_i^2\Big\}^{1/2}\bigg)^n\,. \end{aligned}\tag{3.1.56}$$

Hence, there exists a constant $C'_n>0$ such that:

$$\big|\langle f_{|c}\rangle_{\bigotimes_1^n\mathcal{L}_N^{(\widetilde{\lambda})}}\big| \;\leq\; C'_n\,N^{(\alpha-1)n/2}\Big(||f_{|c}||_{H_{n/2}(\mathbb{R}^n)}\;+\;||\mathcal{K}_\kappa[f]||_{W_0^\infty(\mathbb{R}^n)}\Big)\,. \tag{3.1.57}$$

where the W_0^∞ norm is nothing but the L^∞ norm. In order to bound $||f_{|c}||_{H_{n/2}(\mathbb{R}^n)}$ by the W_n^∞ norms (c.f. their definition (1.6.7)), we observe that:

$$||f_{|c}||^2_{H_{n/2}(\mathbb{R}^n)}\;\leq\;||f_{|c}||_{H_n(\mathbb{R}^n)}\cdot||f_{|c}||_{L^2(\mathbb{R}^n)}\,. \tag{3.1.58}$$

The $L^2(\mathbb{R}^n)$ norm is bounded directly as:

$$||f_{|c}||_{L^2(\mathbb{R}^n)}\;\leq\;C'(2t+2)^{\frac{n}{2}}\cdot||\mathcal{K}_\kappa[f]||_{W_0^\infty(\mathbb{R}^n)}\,. \tag{3.1.59}$$

Finally, in order to bound $||f_{|c}||_{H_n(\mathbb{R}^n)}$, we remark that $(1+|t|)^{2n}\;\leq\;4^n(1+t^2)^n$, so that:

$$\bigg(1+\Big\{\sum_{a=1}^{n}\varphi_a^2\Big\}^{1/2}\bigg)^{2n}\;\leq\;C\sum_{k=0}^{n}P_k(\varphi_1^2,\dots,\varphi_n^2) \tag{3.1.60}$$

for some symmetric homogeneous polynomial of degree k which has the expansion:

$$P_k(\varphi_1^2,\dots,\varphi_n^2)\;=\;\sum_{k_1+\dots+k_n=k}p_{\{k_a\}}\cdot\varphi_1^{2k_1}\cdots\varphi_n^{2k_n}\qquad\text{with}\qquad p_{\{k_a\}}\geq0\,. \tag{3.1.61}$$

This ensures that

$$||f_{|c}||^2_{H_n(\mathbb{R}^n)} \leq C \sum_{k=0}^{n} \sum_{k_1+\dots+k_n=k} p_{\{k_a\}} \int \Big| \prod_{a=1}^{n} \partial_{\xi_a}^{k_a} \cdot f_{|c}(\xi_1, \dots, \xi_n) \Big|^2 \cdot \mathrm{d}^n \boldsymbol{\xi}$$
$$\leq C' \cdot (2+2t)^n \cdot ||\mathcal{K}_\kappa[f]||^2_{W_n^\infty(\mathbb{R}^n)} \cdot \tag{3.1.62}$$

To get the last line, we have repeatedly used the sub-exponential hypothesis (2.2.4). As a consequence, for some constant C'

$$||f_{|c}||_{H_{n/2}(\mathbb{R}^n)} \;\leq\; C' \cdot ||\mathcal{K}_\kappa[f]||^{\frac{1}{2}}_{W_n^\infty(\mathbb{R}^n)} \cdot ||\mathcal{K}_\kappa[f]||^{\frac{1}{2}}_{W_0^\infty(\mathbb{R}^n)} \cdot \tag{3.1.63}$$

Inserting the above bound in (3.1.57), we obtain

$$\big|\langle f_{|c}\rangle_{\otimes_1^n \mathcal{L}_N^{(\tilde{\lambda})}}\big| \;\leq\; C_n''\, N^{(\alpha-1)n/2}\, ||\mathcal{K}_\kappa[f]||^{\frac{1}{2}}_{W_n^\infty(\mathbb{R}^n)} \cdot ||\mathcal{K}_\kappa[f]||^{\frac{1}{2}}_{W_0^\infty(\mathbb{R}^n)} \,, \tag{3.1.64}$$

what leads to the claimed form of the bound on the average $\langle f\rangle_{\otimes_1^n \mathcal{L}_N^{(\lambda)}}$. ■

3.2 The Schwinger–Dyson Equations

In the present section, we derive the system of Schwinger–Dyson equations in our model. The operator

$$\mathcal{U}_N[\phi](\xi) \;=\; \phi(\xi) \cdot \big\{ V'(\xi) \;-\; \mathcal{S}_N[\rho_{\mathrm{eq}}^{(N)}](\xi) \big\} + \mathcal{S}_N[\phi \cdot \rho_{\mathrm{eq}}^{(N)}](\xi) \,, \tag{3.2.1}$$

with $\mathcal{S}_N$ defined in (2.4.15) will arise in their expression, and play a crucial role in the large-N analysis. It will be shown in Proposition 4.3.8 that the operator $\mathcal{S}_N$ is invertible and in Proposition 5.2.1 that the operator $\mathcal{U}_N$ is invertible as well. We will build on this information until the end of this chapter. At a later stage, we shall use as well fine bounds on the $W_\ell^\infty(\mathbb{R})$ norms of functions $\mathcal{K}_\kappa\big[\mathcal{U}_N^{-1}[\phi]\big]$ which will be obtained later in Proposition 5.2.2. We do stress that these results on the invertibility of $\mathcal{S}_N$ and $\mathcal{U}_N$ as well as those relative to estimates involving $\mathcal{U}_N^{-1}$ will be obtained independently of the results obtained in the present chapter. By presenting this technical result only in a later Chapter 5, and using it as a tool in the present chapter, we hope to make the principles of analysis of Schwinger–Dyson equations more transparent.

Since we will be dealing with operators initially defined on functions in one variable but acting on one of the variables of a function in many variables, it is useful to introduce the

Definition 3.2.1 Given an operator $\mathcal{O} : W_p^\infty(\mathbb{R}) \to W_p^\infty(\mathbb{R}^\ell)$ acting on functions of one variable and $\phi \in W_p^\infty(\mathbb{R}^n)$, $\mathcal{O}_k[\phi]$ refers to the function

$$\mathcal{O}_k[\phi](\xi_1,\ldots,\xi_{n+\ell-1}) = \mathcal{O}\big[\phi(\xi_1,\ldots,\xi_{k-1},*,\xi_{k+\ell},\ldots,\xi_{n+\ell-1})\big](\xi_k,\ldots,\xi_{k+\ell-1})\,, \tag{3.2.2}$$

in which $*$ denotes the variable of ϕ on which the operator $\mathcal{O}_k$ acts.

For instance, according to the above definition, we have

$$\mathcal{U}_{N;1}\big[\phi\big](\xi_1,\ldots,\xi_n) \;=\; \mathcal{U}_N\big[\phi(*,\xi_2,\ldots,\xi_n)\big](\xi_1).$$

Definition 3.2.2 If ϕ is a function in $n \geq 1$ variables, we denote ∂_p the differentiation with respect to the pth variable. We also define an operator

$$\Xi^{(p)} \,:\, W_\ell^\infty(\mathbb{R}^n) \to W_\ell^\infty(\mathbb{R}^{n-1})$$

by:

$$\Xi^{(p)}[\phi](\xi_1,\ldots,\xi_n) = \phi(\xi_1,\ldots,\xi_{p-1},\xi_1,\xi_p,\ldots,\xi_{n-1})\,.$$

Proposition 3.2.3 *Let ϕ_n be a function in n real variables such that $\mathcal{K}_\kappa[\phi_n] \in W_1^\infty(\mathbb{R}^n)$, cf.* (3.1.27), *for some $\kappa \geq 0$ that can depend on n. Then, all expectation values appearing below are well-defined. Furthermore, the rank* 1 *Schwinger–Dyson equation takes the form:*

$$-\big\langle\phi_1\big\rangle_{\mathcal{L}_N^{(\lambda)}} + \frac{1}{2}\Big\langle\mathcal{D}_N \circ \mathcal{U}_N^{-1}[\phi_1]\Big\rangle_{\mathcal{L}_N^{(\lambda)}\otimes\mathcal{L}_N^{(\lambda)}} + \frac{(1-\beta)}{N^{1+\alpha}}\big\langle\partial_1\mathcal{U}_N^{-1}[\phi_1]\big\rangle_{\mu_{\mathrm{eq}}^{(N)}}$$
$$+\frac{(1-\beta)}{N^{1+\alpha}}\big\langle\partial_1\mathcal{U}_N^{-1}[\phi_1]\big\rangle_{\mathcal{L}_N^{(\lambda)}} = 0\,. \tag{3.2.3}$$

There, $\mathcal{D}_N$ corresponds to the non-commutative derivative

$$\mathcal{D}_N[\phi](\xi,\eta) = \Big\{\sum_{p=1}^{2}\beta\pi\omega_p\mathrm{cotanh}\big[\pi\omega_p N^\alpha(\xi-\eta)\big]\Big\}\cdot\big(\phi(\xi)-\phi(\eta)\big)\,. \tag{3.2.4}$$

In their turn, the Schwinger–Dyson equation at rank n takes the form:

$$\big\langle\phi_n\big\rangle_{\overset{n}{\otimes}\mathcal{L}_N^{(\lambda)}} = \frac{1}{N^{2+\alpha}}\sum_{p=2}^{n}\Big\langle\Xi^{(p)}\circ\mathcal{U}_{N;1}^{-1}[\partial_p\phi_n]\Big\rangle_{\overset{n-1}{\otimes}\mathcal{L}_N^{(\lambda)}} + \frac{1}{2}\Big\langle\mathcal{D}_{N;1}\circ\mathcal{U}_{N;1}^{-1}[\phi_n]\Big\rangle_{\overset{n+1}{\otimes}\mathcal{L}_N^{(\lambda)}}$$
$$+\frac{(1-\beta)}{N^{1+\alpha}}\Big\langle\partial_1\mathcal{U}_{N;1}^{-1}[\phi_n]\Big\rangle_{\mu_{\mathrm{eq}}^{(N)}\overset{n-1}{\otimes}\mathcal{L}_N^{(\lambda)}}$$
$$+\frac{1}{N^{2+\alpha}}\sum_{p=2}^{n}\Big\langle\Xi^{(p)}\circ\mathcal{U}_{N;1}^{-1}[\partial_p\phi_n]\Big\rangle_{\mu_{\mathrm{eq}}^{(N)}\overset{n-2}{\otimes}\mathcal{L}_N^{(\lambda)}}$$
$$+\frac{(1-\beta)}{N^{1+\alpha}}\Big\langle\partial_1\mathcal{U}_{N;1}^{-1}[\phi_n]\Big\rangle_{\overset{n}{\otimes}\mathcal{L}_N^{(\lambda)}}\,. \tag{3.2.5}$$

Proof Schwinger–Dyson equations express the invariance of an integral under change of variables, or equivalently, integration by parts. Although the principle of derivation is well-known, we include the proof to be self-contained, following the route of infinitesimal change of variables. Let $\phi^{(a)}$, $a = 1, \dots, n+1$ be a collection of smooth and compactly supported functions. We introduce an ϵ-deformation of the probability density p_N given in (2.5.1) by setting:

$$p_N^{(\{\epsilon_a\}_1^n)}(\boldsymbol{\lambda}) = \frac{1}{Z_N(\{\epsilon_a\})} \prod_{a<b}^{N} \Big\{ \sinh\big[\pi\omega_1 N^{\alpha}(\lambda_a - \lambda_b)\big] \sinh\big[\pi\omega_2 N^{\alpha}(\lambda_a - \lambda_b)\big] \Big\}^{\beta} \times \prod_{a=1}^{N} \mathrm{e}^{-N^{1+\alpha} V_{(\{\epsilon_a\})}(\lambda_a)} , \tag{3.2.6}$$

where:

$$V_{(\{\epsilon_a\})}(\lambda) = V(\lambda) + \sum_{a=2}^{n+1} \epsilon_a \bigg(\phi^{(a)}(\xi) - \int \phi^{(a)}(\eta)\, \mathrm{d}\mu_{\mathrm{eq}}^{(N)}(\eta) \bigg) . \tag{3.2.7}$$

the new normalisation constant $Z_N(\{\epsilon_a\})$ in (3.2.6) is such that $p_N^{(\{\epsilon_a\})}$ is still a probability density on $\mathbb{R}^N$.

We then define $G_t(\mu) = \mu + t\phi^{(1)}(\mu)$. Since $\partial_\xi \phi^{(1)}(\xi)$ is bounded from below, for t small enough G_t is a diffeomorphism of $\mathbb{R}$. Let us carry out the change of variables $\lambda_a = G_t(\mu_a)$ and translate the fact that $p_N^{(\{\epsilon_a\})}$ is a probability measure. This yields

$$1 = \int_{\mathbb{R}^N} p_N^{(\{\epsilon_a\})}(\boldsymbol{\lambda}) \prod_{a=1}^{N} \mathrm{d}\lambda_a = \int_{\mathbb{R}^N} p_N^{(\{\epsilon_a\})}\big(G_t(\lambda_1), \dots, G_t(\lambda_N)\big) \prod_{a=1}^{N} G_t'(\lambda_a)\, \mathrm{d}\lambda_a . \tag{3.2.8}$$

As a consequence, the change of variables yields, to the first order in t:

$$1 = \int_{\mathbb{R}^N} \mathrm{d}^N\boldsymbol{\lambda} \Big\{ 1 + t \sum_{a=1}^{N} \partial_{\lambda_a} \phi^{(1)}(\lambda_a) \Big\} \Big\{ 1 - t N^{1+\alpha} \sum_{a=1}^{N} \big(V_{(\{\epsilon_a\})}\big)'(\lambda_a)\, \phi^{(1)}(\lambda_a) \Big\} \Big\{ 1 + t N^{\alpha} \sum_{a<b}^{N} \Big[\sum_{p=1}^{2} \beta\pi\omega_p \mathrm{cotanh}[\pi\omega_p N^{\alpha}(\lambda_a - \lambda_b)] \Big] \Big[\phi^{(1)}(\lambda_a) - \phi^{(1)}(\lambda_b) \Big] \Big\} \times p_N^{(\{\epsilon_a\})}(\boldsymbol{\lambda}) + O(t^2) . \tag{3.2.9}$$

Identifying the terms linear in t leads to:

$$- \Big\langle \phi^{(1)} \partial_1 \big[V_{(\{\epsilon_a\})} \big] \Big\rangle_{L_N^{(\lambda)}}^{(\{\epsilon_a\})} + \frac{1}{2} \Big\langle \mathcal{D}_N[\phi^{(1)}] \Big\rangle_{L_N^{(\lambda)} \otimes L_N^{(\lambda)}}^{(\{\epsilon_a\})} + \frac{(1-\beta)}{N^{1+\alpha}} \Big\langle \partial_1 \phi^{(1)} \Big\rangle_{L_N^{(\lambda)}}^{(\{\epsilon_a\})} = 0 . \tag{3.2.10}$$

The superscript $(\{\epsilon_a\})$ is there to emphasise that the averages should be taken with respect to the probability measure associated with the ϵ-deformed density (3.2.6). We then centralise the empirical measures with respect to $\mu_{\mathrm{eq}}^{(N)}$. By using the integral equation satisfied by the density of the equilibrium measure $V'(\xi) = \mathcal{S}_N\big[\rho_{\mathrm{eq}}^{(N)}\big](\xi)$ for $\xi \in [a_N ; b_N]$, we obtain:

$$
- \Big\langle \mathcal{U}_N[\phi^{(1)}] \Big\rangle^{(\{\epsilon_a\})}_{\mathcal{L}_N^{(\lambda)}} - \sum_{p=2}^{n+1} \epsilon_a \bigg(\big\langle \phi^{(1)} \, \partial_1 \phi^{(p)} \big\rangle_{\mu_{\mathrm{eq}}^{(N)}} + \big\langle \phi^{(1)} \, \partial_1 \phi^{(p)} \big\rangle^{(\{\epsilon_a\})}_{\mathcal{L}_N^{(\lambda)}} \bigg)
$$
$$
+ \frac{1}{2} \Big\langle \mathcal{D}_N[\phi^{(1)}] \Big\rangle^{(\{\epsilon_a\})}_{\mathcal{L}_N^{(\lambda)} \otimes \mathcal{L}_N^{(\lambda)}} + \frac{(1-\beta)}{N^{1+\alpha}} \bigg(\big\langle \partial_1 \phi^{(1)} \big\rangle_{\mu_{\mathrm{eq}}^{(N)}} + \big\langle \partial_1 \phi^{(1)} \big\rangle^{(\{\epsilon_a\})}_{\mathcal{L}_N^{(\lambda)}} \bigg) = 0 \, . \tag{3.2.11}
$$

Sending ϵ_a's to zero in this equation leads to the desired form of the Schwinger–Dyson equation at rank 1. In order to get the Schwinger–Dyson equation at rank n, we should compute the ϵ_a derivatives of (3.2.11) evaluated at $\epsilon_a \equiv 0$. However, first, it is convenient to multiply the above equation by $Z_N(\{\epsilon_a\})/Z_N[V]$ so as to avoid differentiating the $\{\epsilon_a\}$-dependent partition function entering in the definition of the density $p_N^{(\{\epsilon_a\})}(\boldsymbol{\lambda})$. Doing so, however, produces additional averages in front of the averages solely involving the non-stochastic measures μ_{eq}:

$$
- \Big\langle \mathcal{U}_N[\phi^{(1)}](\xi_1) \prod_{a=2}^{n} \phi^{(a)}(\xi_a) \Big\rangle_{\overset{n+1}{\otimes} \mathcal{L}_N^{(\lambda)}} + \frac{1}{N^{2+\alpha}} \sum_{p=2}^{n+1} \Big\langle \phi^{(1)}(\xi_1) \, \partial_1 \phi^{(p)}(\xi_1) \prod_{\substack{a=2 \\ \neq p}}^{n+1} \phi^{(a)}(\xi_a^{(p)}) \Big\rangle_{\overset{n}{\otimes} \mathcal{L}_N^{(\lambda)}}
$$
$$
+ \frac{1}{2} \Big\langle \mathcal{D}_N[\phi^{(1)}](\xi_1, \xi_2) \prod_{a=2}^{n+1} \phi^{(a)}(\xi_{a+1}) \Big\rangle_{\overset{n+2}{\otimes} \mathcal{L}_N^{(\lambda)}} + \frac{(1-\beta)}{N^{1+\alpha}} \Big\langle \partial_1 \phi^{(1)}(\xi_1) \Big\rangle_{\mu_{\mathrm{eq}}^{(N)}}
$$
$$
\times \Big\langle \prod_{a=2}^{n+1} \phi^{(a)}(\xi_{a-1}) \Big\rangle_{\overset{n}{\otimes} \mathcal{L}_N^{(\lambda)}} + \frac{1}{N^{2+\alpha}} \sum_{p=2}^{n+1} \Big\langle \phi^{(1)}(\xi_1) \partial_1 \phi^{(p)}(\xi_1) \Big\rangle_{\mu_{\mathrm{eq}}^{(N)}} \Big\langle \prod_{\substack{a=2 \\ \neq p}}^{n+1} \phi^{(a)}(\xi_{a-1}^{(p)}) \Big\rangle_{\overset{n-1}{\otimes} \mathcal{L}_N^{(\lambda)}}
$$
$$
+ \frac{(1-\beta)}{N^{1+\alpha}} \Big\langle \partial_1 \phi^{(1)}(\xi_1) \prod_{a=2}^{n+1} \phi^{(a)}(\xi_a) \Big\rangle_{\overset{n+1}{\otimes} \mathcal{L}_N^{(\lambda)}} = 0 \, . \tag{3.2.12}
$$

To any $\boldsymbol{\xi} \in \mathbb{R}^{n-1}$, we associate the vector $\boldsymbol{\xi}^{(p)} \in \mathbb{R}^n$ by

$$
\boldsymbol{\xi}^{(p)} = (\xi_1, \ldots, \xi_{p-1}, \xi_1, \xi_p, \ldots, \xi_{n-1}),
$$

whose components arise in products of the type $\prod_{\substack{a=2 \\ \neq p}}^{n+1} \phi^{(a)}(\xi_a^{(p)})$. The representation

$$
\mathcal{U}_N[\phi](\xi) = \phi(\xi) V'(\xi)
$$
$$
+ \int_{a_N}^{b_N} \bigg\{ \sum_{p=1}^{2} \beta \pi \omega_p \mathrm{cotanh}\big[\pi \omega_p N^{\alpha} (\xi - \eta) \big] \bigg\} \big(\phi(\eta) - \phi(\xi) \big) \, \rho_{\mathrm{eq}}^{(N)}(\eta) \, \mathrm{d}\eta \tag{3.2.13}
$$

readily shows that the operators $\mathcal{U}_N$ and $\mathcal{D}_N$ are both continuous as operators $W_1^{\infty}(K) \to W_0^{\infty}(K)$ for any compact $K \subseteq \mathbb{R}$. This continuity along with the finiteness of the measure $\mathbb{P}_N$ is then enough to conclude, by density of $\mathcal{C}_c^{\infty}(\mathbb{R}) \otimes \cdots \otimes \mathcal{C}_c^{\infty}(\mathbb{R})$ in $\mathcal{C}_c^{\infty}(\mathbb{R}^n)$, that equation (3.2.5) holds for all functions $\phi_n \in \mathcal{C}_c^{\infty}(\mathbb{R}^n)$. Eventually, the assumption of compact support can be dropped. Indeed, given any $\phi_n \in \mathcal{C}_c^{\infty}(\mathbb{R}^n)$, the Schwinger–Dyson equation at level n can be presented as

$$\begin{aligned} \Big\langle \mathcal{U}_{N;1}[\phi_n] \Big\rangle_{\otimes^n \mathcal{L}_N^{(\lambda)}} = & \frac{1}{N^{2+\alpha}} \sum_{p=2}^{n} \Big\langle \Xi^{(p)}[\partial_p \phi_n] \Big\rangle_{\otimes^{n-1} \mathcal{L}_N^{(\lambda)}} + \frac{1}{2} \Big\langle \mathcal{D}_{N;1}[\phi_n] \Big\rangle_{\otimes^{n+1} \mathcal{L}_N^{(\lambda)}} \\ & + \frac{(1-\beta)}{N^{1+\alpha}} \Big\langle \partial_1 \phi_n \Big\rangle_{\mu_{\mathrm{eq}}^{(N)} \otimes^{n-1} \mathcal{L}_N^{(\lambda)}} + \frac{1}{N^{2+\alpha}} \sum_{p=2}^{n} \Big\langle \Xi^{(p)}[\partial_p \phi_n] \Big\rangle_{\mu_{\mathrm{eq}}^{(N)} \otimes^{n-2} \mathcal{L}_N^{(\lambda)}} \\ & + \frac{(1-\beta)}{N^{1+\alpha}} \Big\langle \partial_1 \phi_n \Big\rangle_{\otimes^n \mathcal{L}_N^{(\lambda)}} . \end{aligned} \qquad (3.2.14)$$

It is readily seen due to the sub-exponentiality hypothesis (2.2.4) that given $0 < \kappa < \kappa'$ and ϕ_n such that $\mathcal{K}_{\kappa}[\phi_n] \in W_1^{\infty}(\mathbb{R}^n)$, we have:

$$||\mathcal{K}_{\kappa'}\big[\mathcal{U}_{N;1}[\phi_n]\big]||_{W_0^{\infty}(\mathbb{R}^n)} \leq C||\mathcal{K}_{\kappa}[\phi_n]]||_{W_0^{\infty}(\mathbb{R}^n)} \qquad (3.2.15)$$

and likewise for $\mathcal{D}_{N;1}$. Thus, since $\mathcal{K}_{\kappa}[\phi_n] \in W_1^{\infty}(\mathbb{R}^n)$ can be approached in $W_1^{\infty}(\mathbb{R}^n)$ norm by functions $\mathcal{K}_{\kappa}[\psi_n]$ with $\psi_n \in \mathcal{C}_c^{\infty}(\mathbb{R}^n)$, it remains to invoke the finiteness of the measures $\mathbb{P}_N$ and the decomposition of the nth order averages obtained in Lemma 3.1.9 so as to get (3.2.5) in full generality. In the announcement of the result, we actually choose to write the Schwinger–Dyson equation (3.2.5) for $\mathcal{U}_{N;1}^{-1}[\phi_n]$ instead of ϕ_n. This rewriting is possible because we construct in Chapter 5 this inverse $\mathcal{U}_{N;1}^{-1}$, and merely anticipate the use we will make of this equation. The Schwinger–Dyson equation we have proved in the form (3.2.5) holds independently of the invertibility of $\mathcal{U}_{N;1}$. ■

It follows from the form taken by the Schwinger–Dyson equations that, if we want to solve these equations perturbatively we should, in the very first place, construct the inverse to the operator $\mathcal{U}_N$. This should be done is such a way that one can control explicitly or at least in a manageable way, its dependence on N and its possible singularities. Indeed, the building blocks of $\mathcal{U}_N^{-1}$ exhibit, for instance, square root like singularities at the endpoints of the support $[a_N\,;\,b_N]$ of the equilibrium measure. In Section 5.2.1, we shall construct a regular representation for $\mathcal{U}_N^{-1}$. By regularity, we mean that the various square root singularities present in its building blocks eventually cancel out, hence showing that $\mathcal{U}_N^{-1}[H]$ is smooth as long as H is. Then, in Section 5.2.2, we shall provide explicit, N-dependent, bounds on the $W_{\ell}^{\infty}(\mathbb{R})$ norms of $\mathcal{U}_N^{-1}[H]$. These will play a crucial role in the large-N analysis of the Schwinger–Dyson equations.

3.3 Asymptotic Analysis of the Schwinger–Dyson Equations

The asymptotic analysis of the Schwinger–Dyson equation builds heavily on a family of N-weighted norms that we introduce below.

Definition 3.3.1 For any $\phi \in W_n^{\infty}(\mathbb{R}^p)$, the N-weighted L^{∞} norm of order ℓ is defined by

$$\mathcal{N}_N^{(\ell)}[\phi] \;=\; \sum_{k=0}^{\ell} \frac{||\phi||_{W_k^{\infty}(\mathbb{R}^p)}}{N^{k\alpha}} \,. \tag{3.3.1}$$

This notation $\mathcal{N}_N^{(\ell)}[\phi]$ does not specify the number of variables of ϕ since this is usually clear from the context.

The weighted norm satisfies the obvious bound:

$$\mathcal{N}_N^{(\ell)}[\phi] \leq (\ell+1)\cdot ||\phi||_{W_\ell^{\infty}(\mathbb{R}^p)}\,, \tag{3.3.2}$$

and, respectively, the operators of differentiation and "repetition of a variable" $\Xi^{(p)}$ are bounded as:

$$\mathcal{N}_N^{(\ell)}[\partial_p\phi] \leq N^{\alpha}\,\mathcal{N}_N^{(\ell+1)}[\phi]\,, \qquad\qquad \mathcal{N}_N^{(\ell)}\big[\Xi^{(p)}[\phi]\big] \leq \mathcal{N}_N^{(\ell)}[\phi]\,. \tag{3.3.3}$$

Also, it is important to introduce a specific function that allows one to control the dependence on the potential in the various bounds that issue from the Schwinger–Dyson equations.

Definition 3.3.2 The order ℓ estimate of the potential V is defined as

$$\mathfrak{n}_\ell[V] = \frac{\max\Big\{\prod_{a=1}^{\ell} ||\mathcal{K}_\kappa[V']||_{W_{k_a}^{\infty}(\mathbb{R}^n)} \,:\, \sum_{a=1}^{\ell} k_a = 2\ell+1\Big\}}{\Big\{\min\Big(1\,,\, \inf_{[a\,;b]}|V''(\xi)|\,,\, |V'(b+\epsilon)-V'(b)|\,,\, |V'(a-\epsilon)-V'(a)|\Big)\Big\}^{\ell+1}}\,, \tag{3.3.4}$$

where $\epsilon > 0$ is small enough and fixed once for all, while $\kappa > 0$. We also remind that $\mathcal{K}_\kappa$ is the exponential regularisation of Definition 3.1.7.

Since κ only plays a minor role due to the sub-exponentiality hypothesis (2.2.4) in the estimates provided by $\mathfrak{n}_\ell[V]$, we chose to keep its dependence implicit. Note also that the constants $\mathfrak{n}_\ell[V]$ satisfy

$$\mathfrak{n}_\ell[V]\cdot\mathfrak{n}_{\ell'}[V] \;\leq\; \mathfrak{n}_{\ell+\ell'+1}[V]\,. \tag{3.3.5}$$

Lemma 3.3.3 *Let $\kappa > 0$. There exist constants $C_{n;\ell}, \widetilde{C}_{n;\ell} > 0$ such that, for any ϕ satisfying*

- $\mathcal{K}_{\kappa/\ell}[\phi] \in W^{\infty}_{2\ell+1}(\mathbb{R}^n)$
- $\xi \mapsto \phi(\xi, \xi_2, \dots, \xi_n) \in \mathfrak{X}_s([a_N\,;\,b_N])$, $0 < s < 1/2$, *that is to say*[1]

$$\int_{\mathbb{R}+\mathrm{i}\epsilon} \frac{\mathrm{d}\mu}{2\mathrm{i}\pi}\, \chi_{11}(\mu) \int_{a_N}^{b_N} \phi(\xi, \xi_2, \dots, \xi_n)\, \mathrm{e}^{\mathrm{i}\mu N^{\alpha}(\xi - b_N)} \mathrm{d}\xi = 0$$

$$\text{almost everywhere in } (\xi_2, \dots, \xi_n) \in \mathbb{R}^{n-1} \tag{3.3.6}$$

we have the bounds:

$$\mathcal{N}_N^{(\ell)}\Big[\mathcal{K}_{\kappa}\big[\mathcal{U}_{N;1}^{-1}[\phi]\big]\Big] \leq C_{n;\ell} \cdot \mathfrak{n}_{\ell}[V] \cdot N^{\alpha} \cdot \big(\ln N\big)^{2\ell+1} \cdot \mathcal{N}_N^{(2\ell+1)}\big[\mathcal{K}_{\kappa}[\phi]\big], \tag{3.3.7}$$

$$\mathcal{N}_N^{(\ell)}\Big[\mathcal{K}_{\kappa}\big[\mathcal{D}_{N;1}[\phi]\big]\Big] \leq \widetilde{C}_{n;\ell} \cdot (\ln N)^2 \cdot \mathcal{N}_N^{(\ell+1)}\big[\mathcal{K}_{\kappa}[\phi]\big]. \tag{3.3.8}$$

Note that the above lemma implies, in particular, a bound on the weighted norm of $\mathcal{D}_{N;1} \circ \mathcal{U}_{N;1}^{-1}$:

$$\mathcal{N}_N^{(\ell)}\Big[\mathcal{K}_{\kappa}\big[\mathcal{D}_{N;1} \circ \mathcal{U}_{N;1}^{-1}[\phi]\big]\Big] \leq C'_{n,\ell} \cdot \mathfrak{n}_{\ell+1}[V] \cdot N^{\alpha} \cdot \big(\ln N\big)^{2\ell+5} \cdot \mathcal{N}_N^{(2\ell+3)}\big[\mathcal{K}_{\kappa}[\phi]\big], \tag{3.3.9}$$

We stress for the last time that the proof of this Lemma, for the part concerning $\mathcal{U}_{N;1}^{-1}$, relies on estimates of this inverse obtained in Chapter 5 independently of the present chapter.

Proof We first focus on the norm of $\mathcal{K}_{\kappa}\big[\mathcal{D}_{N;1}[\phi]\big]$. In order to obtain (3.3.8), we bound

$$\mathcal{O}_{\mathbf{k}_{n+1}}(\boldsymbol{\xi}_{n+1}) = \prod_{a=1}^{n+1} \partial_{\xi_a}^{k_a} \mathcal{K}_{\kappa}\big[\mathcal{D}_{N;1}[\phi]\big](\xi_1, \dots, \xi_{n+1}) \quad \text{with } \sum_{a=1}^{n+1} k_a \leq \ell \quad k_a \in \mathbb{N} \tag{3.3.10}$$

by different means in the two cases of interest, *viz.* $N^{\alpha}|\xi_1 - \xi_2| \geq (\ln N)^2$ and $N^{\alpha}|\xi_1 - \xi_2| < (\ln N)^2$.

We first treat the case $N^{\alpha}|\xi_1 - \xi_2| \geq (\ln N)^2$. Observe that for $|N^{\alpha}\xi| \geq (\ln N)^2$, we have:

---

[1]It is straightforward to check by carrying out contour deformations that, for functions ψ decaying sufficiently fast at infinity with respect to its first variable, the condition (3.3.6) is equivalent to belonging to $\mathfrak{X}_s(\mathbb{R})$.

$$\forall \ell \geq 0, \qquad \left|\partial_{\xi}^{\ell}\{S(N^{\alpha}\xi)\}\right| \leq \delta_{\ell,0}\, c_0' \,+\, (1-\delta_{\ell,0})\, c_{\ell}'\, N^{\ell\alpha} \mathrm{e}^{-c''\ln^2 N} \,\leq\, c_{\ell}\big(\ln N\big)^2 \tag{3.3.11}$$

for some constants c_ℓ, where S is defined in (2.4.15) and $\delta_{\ell,0}$ being the Kronecker symbol. Therefore:

$$\begin{aligned}
\left|\mathcal{O}_{\boldsymbol{k}_{n+1}}(\boldsymbol{\xi}_{n+1})\right| \leq & \sum_{\substack{p_a+\ell_a=k_a \\ a=1,2}} \prod_{a=1}^{2} \binom{k_a}{p_a} \\
& \times \left|\partial_{\xi_1}^{p_1}\partial_{\xi_2}^{p_2}\big[\phi_{\{k_a\}}(\xi_1,\xi_3,\ldots,\xi_{n+1}) - \phi_{\{k_a\}}(\xi_2,\xi_3,\ldots,\xi_{n+1})\big]\right| \cdot c_{\ell_1+\ell_2} \cdot (\ln N)^2 \\
\leq & \; C \cdot N^{[\max(k_1,k_2)]\alpha} \cdot (\ln N)^2 \sum_{s=0}^{\max(k_1,k_2)} N^{-s\alpha} \max_{\eta\in\{\xi_1,\xi_2\}} \left|\partial_1^s \phi_{\{k_a\}}(\eta,\xi_3,\ldots,\xi_{n+1})\right| \\
\leq & \; C \cdot N^{\ell\alpha} \cdot (\ln N)^2 \cdot \mathcal{N}_N^{(\ell)}\big[\mathcal{K}_{\kappa}[\phi]\big] \,,
\end{aligned} \tag{3.3.12}$$

where, in the intermediate calculations, we have used:

$$\phi_{\{k_a\}}(\xi_1,\xi_2,\ldots,\xi_n) \equiv \prod_{a=3}^{n+1} \partial_{\xi_a}^{k_a}\Big\{\mathcal{K}_{\kappa}[\phi](\xi_1,\xi_2,\ldots,\xi_n)\Big\}\,. \tag{3.3.13}$$

We now turn to the case when $N^{\alpha}|\xi_1-\xi_2| < (\ln N)^2$. Observe that for any $\ell \in \mathbb{N}$ and $|N^{\alpha}\xi| \leq (\ln N)^2$, the function $\widetilde{S}$, with $\widetilde{S}(x) = xS(x)$, satisfies

$$\begin{aligned}
\forall \ell \geq 0, \quad \left|\partial_{\xi}^{\ell}\{\widetilde{S}(N^{\alpha}\xi)\}\right| \leq & \; \delta_{\ell,0}\left|N^{\alpha}\xi\big[S(N^{\alpha}\xi) - \frac{2\beta}{N^{\alpha}\xi}\big] + 2\beta\right| \\
& + \, (1-\delta_{\ell,0})\, N^{\ell\alpha}||\widetilde{S}||_{W_{\ell}^{\infty}(\mathbb{R})} \,\leq\, c_{\ell}N^{\alpha\ell}\big(\ln N\big)^2
\end{aligned} \tag{3.3.14}$$

for some constants c_ℓ. Starting from the integral representation

$$\begin{aligned}
& \mathcal{O}_{\boldsymbol{k}_{n+1}}(\boldsymbol{x}_{n+1}) \\
& = \int_0^1 \frac{\mathrm{d}t}{N^{\alpha}}\partial_{\xi_1}^{k_1}\partial_{\xi_2}^{k_2}\Big\{\partial_1\phi_{\{k_a\}}(\xi_1+t(\xi_2-\xi_1),\xi_3,\ldots,\xi_{n+1})\cdot\widetilde{S}\big(N^{\alpha}(\xi_1-\xi_2)\big)\Big\}\,,
\end{aligned} \tag{3.3.15}$$

we obtain:

$$\left|\mathcal{O}_{k_{n+1}}(\boldsymbol{x}_{n+1})\right| \leq \sum_{\substack{p_a+\ell_a=k_a \\ a=1,2}} \frac{\binom{k_1}{p_1}\binom{k_2}{p_2} c_{\ell_1+\ell_2}}{N^{\alpha(1-\ell_1-\ell_2)}}$$

$$\times \int_0^1 (1-t)^{p_1} t^{p_2} (\partial_1^{p_1+p_2+1} \phi_{\{k_a\}})\big(\xi_1 + t(\xi_2-\xi_1), \xi_3, \ldots, \xi_{n+1}\big)$$

$$\cdot (\ln N)^2 \cdot \mathrm{d}t$$

$$\leq C N^{(k_1+k_2)\alpha} (\ln N)^2 \sum_{s=1}^{k_1+k_2+1} N^{-s\alpha} \max_{\eta \in [\xi_1\,;\xi_2]} \left|(\partial_1^s \phi_{\{k_a\}})(\eta, \xi_3, \ldots, \xi_{n+1})\right|$$

$$\leq C N^{\ell\alpha} (\ln N)^2 \cdot \mathcal{N}_N^{(\ell+1)}\big[\mathcal{K}_\kappa[\phi]\big]\,. \tag{3.3.16}$$

Putting together (3.3.12) and (3.3.16) and taking the supremum over $\{k_a\}$ such that $\sum_a k_a \leq \ell$, we deduce the desired bound (3.3.8) for the weighted norm of $\mathcal{D}_N$.

The bounds for the weighted norm of $\mathcal{K}_\kappa\big[\mathcal{U}_{N;1}^{-1}[\phi]\big]$ are obtained quite straightforwardly by using the $W_\ell^\infty(\mathbb{R})$ bounds on $\mathcal{K}_\kappa\big[\mathcal{U}_{N;1}^{-1}[\phi]\big]$ as derived in Proposition 5.2.2. ■

With the bounds on the action of the operators $\mathcal{U}_{N;1}^{-1}$ and $\mathcal{D}_{N;1}$, we can improve the a priori bounds on the centred expectation values of the correlators through a bootstrap procedure.

Proposition 3.3.4 *Let $\alpha < 1/4$ and pick $\kappa > 0$. There exist an increasing sequence of integers $(m_n)_n$ and positive constants $(C_n)_n$, such that, for any $n \geq 1$ and $\phi \in \mathfrak{X}_s([a_N\,;b_N])$ in the sense of* (3.3.6) *and satisfying $\mathcal{K}_\kappa[\phi] \in W_{m_n}^\infty(\mathbb{R}^n)$, cf.* (3.1.27), *we have:*

$$\left|\langle\phi\rangle_{\bigotimes_1^n \mathcal{L}_N^{(\lambda)}}\right| \leq C_n \cdot \mathfrak{n}_{m_n}[V] \cdot \mathcal{N}_N^{(m_n)}\big[\mathcal{K}_\kappa[\phi]\big]\, N^{(\alpha-1)n}\,. \tag{3.3.17}$$

The whole dependence of the upper bound on the potential V is contained in the constant $\mathfrak{n}_{\ell_n}[V]$, and we can take:

$$m_n = \ell_n^{(q_n)}, \qquad q_n = 1 + \left\lfloor \frac{n}{1-4\alpha} \right\rfloor, \qquad \ell_n^{(q)} = 2^q(n+q) + 3(2^q-1)\,. \tag{3.3.18}$$

Proof The proof utilises a bootstrap-based improvement of the a priori bounds given in Corollary 3.1.10. Namely, assume the existence of sequences $\eta_N \to 0$, $\varkappa_N \in [0\,;1]$, and constants $C_n > 0$ independent of N, and integers ℓ_n increasing with n, such that, for any ϕ such that $\mathcal{K}_\kappa[\phi] \in W_{\ell_n}^\infty(\mathbb{R}^n)$:

$$\left|\langle\phi\rangle_{\bigotimes_1^n \mathcal{L}_N^{(\lambda)}}\right| \leq C_n \cdot \mathfrak{n}_{\ell_n}[V] \cdot \mathcal{N}_N^{(\ell_n)}\big[\mathcal{K}_\kappa[\phi]\big] \cdot \Big(\eta_N^n \cdot \varkappa_N + N^{n(\alpha-1)}\Big)\,. \tag{3.3.19}$$

We will establish that there exists a new constant $C'_n > 0$ and integers $\ell'_n = 2\ell_{n+1} + 3$ such that, for $\mathcal{K}_\kappa[\phi] \in W^\infty_{\ell'_n}(\mathbb{R}^n)$:

$$\left|\langle\phi\rangle_{\bigotimes_1^n \mathcal{L}_N^{(\lambda)}}\right| \leq C_n \cdot \mathfrak{n}_{\ell'_n}[V] \cdot \mathcal{N}_N^{(\ell'_n)}\big[\mathcal{K}_\kappa[\phi]\big] \cdot \left(\eta_N^n \cdot \varkappa'_N + N^{n(\alpha-1)}\right), \tag{3.3.20}$$

where

$$\varkappa'_N = \varkappa_N \left(\ln N\right)^{\ell'_n+2} \max\left(N^\alpha \eta_N \, ; \, N^{\alpha-2}\eta_N^{-2} ; \, N^{\alpha-1}\eta_N^{-1}\right). \tag{3.3.21}$$

Before justifying (3.3.21), let us examine its consequences. The bootstrap approach can be settled if

$$\varkappa'_N = N^{-\gamma_\varkappa} \varkappa_N \tag{3.3.22}$$

for some $\gamma_\varkappa \geq 0$. Assuming momentarily that $\eta_N = N^{-\gamma}$, when $0 < \alpha < 1$, the range of α and γ ensuring (3.3.22) is:

$$\alpha < \gamma < 1 - \alpha \qquad \textit{a fortiori} \ \ \alpha < 1/2 \,. \tag{3.3.23}$$

The rate $\gamma_\varkappa$ at which $\varkappa'_N/\varkappa_N$ goes to zero increases when γ runs from α to $1/2$, is maximal and equal to $1/2 - \alpha$ when $\gamma = 1/2$, and then decreases when γ increases between $1/2$ and $1 - \alpha$.

The *a priori* estimate proved in Corollary 3.1.10 gives:

$$\begin{aligned}\left|\langle\phi\rangle_{\bigotimes_1^n \mathcal{L}_N^{(\lambda)}}\right| &\leq C'_n \cdot ||\mathcal{K}_\kappa[\phi]||^{\frac{1}{2}}_{W_n^\infty(\mathbb{R}^n)} \cdot ||\mathcal{K}_\kappa[\phi]||^{\frac{1}{2}}_{W_0^\infty(\mathbb{R}^n)} \cdot N^{(\alpha-1)n/2} \\ &\leq C'_n \cdot \mathcal{N}_N^{(n)}\big[\mathcal{K}_\kappa[\phi]\big]\, N^{(\alpha-1/2)n} \,. \end{aligned} \tag{3.3.24}$$

Therefore, the assumption (3.3.20) is satisfied with $\eta_N = N^{-\gamma}$ for $\gamma = 1/2 - \alpha$, and the order $\ell_n = n$ for the weighted norm. The bootstrap condition (3.3.23) then implies $\alpha < 1/4$, and in this case, we find:

$$\varkappa'_N \leq \varkappa_N \, (\ln N)^{\ell'_n} \, N^{-\frac{(1-4\alpha)}{2}} \,. \tag{3.3.25}$$

Now, we can iterate the bootstrap until the first term in (3.3.20) becomes less or equal than the second term $N^{(\alpha-1)n}$. This is obtained in a number of steps q_n determined by the equation $N^{-(1/2-\alpha)n} N^{-(1-4\alpha)q_n/2} \ll N^{(\alpha-1)n}$, therefore:

$$q_n = 1 + \left\lfloor \frac{n}{1-4\alpha} \right\rfloor. \tag{3.3.26}$$

The order of the weighted norm appearing in the bound of the n point correlations at step q of the recursion satisfies $\ell_n^{(q)} = 2\ell_{n+1}^{(q-1)} + 3$, with initial condition $\ell_n^{(0)} = n$. The solution is

$$\ell_n^{(q)} = 2^q(n+q) + 3(2^q - 1) \,. \tag{3.3.27}$$

Therefore, we get at the end of the recursion:

$$\left|\langle\phi\rangle_{\bigotimes_1^n \mathcal{L}_N^{(\lambda)}}\right| \leq C_n \cdot N^{(\alpha-1)n} \cdot \mathcal{N}_N^{(m_n)}\big[\mathcal{K}_\kappa[\phi]\big], \qquad m_n = \ell_n^{(q_n)}\,. \tag{3.3.28}$$

We shall now justify the claim (3.3.21). Starting from (3.3.19), we bound $\langle\phi\rangle_{\bigotimes_1^n \mathcal{L}_N^{(\lambda)}}$ given by the Schwinger–Dyson equations of Proposition 3.2.3, using the norms of the operators $\mathcal{U}_{N;1}$ and $\mathcal{D}_N$ obtained in Lemma 3.3.3. We stress that it is indeed licit to apply the bound (3.3.7) for $\mathcal{U}_N^{-1}$ for, if ϕ satisfies the condition (3.3.6), then so do the functions $\partial_p\phi$ with $p = 2, \ldots, n$. Respecting the order of appearance of terms in (3.2.5), we get[2]:

$$\begin{aligned}
\left|\langle\phi\rangle_{\bigotimes_1^n \mathcal{L}_N^{(\lambda)}}\right| \leq\ & C\mathfrak{n}^2_{\ell_{n-1}}[V]\frac{N^{2\alpha}}{N^{2+\alpha}}(\ln N)^{2\ell_{n-1}+1}\mathcal{N}_N^{(2\ell_{n-1}+2)}\big[\mathcal{K}_\kappa[\phi]\big] \\
& \cdot\Big(\eta_N^{n-1}\cdot\varkappa_N + N^{(n-1)(\alpha-1)}\Big) \\
& + C\mathfrak{n}_{\ell_{n+1}}[V]\mathfrak{n}_{\ell_{n+1}+1}[V]N^{\alpha}(\ln N)^{2\ell_{n+1}+5}\cdot\mathcal{N}_N^{(2\ell_{n+1}+3)}\big[\mathcal{K}_\kappa[\phi]\big] \\
& \cdot\Big(\eta_N^{n+1}\cdot\varkappa_N + N^{(n+1)(\alpha-1)}\Big) \\
& + C\,\mathfrak{n}_{\ell_{n-1}}[V]\mathfrak{n}_{\ell_{n-1}+1}[V]\frac{N^{2\alpha}}{N^{1+\alpha}}(\ln N)^{2\ell_{n-1}+3}\cdot\mathcal{N}_N^{(2\ell_{n-1}+3)}\big[\mathcal{K}_\kappa[\phi]\big] \\
& \cdot\Big(\eta_N^{n-1}\cdot\varkappa_N + N^{(n-1)(\alpha-1)}\Big) \\
& + C(\mathfrak{n}_{\ell_{n-2}}[V])^2\frac{N^{2\alpha}}{N^{2+\alpha}}(\ln N)^{2\ell_{n-2}+2}\cdot\mathcal{N}_N^{(2\ell_{n-2}+1)}\big[\mathcal{K}_\kappa[\phi]\big] \\
& \cdot\Big(\eta_N^{n-2}\cdot\varkappa_N + N^{(n-2)(\alpha-1)}\Big) \\
& + C\,\mathfrak{n}_{\ell_n}[V]\mathfrak{n}_{\ell_n+1}[V]\frac{N^{2\alpha}}{N^{1+\alpha}}(\ln N)^{2\ell_{n-1}+3}\cdot\mathcal{N}_N^{(2\ell_n+3)}\big[\mathcal{K}_\kappa[\phi]\big] \\
& \cdot\Big(\eta_N^{n}\cdot\varkappa_N + N^{n(\alpha-1)}\Big)\,,
\end{aligned} \tag{3.3.29}$$

for some constant $C > 0$ depending on n and κ only. Note that terms integrated against the probability measure $\mu_{\mathrm{eq}}^{(N)}$ have been bounded by means of sup norms. The maximal powers of N are exactly as in (3.3.21)—since we assume $\eta_N \to 0$, the powers arising in the first line are negligible compared to those in the fourth line. We can then use (3.3.5) to bound the products of $\mathfrak{n}_\ell[V]$'s in terms of $\mathfrak{n}_{\ell'_n}[V]$ for a choice:

$$\ell'_n \geq \max\big(2\ell_{n-1}+2,\, 2\ell_{n+1}+3,\, 2\ell_{n-1}+3,\, 2\ell_{n-2}+2,\, 2\ell_n+3\big)\,. \tag{3.3.30}$$

Since $(\ell_n)_n$ is increasing, we can take $\ell'_n = 2\ell_{n+1}+3$, and we indeed find (3.3.20) for N large enough. Note that, the new sequence $(\ell'_n)_n$ is, again, increasing. Then, the

---

[2]The third and fifth line are absent in the case $\beta = 1$, and it gives a larger range of $\alpha > 0$ for which η_N can be chosen so that the bootstrap works. But, eventually, this does not lead to a stronger bound because we can only initialize the bootstrap with the concentration bound (3.1.10) i.e. $\eta_N = N^{-(1/2-\alpha)}$.

maximal power of $\ln N$ occurs in the second line, and is $(\ln N)^{2\ell_{n+1}+5} = (\ln N)^{\ell'_n+2}$. So, we have fully justified (3.3.20). ■

The improved estimates on the multi-point correlators are almost all that is needed for obtaining the large N asymptotic expansion of general one-point functions up to $o(N^{-(2+\alpha)})$ corrections. Prior to deriving such results, we still need to introduce an operator $\widetilde{\mathcal{X}}_N$ mapping any function in $W_p^{\infty}(O)$, O a bounded open subset in $\mathbb{R}^n$, onto a function belonging to $\mathfrak{X}_s([a_N\,;\,b_N])$ in the sense of (3.3.6).

Definition 3.3.5 Let $\mathcal{X}_N$ be the linear form on $W_1^{\infty}([a_N\,;\,b_N])$:

$$\mathcal{X}_N[\phi] \;=\; \frac{\mathrm{i}N^{\alpha}}{\chi_{11;+}(0)} \int\limits_{\mathbb{R}+\mathrm{i}\epsilon} \frac{\mathrm{d}\mu}{2\mathrm{i}\pi}\,\chi_{11}(\mu) \int\limits_{a_N}^{b_N} \mathrm{e}^{\mathrm{i}\mu N^{\alpha}(\xi-b_N)}\,\phi(\xi)\,\mathrm{d}\xi \quad . \tag{3.3.31}$$

Then, we denote by $\widetilde{\mathcal{X}}_N$ the operator

$$\widetilde{\mathcal{X}}_N[\phi](\xi) \;=\; \phi(\xi) \;-\; \mathcal{X}_N[\phi] \tag{3.3.32}$$

and also define:

$$\widetilde{\mathcal{U}}_N^{-1} = \mathcal{U}_N^{-1}\circ\widetilde{\mathcal{X}}_N\,, \qquad \widetilde{\mathcal{W}}_N = \mathcal{W}_N\circ\widetilde{\mathcal{X}}_N\;. \tag{3.3.33}$$

It follows readily from the identity

$$\int\limits_{\mathbb{R}+\mathrm{i}\epsilon} \chi_{11}(\mu)\cdot\frac{1-\mathrm{e}^{-\mathrm{i}\mu\overline{x}_N}}{\mu}\cdot\frac{\mathrm{d}\mu}{2\mathrm{i}\pi} \;=\; \chi_{11;+}(0) \qquad \text{with} \qquad \overline{x}_N \;=\; N^{\alpha}(b_N-a_N)\,, \tag{3.3.34}$$

that $\widetilde{\mathcal{X}}_N[\phi]\in\mathfrak{X}_s([a_N\,;\,b_N])$ in the sense of (3.3.6). The proof of (3.3.34) follows from the use of the boundary conditions $\mathrm{e}^{-\mathrm{i}\lambda N^{\alpha}(b_N-a_N)}\chi_{11;+}(\lambda)=\chi_{11;-}(\lambda)$, $\lambda\in\mathbb{R}$ the fact that $\chi_{11}\in\mathcal{O}(\mathbb{C}\setminus\mathbb{R})$ and that $\chi_{11}(\lambda)=\mathrm{O}\big(|\lambda|^{-1/2}\big)$ at infinity. Likewise, by using the bounds (6.1.23) obtained in Corollary 6.1.3 it is readily seen that

$$\mathcal{N}_N^{(p)}\Big[\mathcal{K}_{\kappa}\big[\widetilde{\mathcal{X}}_N[\phi]\big]\Big] \;\le\; C\cdot\mathcal{N}_N^{(p)}\big[\mathcal{K}_{\kappa}[\phi]\big]\,. \tag{3.3.35}$$

Proposition 3.3.6 *Given any $\kappa>0$, and any ϕ satisfying $\mathcal{K}_{\kappa}[\phi]\in W_{\ell}^{\infty}(\mathbb{R})$, we have:*

$$\begin{aligned}\langle\phi\rangle_{\mathcal{L}_N^{(\lambda)}} = {} & \frac{(1-\beta)}{N^{1+\alpha}}\cdot\Big\langle\partial_1\widetilde{\mathcal{U}}_N^{-1}[\phi]\Big\rangle_{\mu_{\mathrm{eq}}^{(N)}} \;+\; \frac{1}{2N^{2+\alpha}}\Big\langle\Xi^{(2)}\Big[\partial_2\widetilde{\mathcal{U}}_{N;1}^{-1}\Big[\mathcal{D}_N\big[\widetilde{\mathcal{U}}_N^{-1}[\phi]\big]\Big]\Big]\Big\rangle_{\mu_{\mathrm{eq}}^{(N)}} \\ & + \frac{(1-\beta)^2}{2N^{2(1+\alpha)}}\Big\langle\partial_1\partial_2\widetilde{\mathcal{U}}_{N;1}^{-1}\widetilde{\mathcal{U}}_{N;2}^{-1}\Big[\mathcal{D}_N\big[\widetilde{\mathcal{U}}_N^{-1}[\phi]\big]\Big]\Big\rangle_{\otimes^2\mu_{\mathrm{eq}}^{(N)}} \\ & + \frac{(1-\beta)^2}{N^{2(1+\alpha)}}\Big\langle\partial_1\widetilde{\mathcal{U}}_N^{-1}\big[\partial_1\widetilde{\mathcal{U}}_N^{-1}[\phi]\big]\Big\rangle_{\mu_{\mathrm{eq}}^{(N)}} \;+\; \frac{\delta_N[\phi,V]}{N^{2+\alpha}}\,. \end{aligned} \tag{3.3.36}$$

The remainder $\delta_N[\phi, V]$ *is bounded as:*

$$\big|\delta_N[\phi, V]\big| \;\leq\; C \cdot \mathfrak{n}_{\ell}[V] \cdot \mathcal{N}_N^{(\ell')}\big[\mathcal{K}_{\kappa}[\phi]\big] \cdot N^{6\alpha-1}\,(\ln N)^{\ell''} \tag{3.3.37}$$

for a constant $C > 0$ *that does not dependent on* ϕ *nor on the potential* V*, and the integers:*

$$\ell = \max(3m_3 + 5, 8m_2 + 18), \quad \ell' = \max(4m_3 + 9, 14m_2 + 37),$$
$$\ell'' = \max(14m_2 + 17, 6m_3 + 16)$$

given in terms of the sequence $(m_n)_n$ *introduced in* (3.3.18).

Proof The strategy is to exploit the Schwinger–Dyson equation and get rid of expectation values of functions integrated against the random measure $L_N^{(\lambda)}$. This can be done by replacing them approximately by integration against a deterministic measure of a transformed function, up to corrections that we can estimate.

Let ϕ be a sufficiently regular function of one variable. Since the signed measure $\mathcal{L}_N^{(\lambda)}$ has zero mass, it follows that $\langle \phi \rangle_{\mathcal{L}_N^{(\lambda)}} = \langle \widetilde{\mathcal{X}}_N[\phi] \rangle_{\mathcal{L}_N^{(\lambda)}}$. We can use the Schwinger–Dyson equation at rank 1 (3.2.3) for the function $\widetilde{\mathcal{X}}_N[\phi]$, and apply the sharp bounds of Proposition 3.3.4 to derive:

$$\Big|\langle \phi \rangle_{\mathcal{L}_N^{(\lambda)}} - \frac{1-\beta}{N^{1+\alpha}}\Big\langle \partial_1 \mathcal{U}_N^{-1}\big[\widetilde{\mathcal{X}}_N[\phi]\big]\Big\rangle_{\mu_{\mathrm{eq}}^{(N)}}\Big| \leq C \cdot \mathfrak{n}_{2m_2+2}[V]$$
$$\times\; \mathcal{N}_N^{(2m_2+3)}\big[\mathcal{K}_{\kappa}[\phi]\big] \cdot N^{3\alpha-2}(\ln N)^{2m_2+5}\,. \tag{3.3.38}$$

Above, we have stressed explicitly the composition of the operator $\mathcal{U}_N^{-1}$ with $\widetilde{\mathcal{X}}_N$. This bound ensures that

$$\Big|\frac{1-\beta}{N^{1+\alpha}}\langle \partial_1 \widetilde{\mathcal{U}}_N^{-1}[\phi]\rangle_{\mathcal{L}_N^{(\lambda)}} - \frac{(1-\beta)^2}{N^{2(1+\alpha)}}\Big\langle \partial_1 \widetilde{\mathcal{U}}_N^{-1}\big[\partial_1 \widetilde{\mathcal{U}}_N^{-1}[\phi]\big]\Big\rangle_{\mu_{\mathrm{eq}}^{(N)}}\Big|$$
$$\leq C' \cdot \mathfrak{n}_{4m_2+7}[V] \cdot \mathcal{N}_N^{(4m_2+9)}\big[\mathcal{K}_{\kappa}[\phi]\big] \cdot N^{4\alpha-3}(\ln N)^{6m_2+14}\,. \tag{3.3.39}$$

where we remind that $\widetilde{\mathcal{U}}_N^{-1} = \mathcal{U}_N^{-1} \circ \widetilde{\mathcal{X}}_N$. Equation (3.3.39) can be used for a substitution of the term proportional to $(1-\beta)$ in the Schwinger–Dyson equation at rank 1 (3.2.3), and we get:

$$\Big|\langle \phi \rangle_{\mathcal{L}_N^{(\lambda)}} - \frac{1-\beta}{N^{1+\alpha}}\langle \partial_1 \widetilde{\mathcal{U}}_N^{-1}[\phi]\rangle_{\mu_{\mathrm{eq}}^{(N)}} - \frac{(1-\beta)^2}{N^{2(1+\alpha)}}\Big\langle \partial_1 \widetilde{\mathcal{U}}_N^{-1}\big[\partial_1 \widetilde{\mathcal{U}}_N^{-1}[\phi]\big]\Big\rangle_{\mu_{\mathrm{eq}}^{(N)}}$$
$$- \frac{1}{2}\langle \mathcal{D}_N \circ \widetilde{\mathcal{U}}_N^{-1}[\phi]\rangle_{\otimes^2 \mathcal{L}_N^{(\lambda)}}\Big|$$
$$\leq C' \cdot \mathfrak{n}_{4m_2+7}[V] \cdot \mathcal{N}_N^{(4m_2+9)}\big[\mathcal{K}_{\kappa}[\phi]\big] \cdot N^{4\alpha-3}(\ln N)^{6m_2+14}\,. \tag{3.3.40}$$

In order to gain a better control on the term involving $\mathcal{D}_N$—which is a two-point correlator—we need to study the Schwinger–Dyson equation at rank $n = 2$ (3.2.5). Given a sufficiently regular function ψ_2 in two variables, using the sharp bounds of Proposition 3.3.4, we find:

$$\begin{aligned}&\Big|\langle\psi_2\rangle_{\otimes^2\mathcal{L}_N^{(\lambda)}} - \frac{1}{N^{2+\alpha}}\Big\langle \Xi^{(2)}\Big[\partial_2\widetilde{\mathcal{U}}_{N;1}^{-1}[\psi_2]\Big]\Big\rangle_{\mu_{\mathrm{eq}}^{(N)}} - \frac{1-\beta}{N^{1+\alpha}}\Big\langle\Big(\partial_1\widetilde{\mathcal{U}}_{N;1}^{-1}[\psi_2]\Big)\Big\rangle_{\mu_{\mathrm{eq}}^{(N)}\otimes\mathcal{L}_N^{(\lambda)}}\Big|\\ &\le C\cdot \mathfrak{n}_{2m_3+2}[V]\cdot\mathcal{N}_N^{(2m_3+3)}\big[\mathcal{K}_\kappa[\psi_2]\big]\cdot N^{4\alpha-3}(\ln N)^{2m_3+5}\,. \qquad (3.3.41)\end{aligned}$$

We apply this estimate to the particular choice:

$$\psi_2(\xi_1,\xi_2) = \mathcal{D}_N\big[\widetilde{\mathcal{U}}_N^{-1}[\phi]\big](\xi_1,\xi_2)\,. \qquad (3.3.42)$$

Thanks to the bound (3.3.9) on the norm of $\mathcal{D}_N\circ\mathcal{U}_N^{-1}$ and the sub-multiplicativity (3.3.5) of the $\mathfrak{n}_\ell[V]$'s, we deduce:

$$\begin{aligned}&\Big|\langle\psi_2\rangle_{\otimes^2\mathcal{L}_N^{(\lambda)}} - \frac{1}{N^{2+\alpha}}\Big\langle \Xi^{(2)}\Big[\partial_2\widetilde{\mathcal{U}}_{N;1}^{-1}[\psi_2]\Big]\Big\rangle_{\mu_{\mathrm{eq}}^{(N)}} - \frac{1-\beta}{N^{1+\alpha}}\Big\langle\Big(\partial_1\widetilde{\mathcal{U}}_{N;1}^{-1}[\psi_2]\Big)\Big\rangle_{\mu_{\mathrm{eq}}^{(N)}\otimes\mathcal{L}_N^{(\lambda)}}\Big|\\ &\le C\cdot \mathfrak{n}_{4m_3+7}[V]\cdot\mathcal{N}_N^{(4m_3+9)}\big[\mathcal{K}_\kappa[\phi]\big]\cdot N^{5\alpha-3}\,(\ln N)^{6m_3+16}\,. \qquad (3.3.43)\end{aligned}$$

This can be used for a substitution of $\langle\psi_2\rangle = \langle\mathcal{D}_N\circ\mathcal{U}_N^{-1}[\phi]\rangle$ in the left-hand side of (3.3.40). Before performing this substitution, we still need to control the term in (3.3.43) which is proportional to $(1-\beta)$. This is a one-point correlator for the function:

$$\psi_1(\xi) = \frac{1-\beta}{N^{1+\alpha}}\int \partial_\eta\widetilde{\mathcal{U}}_{N;1}^{-1}\big[\psi_2(*,\xi)\big](\eta)\,\mathrm{d}\mu_{\mathrm{eq}}^{(N)}(\eta)\,. \qquad (3.3.44)$$

Applying the one-point estimate (3.3.38) to the function ψ_1, along with the bounds (3.3.7) and (3.3.8) for the norms of $\mathcal{U}_N^{-1}$ and $\mathcal{D}_N$, we find:

$$\begin{aligned}&\Big|\langle\psi_1\rangle_{\mathcal{L}_N^{(\lambda)}} - \frac{1-\beta}{N^{1+\alpha}}\Big\langle\partial_1\widetilde{\mathcal{U}}_N^{-1}[\psi_1]\Big\rangle_{\mu_{\mathrm{eq}}^{(N)}}\Big|\\ &\le C\cdot\mathfrak{n}_{8m_2+18}[V]\cdot\mathcal{N}_N^{(8m_2+21)}\big[\mathcal{K}_\kappa[\phi]\big]\cdot N^{5\alpha-3}(\ln N)^{14m_2+37}\,. \qquad (3.3.45)\end{aligned}$$

This leads to:

$$\begin{aligned}&\Big|\langle\psi_2\rangle_{\otimes^2\mathcal{L}_N^{(\lambda)}} - \frac{1}{N^{2+\alpha}}\Big\langle\Xi^{(2)}\circ\partial_2\widetilde{\mathcal{U}}_{N;1}^{-1}[\psi_2]\Big\rangle_{\mu_{\mathrm{eq}}^{(N)}}\\ &\quad - \frac{(1-\beta)^2}{N^{2(1+\alpha)}}\Big\langle\partial_1\widetilde{\mathcal{U}}_{N;1}^{-1}\partial_2\big[\widetilde{\mathcal{U}}_{N;2}^{-1}[\psi_2]\big]\Big\rangle_{\mu_{\mathrm{eq}}^{(N)}\otimes\mu_{\mathrm{eq}}^{(N)}}\Big|\\ &\le C\cdot\mathfrak{n}_{8m_2+18}[V]\cdot\mathcal{N}_N^{(8m_3+21)}\big[\mathcal{K}_\kappa[\phi]\big]\cdot N^{5\alpha-3}(\ln N)^{14m_2+37}\,. \qquad (3.3.46)\end{aligned}$$

The result follows by substituting this inequality in (3.3.40). ■

3.4 The Large-N Asymptotic Expansion of ln $Z_N[V]$ up to o(1)

We can use the large-N analysis of the Schwinger–Dyson equations to establish the existence of an asymptotic expansion up to o(1) of ln $Z_N[V]$. The coefficients in this asymptotic expansion are single and double integrals whose integrand depends on N. We will work out the large-N asymptotic expansion of these coefficients in Sections 6.1–6.3. Prior to writing down this large-N asymptotic expansion, we need to introduce those single and double integrals that will enter in the description of the result. We also remind the notation $\widetilde{\mathcal{W}}_N = \mathcal{W}_N \circ \widetilde{\mathcal{X}}_N$ where $\mathcal{W}_N$ is the inverse of $\mathcal{S}_N$ (*cf.* (2.4.15)), studied in Section 4.3.4, and $\widetilde{\mathcal{X}}_N$ is the operator introduced in Definition 3.3.5. Given G, H sufficiently regular on $[a_N\,;b_N]$, we define the one-dimensional integrals:

$$\begin{aligned}\mathfrak{I}_{\mathrm{s}}\big[H,G\big] &= \int_{a_N}^{b_N} H(\xi)\cdot\mathcal{W}_N[G](\xi)\cdot \mathrm{d}\xi\;,\\ \mathfrak{I}^{(1)}_{\mathrm{s};\beta}\big[H,G\big] &= \int_{a_N}^{b_N} \mathcal{W}_N[G'](\xi)\,\partial_\xi\bigg\{\frac{\widetilde{\mathcal{W}}_N[H](\xi)}{\mathcal{W}_N[G'](\xi)}\bigg\}\,\mathrm{d}\xi\end{aligned} \tag{3.4.1}$$

and

$$\mathfrak{I}^{(2)}_{\mathrm{s};\beta}\big[H,G\big] = \int_{a_N}^{b_N} \mathcal{W}_N[G'](\xi)\,\partial_\xi\left\{\frac{\widetilde{\mathcal{W}}_N\Big[\partial_1\Big(\frac{\widetilde{\mathcal{W}}_N[H]}{\mathcal{W}_N[G']}\Big)\Big](\xi)}{\mathcal{W}_N[G'](\xi)}\right\}\mathrm{d}\xi\;. \tag{3.4.2}$$

We also define the two-dimensional integrals:

$$\mathfrak{I}_{\mathrm{d}}\big[H,G\big] = \int_{a_N}^{b_N} \widetilde{\mathcal{W}}_N\bigg[\partial_\xi\Big\{S\big(N^\alpha(\xi-*)\big)\Big(\frac{\widetilde{\mathcal{W}}_N[H](\xi)}{\mathcal{W}_N[G'](\xi)} - \frac{\widetilde{\mathcal{W}}_N[H](*)}{\mathcal{W}_N[G'](*)}\Big)\Big\}\bigg](\xi)\,\mathrm{d}\xi \tag{3.4.3}$$

and

$$\mathfrak{I}_{\mathrm{d};\beta}\big[H, G\big] = \frac{1}{2}\int\limits_{a_N}^{b_N} \mathrm{d}\xi \mathrm{d}\eta\, \mathcal{W}_N[G'](\xi)\cdot\mathcal{W}_N[G'](\eta)$$

$$\times\, \partial_\xi\partial_\eta\left(\frac{1}{\mathcal{W}_N[G'](\xi)\cdot\mathcal{W}_N[G'](\eta)}\right.$$

$$\left.\widetilde{\mathcal{W}}_{N;1}\circ\widetilde{\mathcal{W}}_{N;2}\left[S\big(N^\alpha(*_1-*_2)\big)\times\left\{\frac{\widetilde{\mathcal{W}}_N[H](*_1)}{\mathcal{W}_N[G'](*_1)}-\frac{\widetilde{\mathcal{W}}_N[H](*_2)}{\mathcal{W}_N[G'](*_2)}\right\}\right](\xi,\eta)\right). \tag{3.4.4}$$

Above, $*$ refers to the variables on which the operators act, $*_1$, *viz.* $*_2$, to the first, respectively second, running variable on which the product of operators $\mathcal{W}_{N;1}\cdot\mathcal{W}_{N;2}$ acts. The subscript β reminds that the terms concerned are absent in the case $\beta = 1$.

Proposition 3.4.1 *Let* $V_{G;N}(\lambda) = g_N\lambda^2 + t_N\lambda$ *be the unique Gaussian potential associated with an equilibrium measure supported on* $[a_N\,;\,b_N]$ *as given in Lemma D.0.18 and assume that* $0 < \alpha < 1/6$. *Then there exists* $\ell \in \mathbb{N}$ *such that one has the large-N asymptotic expansion*

$$\ln\left(\frac{Z_N[V]}{Z_N\big[V_{G;N}\big]}\right) = -N^{2+\alpha}\int\limits_0^1 \mathfrak{I}_{\mathrm{s}}\big[\partial_t V_t, V_t'\big]\cdot\mathrm{d}t \;-\; N(1-\beta)\int\limits_0^1 \mathfrak{I}^{(1)}_{\mathrm{s};\beta}\big[\partial_t V_t, V_t\big]\cdot\mathrm{d}t$$

$$-\frac{1}{2}\int\limits_0^1 \mathfrak{I}_{\mathrm{d}}\big[\partial_t V_t, V_t\big]\cdot\mathrm{d}t - \frac{(1-\beta)^2}{N^\alpha}\int\limits_0^1\Big\{\mathfrak{I}^{(2)}_{\mathrm{s};\beta}\big[\partial_t V_t, V_t\big]$$

$$+\,\mathfrak{I}_{\mathrm{d};\beta}\big[\partial_t V_t, V_t\big]\Big\}\cdot\mathrm{d}t \;+\; \mathrm{O}\big(N^{6\alpha-1}(\ln N)^{2\ell}\big)\,. \tag{3.4.5}$$

Here $V_t = (1-t)V_{G;N} + tV$.

Proof The result follows from (2.5.4). Indeed, the remarks above (2.5.8) allow to identify the equilibrium measures $\mu^{(N)}_{\mathrm{eq};V_t} = (1-t)\mu^{(N)}_{\mathrm{eq};V_{G;N}} + t\mu^{(N)}_{\mathrm{eq};V}$ for all $t\in[0,1]$. One can then use Proposition 3.3.6 to expand $\langle\partial_t V_t\rangle^{V_t}_{L_N^{(\lambda)}}$, along with the representation for $\mathcal{U}_N^{-1}$ on the support of the equilibrium measure which reads

$$\widetilde{\mathcal{U}}_N^{-1}[H](\xi) \;=\; \frac{\widetilde{\mathcal{W}}_N[H](\xi)}{\mathcal{W}_N[V'](\xi)}\,. \tag{3.4.6}$$

Taking these data into account, it solely remains to write down explicitly the one and two-dimensional integrals arising in Proposition 3.3.6. ■

Note that the factors $\mathfrak{I}^{(2)}_{\mathrm{s};\beta}\big[\partial_t V_t, V_t\big]$ and $\mathfrak{I}_{\mathrm{d};\beta}\big[\partial_t V_t, V_t\big]$ are preceded by the negative power of $N^{-\alpha}$. Still, it does not mean that these do not contribute to the leading contribution, *i.e.* up to o(1), to the asymptotics of the partition function. Indeed, the presence of derivatives in their associated integrands generates additional powers of N^α.

Reference

1. Maïda, M., Maurel-Segala, E.: Free transport-entropy inequalities for non-convex potentials and application to concentration for random matrices. Probab. Theory Relat. Fields **159** no. 1–2, 329–356 (2014). arXiv:math.PR/1204.3208

Chapter 4
The Riemann–Hilbert Approach to the Inversion of $\mathcal{S}_N$

Abstract In the present chapter we focus on a class of singular integral equation driven by a one parameter γ-regularisation of the operator $\mathcal{S}_N$. More precisely, we introduce the singular integral operator $\mathcal{S}_{N;\gamma}$

$$\mathcal{S}_{N;\gamma}[\phi](\xi) = \fint_{a_N}^{b_N} S_\gamma\big(N^\alpha(\xi-\eta)\big)\phi(\eta)\cdot \mathrm{d}\eta \quad where \quad \left\{ \begin{array}{l} S_\gamma(\xi) \; = \; S(\xi)\cdot \mathbf{1}_{[-\gamma\overline{x}_N\,;\,\gamma\overline{x}_N]} \\ \overline{x}_N = N^\alpha\cdot(b_N - a_N) \end{array} \right. . \tag{4.0.1}$$

This operator is a regularisation of the operator $\mathcal{S}_N$ in the sense that, formally, $\mathcal{S}_{N;\infty} = \mathcal{S}_N$. This regularisation enables to set a well defined associated Riemann–Hilbert problem, and is such that, once all calculations have been done and the inverse of $\mathcal{S}_{N;\gamma}$ constructed, we can send $\gamma \to +\infty$ at the level of the obtained answer. It is then not a problem to check that this limiting operator does indeed provide one with the inverse of $\mathcal{S}_N$. We start this analysis by, first, recasting the singular integral equation into a form where the variables have been re-scaled. Then, we put the problem of inverting the re-scaled operator associated with $\mathcal{S}_{N;\gamma}$ in correspondence with a vector valued Riemann-Hilbert problem. The resolution of this vector problem demands the resolution of a 2×2 matrix Riemann–Hilbert problem for an auxiliary matrix χ. We construct the solution to this problem, for N-large enough, in Section 4.2.2 and then exhibit some of the overall properties of the solution χ in Section 4.2.3. We shall build on these results so as to invert $\mathcal{S}_{N;\gamma}$ and then $\mathcal{S}_N$ in subsequent sections.

4.1 A Re-writing of the Problem

4.1.1 A Vector Riemann–Hilbert Problem

In the handlings that will follow, it will appear more convenient to consider a properly rescaled problem. Namely define

G. Borot et al., *Asymptotic Expansion of a Partition Function Related to the Sinh-model*, Mathematical Physics Studies,
DOI 10.1007/978-3-319-33379-3_4

$$\varphi(\xi) = \phi\big((\xi + N^{\alpha} a_N)N^{-\alpha}\big) \qquad \text{and} \qquad h(\xi) = \frac{N^{\alpha}}{2\mathrm{i}\pi\beta} H\big((\xi + N^{\alpha} a_N)N^{-\alpha}\big) . \tag{4.1.1}$$

It is then clear that solutions to $\mathcal{S}_{N;\gamma}\big[\phi\big](\xi) = H(\xi)$ are in a one-to-one correspondence with those of

$$\mathscr{S}_{N;\gamma}[\varphi](\xi) = \int\limits_{0}^{\overline{x}_N} S_{\gamma}(\xi - \eta)\varphi(\eta) \cdot \frac{\mathrm{d}\eta}{2\mathrm{i}\pi\beta} = h(\xi) . \tag{4.1.2}$$

For any N and $\gamma \geq 0$ and s $\in \mathbb{R}$, the operator $\mathscr{S}_{N;\gamma}$ is continuous as an operator

$$\mathscr{S}_{N;\gamma} : H_s\big([0\,;\overline{x}_N]\big) \longrightarrow H_s\big([-\gamma\overline{x}_N\,;\gamma\overline{x}_N]\big) \subseteq H_s\big(\mathbb{R}\big) . \tag{4.1.3}$$

Indeed, this continuity follows readily from the boundedness of the Fourier transform $\mathcal{F}[S_{\gamma}]$ of the operator's integral kernel, *c.f.* Lemma 4.1.2 to come.

First, we shall start by focusing on spaces with a negative index $s < 0$ and going to construct a class of its inverses

$$\mathscr{S}_{N;\gamma}^{-1} : H_s\big([-\gamma\overline{x}_N\,;\gamma\overline{x}_N]\big) \longrightarrow H_s\big([0\,;\overline{x}_N]\big) . \tag{4.1.4}$$

What we mean here is that, *per se*, the operator is non-invertible in that, as will be inferred from our analysis, for $-k < s < -(k-1)$

$$\dim\ker \mathscr{S}_{N;\gamma} = k . \tag{4.1.5}$$

In fact, the analysis that will follow, provides one with a thorough characterisation of its kernel. Furthermore, when restricting the operator $\mathscr{S}_{N;\gamma}$ to spaces of more regular functions like $H_s\big([0\,;\overline{x}_N]\big)$ with $s > 0$, we get that the image $\mathscr{S}_{N;\gamma}\big[H_s\big([0\,;\overline{x}_N]\big)\big]$ is a closed, explicitly characterisable subspace of $H_s\big([-\gamma\overline{x}_N\,;(\gamma+1)\overline{x}_N]\big)$, and that the operator becomes continuously invertible on it.

In the following, we shall invert the operator $\mathscr{S}_{N;\gamma}$ by means of the resolution of an auxiliary 2×2 Riemann–Hilbert problem and then by implementing a Wiener–Hopf factorisation. The analysis is inspired by the paper of Novokshenov [1] where a correspondence has been built between singular integral equations on a finite segment subordinate to integral kernels depending on the difference on the one hand and Riemann–Hilbert problems on the other one. The large parameter analysis is, however, new.

In fact the very setting of the Riemann–Hilbert problem-based analysis enables one to naturally construct the pseudo-inverse of $\mathscr{S}_{N;\gamma}$—*i.e.* modulo elements of $\ker\big[\mathscr{S}_{N;\gamma}\big]$—when the operator is understood to act on H_s spaces with *negative* index $s < 0$. The inversion of $\mathscr{S}_{N;\gamma}$ understood as an operator on H_s spaces with *positive* index $s \geq 0$ goes, however, beyond, the "crude" construction issuing from the Riemann–Hilbert problem-based analysis. It is, in particular, based on an explicit characterisation, through linear constraints, of the image space $\mathscr{S}_{N;\gamma}\big[H_s([0\,;\overline{x}_N]])\big]$, $s \geq 0$. For $0 < s < 1/2$, which is the case of interest for us, we show that $\mathscr{S}_{N;\gamma}\big[H_s([0\,;\overline{x}_N])\big]$ coincides with $\mathfrak{X}_s([-\gamma\overline{x}_N\,;(\gamma+1)\overline{x}_N])$.

Lemma 4.1.1 *Let $h \in H_s\big([0\,;\overline{x}_N]\big)$, $s < 0$. For any solution $\varphi \in H_s\big([0\,;\overline{x}_N]\big)$ of* (4.1.2)*, there exists a two-dimensional vector function $\Phi \in \mathcal{O}(\mathbb{C}\setminus\mathbb{R})$ such that $\varphi = \mathcal{F}^{-1}[(\Phi_1)_+]$ and Φ is a solution to the boundary value problem:*

- $\big(\Phi_a\big)_\pm \in \mathcal{F}\big[H_s\big(\mathbb{R}^\pm\big)\big]$ *for $a \in \{1,2\}$, and there exists $C > 0$ such that:*

$$\forall \mu > 0, \quad \forall a \in \{1,2\}, \qquad \int_{\mathbb{R}} \big|\Phi_a(\lambda \pm \mathrm{i}\mu)\big|^2 \cdot \big(1+|\lambda|+|\mu|\big)^{2s} \cdot \mathrm{d}\lambda \; < \; C\,. \tag{4.1.6}$$

- *We have the jump equation for $\Phi_+(\lambda) = G_\chi(\lambda)\cdot\Phi_-(\lambda) + \boldsymbol{H}(\lambda)$ for $\lambda \in \mathbb{R}$, with:*

$$G_\chi(\lambda) = \begin{pmatrix} \mathrm{e}^{\mathrm{i}\lambda\overline{x}_N} & 0 \\ \frac{1}{2\mathrm{i}\pi\beta}\cdot\mathcal{F}\big[S_\gamma\big](\lambda) & -\mathrm{e}^{-\mathrm{i}\lambda\overline{x}_N} \end{pmatrix} \quad \textit{and} \quad \boldsymbol{H}(\lambda) = \begin{pmatrix} 0 \\ -\mathrm{e}^{-\mathrm{i}\lambda\overline{x}_N}\mathcal{F}\big[h_{\mathfrak{e}}\big](\lambda) \end{pmatrix}. \tag{4.1.7}$$

Conversely, for any solution $\Phi \in \mathcal{O}(\mathbb{C}\setminus\mathbb{R})$ of the above boundary value problem, $\varphi = \mathcal{F}^{-1}\big[\,\big(\Phi_1\big)_+\big]$ is a solution of (4.1.2).

We do remind that $\pm$ denotes the upper/lower boundary values on $\mathbb{R}$ with the latter being oriented from $-\infty$ to $+\infty$; $h_{\mathfrak{e}}$ denotes any extension of h to $H_s(\mathbb{R})$; $\mathcal{F}\big[S_\gamma\big](\lambda)$ refers to the Fourier transform of the principal value distribution induced by S_γ:

$$\mathcal{F}\big[S_\gamma\big](\lambda) \;=\; \fint_{-\gamma\overline{x}_N}^{\gamma\overline{x}_N} S(\xi)\,\mathrm{e}^{\mathrm{i}\lambda\xi}\mathrm{d}\xi\,. \tag{4.1.8}$$

Proof Assume that one is given a solution φ in $H_s\big([0\,;\overline{x}_N]\big)$ to (4.1.2). Then, let ψ_L, ψ_R be two functions such that

$$\operatorname{supp}(\psi_R) \;=\; [\overline{x}_N\,;+\infty[, \qquad \operatorname{supp}(\psi_L) \;=\,]-\infty\,;0] \tag{4.1.9}$$

and

$$\fint_0^{\overline{x}_N} S_\gamma(\xi-\eta)\varphi(\eta)\cdot\frac{\mathrm{d}\eta}{2\mathrm{i}\pi\beta} \;-\; h_{\mathfrak{e}}(\xi) \;=\; \psi_L(\xi) \;+\; \psi_R(\xi)\,. \tag{4.1.10}$$

Then, by going to the Fourier space, we get:

$$\frac{1}{2\mathrm{i}\pi\beta}\cdot\mathcal{F}\big[S_\gamma\big](\lambda)\cdot\mathcal{F}[\varphi](\lambda) \;-\; \mathcal{F}[h_{\mathfrak{e}}](\lambda) \;=\; \mathcal{F}[\psi_L](\lambda) \;+\; \mathcal{F}[\psi_R](\lambda)\,. \tag{4.1.11}$$

By Lemma 4.1.2 that will be proved below, $\mathcal{F}\big[S_\gamma\big] \in L^\infty(\mathbb{R})$. Hence $\psi_R \in H_s(\mathbb{R}^+)$ whereas $\psi_L \in H_s(\mathbb{R}^-)$. Then, we introduce the vectors

$$\boldsymbol{F}_\uparrow(\lambda) \;=\; \begin{pmatrix} \mathcal{F}[\varphi](\lambda) \\ \mathrm{e}^{-\mathrm{i}\lambda\overline{x}_N}\mathcal{F}[\psi_R](\lambda) \end{pmatrix} \quad \text{and} \quad \boldsymbol{F}_\downarrow(\lambda) \;=\; \begin{pmatrix} \mathcal{F}[\varphi_{\overline{x}_N}](\lambda) \\ \mathcal{F}[\psi_L](\lambda) \end{pmatrix} \tag{4.1.12}$$

where we agree upon $\varphi_{\overline{x}_N}(\xi) = \varphi(\xi + \overline{x}_N)$. Since $\big[\boldsymbol{F}_\uparrow\big]_a \in \mathcal{F}\big[H_s(\mathbb{R}^+)\big]$, respectively $\big[\boldsymbol{F}_\downarrow\big]_a \in \mathcal{F}\big[H_s(\mathbb{R}^-)\big]$, it is readily seen that

$$\widetilde{\boldsymbol{F}}_{\uparrow;a}(\lambda) \;=\; \big(1-\mathrm{i}\lambda\big)^s \cdot \big[\boldsymbol{F}_\uparrow\big]_a(\lambda) \qquad \text{resp.} \qquad \widetilde{\boldsymbol{F}}_{\downarrow;a}(\lambda) \;=\; \big(1+\mathrm{i}\lambda\big)^s \cdot \big[\boldsymbol{F}_\downarrow\big]_a(\lambda) \tag{4.1.13}$$

defines a holomorphic function on $\mathbb{H}^+$, respectively $\mathbb{H}^-$, with $L^2(\mathbb{R})$ $+$, respectively $-$, boundary values on $\mathbb{R}$. The Paley–Wiener Theorem A.0.14 then shows the existence of $C > 0$ such that:

$$\forall \mu > 0, \quad \forall a \in \{1,2\}, \quad \int\limits_{\mathbb{R}} \big|\big[\boldsymbol{F}_{\uparrow/\downarrow}\big]_a(\lambda \pm \mathrm{i}\mu)\big|^2 \cdot \big(1+|\lambda|+|\mu|\big)^{2s} \cdot \mathrm{d}\lambda \;<\; C\,. \tag{4.1.14}$$

In other words the function:

$$\Phi \;=\; \boldsymbol{F}_\uparrow \cdot \mathbf{1}_{\mathbb{H}^+} \;+\; \boldsymbol{F}_\downarrow \cdot \mathbf{1}_{\mathbb{H}^-} \tag{4.1.15}$$

solves the vector valued Riemann–Hilbert problem.

Reciprocally, suppose that one is given a solution Φ to the vector-valued Riemann–Hilbert problem in question. Then, set $\varphi = \mathcal{F}^{-1}\big[\big(\Phi_1\big)_+\big]$. We clearly have $\varphi \in H_s(\mathbb{R}^+)$, but we now show that the support of φ is actually included in $[0, \overline{x}_N]$. Let $(\cdot,\cdot)$ be the canonical scalar product on $L^2(\mathbb{R},\mathbb{C})$. If ρ_R is a $\mathscr{C}^\infty$ function with compact support included in $]\overline{x}_N, +\infty[$, we have:

$$2\pi \cdot (\rho_R, \varphi) = \big(\mathcal{F}[\rho_R]\,,\, \mathcal{F}[\varphi]\big) = (e^{-\mathrm{i}\overline{x}_N *}\mathcal{F}[\rho_R](1-\mathrm{i}*)^{-s}\,,\, (1+\mathrm{i}*)^s(\Phi_1)_-)\,, \tag{4.1.16}$$

where $*$ denotes the running variable. But this is zero since $(1+\mathrm{i}*)^s(\Phi_1)_- \in \mathcal{F}[L^2(\mathbb{R}^-)]$, whereas, by the Paley–Wiener Theorem A.0.14, $\mathrm{e}^{-\mathrm{i}\overline{x}_N *}\mathcal{F}[\rho_R](1-\mathrm{i}*)^{-s} \in \mathcal{F}[L^2(\mathbb{R}^-)]$. Finally, the fact that $\varphi \in H_s([0\,;\overline{x}_N])$ satisfies (4.1.2) follows from taking the Fourier transform of the second line of the jump equation (4.1.7) for Φ. ■

For further handlings, it is useful to characterise the distributional Fourier transform $\mathcal{F}[S_\gamma]$ slightly better.

Lemma 4.1.2 *The distributional Fourier transform* $\mathcal{F}[S_\gamma](\lambda)$ *defined by* (4.1.8) *admits the representation*

$$\frac{\mathcal{F}[S_\gamma](\lambda)}{2\mathrm{i}\pi\beta} = R(\lambda) + \left(\mathrm{e}^{\mathrm{i}\lambda\gamma\overline{x}_N} + \mathrm{e}^{-\mathrm{i}\lambda\gamma\overline{x}_N}\right)\frac{\kappa_N}{\lambda} + r_N(\lambda)$$

$$\textit{where}\quad \kappa_N = -\sum_{p=1}^{2}\frac{\omega_p}{2}\operatorname{cotanh}[\pi\omega_p\gamma\overline{x}_N] \tag{4.1.17}$$

$$R(\lambda) = \frac{\sinh\left[\frac{\lambda(\omega_1+\omega_2)}{2\omega_1\omega_2}\right]}{2\sinh\left[\frac{\lambda}{2\omega_1}\right]\sinh\left[\frac{\lambda}{2\omega_2}\right]}, \tag{4.1.18}$$

and

$$r_N(\lambda) = \sum_{p=1}^{2}\frac{(\pi\omega_p)^2}{\mathrm{i}\lambda(1-\mathrm{e}^{-\lambda/\omega_p})}\int_0^{\mathrm{i}/\omega_p}\left\{\frac{\mathrm{e}^{-\mathrm{i}\lambda\gamma\overline{x}_N}}{\sinh^2[\pi\omega_p(\xi-\gamma\overline{x}_N)]} - \frac{\mathrm{e}^{\mathrm{i}\lambda\gamma\overline{x}_N}}{\sinh^2[\pi\omega_p(\xi+\gamma\overline{x}_N)]}\right\}\cdot\frac{\mathrm{e}^{\mathrm{i}\lambda\xi}\,\mathrm{d}\xi}{2\mathrm{i}\pi}. \tag{4.1.19}$$

Besides, for $\operatorname{Im}\lambda = \epsilon > 0$ *small enough, there exists* $C_\epsilon > 0$ *independent of* N *such that, uniformly in* $\operatorname{Re}\lambda \in \mathbb{R}$*:*

$$|r_N(\lambda)| \le C_\epsilon\,|\lambda|^{-2}\cdot\exp\left\{-\gamma\overline{x}_N(2\pi\min[\omega_1,\omega_2]-\epsilon)\right\}. \tag{4.1.20}$$

Proof One has that

$$\begin{aligned}\frac{\mathcal{F}[S_\gamma](\lambda)}{2\mathrm{i}\pi\beta} &= \frac{1}{2}\sum_{p=1}^{2}\lim_{t\to0^+}\sum_{\epsilon\in\{\pm1\}}\int_{-\gamma\overline{x}_N}^{\overline{x}_N}\pi\omega_p\operatorname{cotanh}[\pi\omega_p(\xi+\mathrm{i}\epsilon t)]\cdot\frac{e^{\mathrm{i}\lambda\xi}\mathrm{d}\xi}{2\mathrm{i}\pi}\\ &= \frac{1}{2}\sum_{\substack{p\in\{1.2\}\\ \epsilon\in\{\pm1\}}}\frac{\pi\omega_p}{1-e^{-\lambda/\omega_p}}\lim_{t\to0^+}\int_{\Gamma_p}\operatorname{cotanh}\left[\pi\omega_p(\xi+\mathrm{i}\epsilon t)\right]\cdot\frac{e^{\mathrm{i}\lambda\xi}\mathrm{d}\xi}{2\mathrm{i}\pi},\end{aligned} \tag{4.1.21}$$

where $\Gamma_p = [-\gamma\overline{x}_N\,;\,\gamma\overline{x}_N] \cup [\gamma\overline{x}_N+\mathrm{i}/\omega_p\,;\,-\gamma\overline{x}_N+\mathrm{i}/\omega_p]$, where the second interval is endowed with an opposite orientation. It then remains to add the counter-term:

$$\begin{aligned}r_N(\lambda) = \sum_{p=1}^{2}\frac{\pi\omega_p}{1-\mathrm{e}^{-\lambda/\omega_p}}\int_0^{\frac{\mathrm{i}}{\omega_p}}\Big\{&\mathrm{e}^{-\mathrm{i}\lambda\gamma\overline{x}_N}\Big(\operatorname{cotanh}[\pi\omega_p\gamma\overline{x}_N] + \operatorname{cotanh}[\pi\omega_p(\xi-\gamma\overline{x}_N)]\Big)\\ &+\mathrm{e}^{\mathrm{i}\lambda\gamma\overline{x}_N}\Big(\operatorname{cotanh}[\pi\omega_p\gamma\overline{x}_N] - \operatorname{cotanh}[\pi\omega_p(\xi+\gamma\overline{x}_N)]\Big)\Big\}\cdot\frac{\mathrm{e}^{\mathrm{i}\lambda\xi}\mathrm{d}\xi}{2\mathrm{i}\pi}.\end{aligned} \tag{4.1.22}$$

to form a closed contour $\widetilde{\Gamma}_p$. Upon integrating by parts, we find the expression (4.1.19) for $r_N(\lambda)$. Then, we pick up the residues surrounded by $\widetilde{\Gamma}_p$, and we also write aside the term behaving as $O(1/\lambda)$ when $\lambda \to \infty$. This leads to the appearence of κ_N in (4.1.17). The bounds on the line $|\mathrm{Im}\,\lambda| = \epsilon > 0$, with ϵ small enough are then obtained through straightforward majorations. ■

The resolution of the vector Riemann–Hilbert problem for Φ can be done with the help of a matrix Wiener–Hopf factorization. In order to apply this method, we first need to obtain a $\pm$-factorization of the matrix G_χ given by (4.1.7). This leads to an 2×2 matrix Riemann–Hilbert problem that we formulate and solve, for N sufficiently large, in the next subsections.

4.1.2 A Scalar Riemann–Hilbert Problem

In order to state the main result regarding to the auxiliary 2×2 matrix Riemann–Hilbert problem, we first need to introduce some objects. To start with, we introduce a factorization of $R(\lambda)$ defined in (4.1.18) that separates contributions from zeroes and poles between the lower and upper half-planes $\lambda \in \mathbb{H}^{\pm}$. In other words, we consider the solution υ to the following scalar Riemann–Hilbert problem, depending on $\epsilon > 0$ small enough and chosen once for all:

- $\upsilon \in \mathcal{O}\big(\mathbb{C} \setminus \{\mathbb{R} + \mathrm{i}\epsilon\}\big)$ and has continuous $\pm$-boundary values on $\mathbb{R} + \mathrm{i}\epsilon$;
- $\upsilon(\lambda) = \begin{cases} (-\mathrm{i}\lambda)^{\frac{1}{2}} \cdot \big(1 + \mathrm{O}(\lambda^{-1})\big) & \text{if } \mathrm{Im}\,\lambda > \epsilon \\ -\mathrm{i}(\mathrm{i}\lambda)^{\frac{1}{2}} \cdot \big(1 + \mathrm{O}(\lambda^{-1})\big) & \text{if } \mathrm{Im}\,\lambda < \epsilon \end{cases}$

  when $\lambda \to \infty$ non-tangentially to $\mathbb{R} + \mathrm{i}\epsilon$;
- $\upsilon_+(\lambda) \cdot R(\lambda) = \upsilon_-(\lambda)$ for $\lambda \in \mathbb{R} + \mathrm{i}\epsilon$.

This problem admits a unique solution given by

$$\upsilon(\lambda) = \begin{cases} R_{\uparrow}^{-1}(\lambda) & \text{if } \mathrm{Im}\,\lambda > \epsilon \\ R_{\downarrow}(\lambda) & \text{if } \mathrm{Im}\,\lambda < \epsilon \end{cases} \tag{4.1.23}$$

where

$$R_{\uparrow}(\lambda) = \frac{\mathrm{i}}{\lambda} \cdot \sqrt{\omega_1 + \omega_2} \cdot \Big(\frac{\omega_2}{\omega_1 + \omega_2}\Big)^{\frac{\mathrm{i}\lambda}{2\pi\omega_1}} \cdot \Big(\frac{\omega_1}{\omega_1 + \omega_2}\Big)^{\frac{\mathrm{i}\lambda}{2\pi\omega_2}} \cdot \frac{\prod_{p=1}^{2} \Gamma\big(1 - \frac{\mathrm{i}\lambda}{2\pi\omega_p}\big)}{\Gamma\big(1 - \frac{\mathrm{i}\lambda(\omega_1+\omega_2)}{2\pi\omega_1\omega_2}\big)} \tag{4.1.24}$$

and

$$R_{\downarrow}(\lambda) = \frac{\lambda}{2\pi\sqrt{\omega_1 + \omega_2}} \cdot \Big(\frac{\omega_2}{\omega_1 + \omega_2}\Big)^{-\frac{\mathrm{i}\lambda}{2\pi\omega_1}} \cdot \Big(\frac{\omega_1}{\omega_1 + \omega_2}\Big)^{-\frac{\mathrm{i}\lambda}{2\pi\omega_2}} \cdot \frac{\prod_{p=1}^{2} \Gamma\big(\frac{\mathrm{i}\lambda}{2\pi\omega_p}\big)}{\Gamma\big(\frac{\mathrm{i}\lambda(\omega_1+\omega_2)}{2\pi\omega_1\omega_2}\big)} \,. \tag{4.1.25}$$

Note that

$$R_{\downarrow}(0) \;=\; -\mathrm{i}\sqrt{\omega_1+\omega_2} \qquad \text{and} \qquad \Big(\lambda R_{\uparrow}(\lambda)\Big)_{|\lambda=0} \;=\; \mathrm{i}\sqrt{\omega_1+\omega_2}\,. \tag{4.1.26}$$

Also, $R_{\uparrow}$ and $R_{\downarrow}$ satisfy to the relations

$$R_{\uparrow}(-\lambda) \;=\; \lambda^{-1}\cdot R_{\downarrow}(\lambda) \qquad \text{and} \qquad \Big(R_{\uparrow}(\lambda^{*})\Big)^{*} \;=\; \lambda^{-1}\cdot R_{\downarrow}(\lambda)\,. \tag{4.1.27}$$

Furthermore, $R_{\uparrow/\downarrow}$ exhibit the asymptotic behaviour

$$R_{\uparrow}(\lambda) = \big(-\mathrm{i}\lambda\big)^{-\frac{1}{2}}\cdot\Big(1\,+\,\mathrm{O}\big(\lambda^{-1}\big)\Big) \qquad \text{for} \;\; \lambda \underset{\lambda\in\mathbb{H}^{+}}{\longrightarrow} \infty \tag{4.1.28}$$

$$R_{\downarrow}(\lambda) = -\mathrm{i}\big(\mathrm{i}\lambda\big)^{\frac{1}{2}}\cdot\Big(1\,+\,\mathrm{O}\big(\lambda^{-1}\big)\Big) \qquad \text{for} \;\; \lambda \underset{\lambda\in\mathbb{H}^{-}}{\longrightarrow} \infty \tag{4.1.29}$$

as it should be. The notation $\uparrow$ and $\downarrow$ indicates the direction in the complex plane where $R_{\uparrow/\downarrow}$ have no pole nor zeroes.

Preliminary definitions

We need a few other definitions before describing the solution to the factorisation problem for G_{χ}. Let:

$$\mathcal{R}_{\uparrow}(\lambda) \;=\; \begin{pmatrix} 0 & -1 \\ 1 & -R(\lambda)\mathrm{e}^{\mathrm{i}\lambda\overline{x}_N} \end{pmatrix} \qquad \text{and} \qquad \mathcal{R}_{\downarrow}(\lambda) \;=\; \begin{pmatrix} -1 & R(\lambda)\mathrm{e}^{-\mathrm{i}\lambda\overline{x}_N} \\ 0 & 1 \end{pmatrix}\,, \tag{4.1.30}$$

as well as their "asymptotic" versions:

$$\mathcal{R}_{\uparrow}^{(\infty)} \;=\; \begin{pmatrix} 0 & -1 \\ 1 & 0 \end{pmatrix} \qquad \text{and} \qquad \mathcal{R}_{\downarrow}^{(\infty)} \;=\; \begin{pmatrix} -1 & 0 \\ 0 & 1 \end{pmatrix}\,. \tag{4.1.31}$$

We also need to introduce

$$M_{\uparrow}(\lambda) = \begin{pmatrix} 1 & 0 \\ -\dfrac{1-R^2(\lambda)}{\upsilon^2(\lambda)\cdot R(\lambda)}\mathrm{e}^{\mathrm{i}\lambda\overline{x}_N} & 1 \end{pmatrix} \quad \text{and} \quad M_{\downarrow}(\lambda) = \begin{pmatrix} 1 & \upsilon^2(\lambda)\cdot\dfrac{1-R^2(\lambda)}{R(\lambda)}\mathrm{e}^{-\mathrm{i}\lambda\overline{x}_N} \\ 0 & 1 \end{pmatrix}, \tag{4.1.32}$$

where υ is given by (4.1.23), and:

$$P_R(\lambda) = I_2 + \frac{\theta_R}{\lambda}\Pi^{-1}(0)\sigma^{-}\Pi(0) \quad \text{and} \quad \left\{\begin{array}{l} P_{L;\uparrow}(\lambda) = I_2 + \kappa_N\,\lambda^{-1}\,\mathrm{e}^{\mathrm{i}(\gamma-1)\lambda\overline{x}_N}\cdot\sigma^{-} \\ P_{L;\downarrow}(\lambda) = I_2 + \kappa_N\,\lambda^{-1}\,\mathrm{e}^{-\mathrm{i}(\gamma-1)\lambda\overline{x}_N}\cdot\sigma^{-} \end{array}\right., \tag{4.1.33}$$

in which $\Pi(0)$ is a constant matrix that will coincide later with the value at 0 of the matrix function Π, *cf.* (4.2.14), and

$$\theta_R = \frac{1}{\upsilon^2(0)} \frac{\kappa_N}{1+\kappa_N/(\omega_1+\omega_2)} . \tag{4.1.34}$$

4.2 The Auxiliary 2 × 2 Matrix Riemann–Hilbert Problem for χ

4.2.1 Formulation and Main Result

The factorisation problem for the jump matrix G_χ corresponds to solving the 2×2 matrix Riemann–Hilbert problem given below. This problem is solvable for N large enough.

Proposition 4.2.1 *There exists N_0 such that, for any $N \geq N_0$, the given below 2×2 Riemann–Hilbert problem has a unique solution. This solution is as given in Figure 4.2*

- *the 2×2 matrix function $\chi \in \mathcal{O}(\mathbb{C} \setminus \mathbb{R})$ has continuous $\pm$-boundary values on $\mathbb{R}$;*
- $\chi(\lambda) = \begin{cases} P_{L;\uparrow}(\lambda) \cdot \begin{pmatrix} -\operatorname{sgn}(\mathrm{Re}\lambda)\cdot \mathrm{e}^{\mathrm{i}\lambda\overline{x}_N} & 1 \\ -1 & 0 \end{pmatrix} \cdot (-\mathrm{i}\lambda)^{-\frac{\sigma_3}{2}} \cdot \Big(I_2 + \frac{\chi_1}{\lambda} + \mathrm{O}(\lambda^{-2})\Big) & \lambda \in \mathbb{H}^+ \\ P_{L;\downarrow}(\lambda) \cdot \begin{pmatrix} -1 & \operatorname{sgn}(\mathrm{Re}\lambda)\cdot \mathrm{e}^{-\mathrm{i}\lambda\overline{x}_N} \\ 0 & 1 \end{pmatrix} \cdot (\mathrm{i}\lambda)^{-\frac{\sigma_3}{2}} \mathrm{e}^{\mathrm{i}\frac{\pi}{2}\sigma_3} \cdot \Big(I_2 + \frac{\chi_1}{\lambda} + \mathrm{O}(\lambda^{-2})\Big) & \lambda \in \mathbb{H}^- \end{cases}$

  for some constant matrix χ_1, when $\lambda \to \infty$ non-tangentially to $\mathbb{R}$;
- $\chi_+(\lambda) = G_\chi(\lambda) \cdot \chi_-(\lambda)$ *for* $\lambda \in \mathbb{R}$.

Furthermore, the unique solution to the above Riemann–Hilbert problem satisfies $\det \chi(\lambda) = \operatorname{sgn}\big(\mathrm{Im}(\lambda)\big)$ *for any* $\lambda \in \mathbb{C} \setminus \mathbb{R}$.

The existence of a solution χ will be established in Section 4.2.2, by a set of transformations:

$$\chi \rightsquigarrow \Psi \rightsquigarrow \Pi \tag{4.2.1}$$

which maps the initial RHP for χ, to a RHP for Π whose jump matrices are uniformly close to the identity when N is large, and thus solvable by perturbative arguments [2]. The structure of χ in terms of the solution Π is summarized in Figure 4.2. The uniqueness of χ follows from standard arguments, see *e.g.* [3], that we now reproduce.

Proof (of uniqueness)
Define, for $\lambda \in \mathbb{C}\setminus\mathbb{R}$,

$$d(\lambda) = \det[\chi(\lambda)]\mathbf{1}_{\mathbb{H}^+}(\lambda) - \det[\chi(\lambda)]\mathbf{1}_{\mathbb{H}^-}(\lambda) . \tag{4.2.2}$$

Since χ has continuous $\pm$-boundary on $\mathbb{R}$, it follows that $d \in \mathcal{O}(\mathbb{C}\setminus\mathbb{R})$ has continuous $\pm$ boundary values on $\mathbb{R}$ as well. Furthermore these satisfy $d_+(\lambda) = d_-(\lambda)$. Finally,

d admits the asymptotic behaviour $d(\lambda) = 1 + \mathrm{O}(\lambda^{-1})$. d can thus be continued to an entire function that is bounded at infinity. Hence, by Liouville theorem, $d \equiv 1$. Let χ_1, χ_2 be two solutions to the Riemann–Hilbert problem for χ. Since χ_2 can be analytically inverted, it follows that $\widetilde{\chi} = \chi_2^{-1} \cdot \chi_1$ solves the Riemann–Hilbert problem:

- $\widetilde{\chi} \in \mathcal{O}(\mathbb{C}\backslash\mathbb{R})$ and has continuous $\pm$-boundary values on $\mathbb{R}$;
- $\widetilde{\chi}(\lambda) = I_2 + \mathrm{O}(\lambda^{-1})$ when $\lambda \to \infty$ non-tangentially to $\mathbb{R}$;
- $\widetilde{\chi}_+(\lambda) = \widetilde{\chi}_-(\lambda)$ for $\lambda \in \mathbb{R}$.

Thus, by analytic continuation through $\mathbb{R}$ and Liouville theorem $\widetilde{\chi} = I_2$, hence ensuring the uniqueness of solutions. ■

4.2.2 *Transformation $\chi \rightsquigarrow \Psi \rightsquigarrow \Pi$ and Solvability of the Riemann–Hilbert Problem*

We construct a piecewise analytic matrix Ψ out of the matrix χ according to Figure 4.1. It is readily checked that the Riemann–Hilbert problem for χ is equivalent to the following Riemann–Hilbert problem for Ψ:

- $\Psi \in \mathcal{O}(\mathbb{C}^*\backslash\Sigma_\Psi)$ and has continuous boundary values on Σ_Ψ;
- The matrix $\begin{pmatrix} -1 & 0 \\ -\kappa_N \lambda^{-1} & 1+\kappa_N/(\omega_1+\omega_2) \end{pmatrix} \cdot \big[\upsilon(0)\big]^{-\sigma_3} \cdot \Psi(\lambda)$ has a limit when $\lambda \to 0$;
- $\Psi(\lambda) = I_2 + \mathrm{O}(\lambda^{-1})$ when $\lambda \to \infty$ non-tangentially to Σ_Ψ;
- $\Psi_+(\lambda) = G_\Psi(\lambda) \cdot \Psi_-(\lambda)$ for $\lambda \in \Sigma_\Psi$;

where the jump matrix G_Ψ takes the form:

$$\text{for } \lambda \in \Gamma_\uparrow \qquad G_\Psi(\lambda) = I_2 + \frac{\mathrm{e}^{\mathrm{i}\lambda\bar{x}_N}}{\upsilon^2(\lambda)R(\lambda)} \cdot \sigma^- , \tag{4.2.3}$$

$$\text{for } \lambda \in \Gamma_\downarrow \qquad G_\Psi(\lambda) = I_2 + \frac{\upsilon^2(\lambda)\,\mathrm{e}^{-\mathrm{i}\lambda\bar{x}_N}}{R(\lambda)} \cdot \sigma^+ , \tag{4.2.4}$$

and for $\lambda \in \mathbb{R} + \mathrm{i}\epsilon$

$$G_\Psi(\lambda) = I_2 + \frac{r_N(\lambda)}{R(\lambda)} \cdot \begin{pmatrix} \dfrac{1}{\dfrac{\mathrm{e}^{\mathrm{i}\lambda\bar{x}_N}}{\upsilon_+(\lambda)\upsilon_-(\lambda)}} & -\upsilon_+(\lambda)\upsilon_-(\lambda)\mathrm{e}^{-\mathrm{i}\lambda\bar{x}_N} \\ & -1 \end{pmatrix} . \tag{4.2.5}$$

The motivation underlying the construction of Ψ is that its jump matrix G_Ψ not only satisfies $G_\Psi - I_2 \in \mathcal{M}_2\Big(\big(L^2 \cap L^\infty\big)\big(\Sigma_\Psi\big)\Big)$, but is, in fact, exponentially in N close to the identity

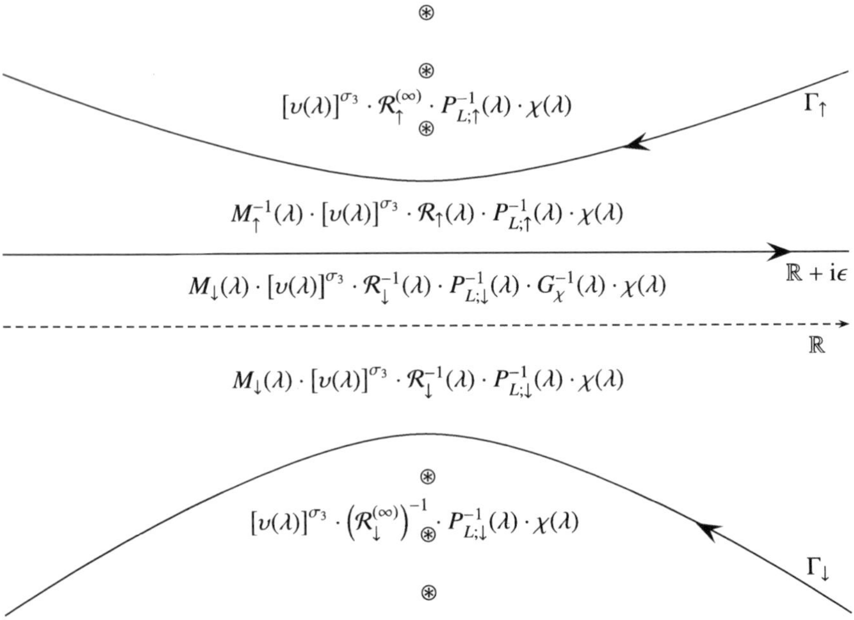

Fig. 4.1 $\Sigma_{\Psi} = \Gamma_{\uparrow} \cup \Gamma_{\downarrow} \cup \{\mathbb{R} + \mathrm{i}\epsilon\}$ is the contour appearing in the Riemann–Hilbert problem for Ψ. $\Gamma_{\uparrow/\downarrow}$ separates all the poles of $R^{-1}(\lambda)$ from $\mathbb{R}$ (they are indicated by $\circledast$), and is such that $\mathrm{dist}(\Gamma_{\uparrow/\downarrow}, \mathbb{R}) > \delta$ for some $\delta > 0$ but sufficiently small

$$||G_{\Psi} - I_2||_{\mathcal{M}_2(L^2(\Sigma_{\Psi}))} + ||G_{\Psi} - I_2||_{\mathcal{M}_2(L^{\infty}(\Sigma_{\Psi}))} = \mathrm{O}\big(\mathrm{e}^{-\varkappa_{\epsilon} N^{\alpha}}\big) , \tag{4.2.6}$$

with

$$\varkappa_{\epsilon} = (b_N - a_N) \cdot \min\Big\{ \inf_{\lambda \in \Gamma_{\uparrow} \cup \Gamma_{\downarrow}} |\mathrm{Im}\, \lambda| \, ; \, 2\gamma\big(\pi \min[\omega_1, \omega_2] - \epsilon\big)\Big\} . \tag{4.2.7}$$

Note that we have a freedom of choice of the curves $\Gamma_{\uparrow/\downarrow}$, provided that these avoid (respectively from below/above) all the poles of $R^{-1}(\lambda)$ in $\mathbb{H}^{+/-}$. As a consequence, we have the natural bound:

$$\inf_{\lambda \in \Gamma_{\uparrow} \cup \Gamma_{\downarrow}} |\mathrm{Im}\, \lambda| \le \frac{2\pi\, \omega_1 \omega_2}{\omega_1 + \omega_2} . \tag{4.2.8}$$

These bounds are enough so as to solve the Riemann–Hilbert problem for Ψ. Indeed, introduce the singular integral operator on the space $\mathcal{M}_2\big(L^2(\Sigma_{\Psi})\big)$ of 2×2 matrix-valued $L^2\big(\Sigma_{\Psi}\big)$ functions by

$$\mathcal{C}_{\Sigma_{\Psi}}^{(-)}\big[\Pi\big](\lambda) = \lim_{\substack{z \to \lambda \\ z \in -\text{side of } \Sigma_{\Psi}}} \int_{\Sigma_{\Psi}} \frac{(G_{\Psi} - I_2)(t) \cdot \Pi(t)}{t - z} \cdot \frac{\mathrm{d}t}{2\mathrm{i}\pi} . \tag{4.2.9}$$

Since $G_{\Psi} - I_2 \in \mathcal{M}_2\Big(\big(L^{\infty} \cap L^2\big)(\Sigma_{\Psi})\Big)$ and Σ_{Ψ} is a Lipschitz curve, it follows from Theorem A.0.13 that $\mathcal{C}^{(-)}_{\Sigma_{\Psi}}$ is a continuous endomorphism on $\mathcal{M}_2(L^2(\Sigma_{\Psi}))$ that furthermore satisfies:

$$|||\mathcal{C}^{(-)}_{\Sigma_{\Psi}}|||_{\mathcal{M}_2(L^2(\Sigma_{\Psi}))} \;\leq\; C\mathrm{e}^{-\varkappa_{\epsilon} N^{\alpha}} \;. \tag{4.2.10}$$

Hence, since

$$G_{\Psi} - I_2 \in \mathcal{M}_2\Big(L^2\big(\Sigma_{\Psi}\big)\Big) \quad \text{and} \quad \mathcal{C}^{(-)}_{\Sigma_{\Psi}}[I_2] \in \mathcal{M}_2\Big(L^2\big(\Sigma_{\Psi}\big)\Big) \tag{4.2.11}$$

provided that N is large enough, it follows that the singular integral equation

$$\Big(I_2 - \mathcal{C}^{(-)}_{\Sigma_{\Psi}}\Big)\big[\Pi_-\big] \;=\; I_2 \tag{4.2.12}$$

admits a unique solution Π_- such that $\Pi_- - I_2 \in \mathcal{M}_2\big(L^2(\Sigma_{\Psi})\big)$. The bound (4.2.6) also implies that:

$$||\Pi_- - I_2||_{\mathcal{M}_2(L^2(\Sigma_{\Psi}))} \leq 1 \tag{4.2.13}$$

for N large enough. It is then a standard fact [2] in the theory of Riemann–Hilbert problems that the matrix

$$\Pi(\lambda) \;=\; I_2 \;+\; \int_{\Sigma_{\Psi}} \frac{(G_{\Psi} - I_2)(t) \cdot \Pi_-(t)}{t-\lambda} \cdot \frac{\mathrm{d}t}{2\mathrm{i}\pi} \tag{4.2.14}$$

is the unique solution to the Riemann–Hilbert problem:

- $\Pi \in \mathcal{O}(\mathbb{C}\setminus\Sigma_{\Psi})$ and has continuous $\pm$ boundary values on Σ_{Ψ};
- $\Pi(\lambda) = I_2 \,+\, \mathrm{O}\big(\lambda^{-1}\big)$ when $\lambda \to \infty$ non-tangentially to Σ_{Ψ};
- $\Pi_+(\lambda) \;=\; G_{\Psi}(\lambda) \cdot \Pi_-(\lambda)$ for $\lambda \in \Sigma_{\Psi}$.

We claim that for any open neighbourhood U of Σ_{Ψ} such that $\mathrm{dist}(\Sigma_{\Psi}, \partial U) > \delta > 0$, there exists a constant $C > 0$ such that:

$$\forall \lambda \in U, \qquad \max_{a,b \in \{1,2\}} \big[\Pi(\lambda) - I_2\big]_{ab} \leq \frac{C\, e^{-\varkappa_{\epsilon} N^{\alpha}}}{1+|\lambda|} \;. \tag{4.2.15}$$

Indeed, we can write:

$$\max_{a,b\in\{1,2\}} \big[\Pi(\lambda) - I_2\big]_{ab} \leq \max_{a,b\in\{1,2\}} \bigg| \int_{\Sigma_{\Psi}} \frac{(G_{\Psi} - I_2)_{ab}(t)}{t-\lambda} \cdot \frac{\mathrm{d}t}{2\mathrm{i}\pi} \bigg| \\ + \sum_{a,b\in\{1,2\}} ||\Pi_- - I_2||_{\mathcal{M}_2(L^2(\Sigma_{\Psi}))} \cdot \bigg(\int_{\Sigma_{\Psi}} \frac{\big|(G_{\Psi} - I_2)_{ab}(t)\big|^2}{|t-\lambda|^2} \cdot \frac{|\mathrm{d}t|}{(2\pi)^2} \bigg)^{1/2} . \tag{4.2.16}$$

The second term is readily bounded with (4.2.13) and the fact (4.2.6) that G_Ψ is exponentially close to the identity matrix. For the first term, we study the asymptotic behaviour of $G_\Psi - I_2$ with help of Section 4.1.2:

$$\text{if } t \in \Gamma_\downarrow \cup \Gamma_\uparrow, \qquad |(G_\Psi - I_2)_{ab}(t)| \leq C\, e^{-|\mathrm{Re}\, t|} \cdot e^{-\varkappa_\epsilon N^\alpha}\,, \tag{4.2.17}$$

$$\text{if } t \in \mathbb{R} + \mathrm{i}\epsilon, \qquad |(G_\Psi - I_2)_{ab}(t)| \leq C\, |t|^{-1} \cdot e^{-\varkappa_\epsilon N^\alpha}\,. \tag{4.2.18}$$

For the contribution on $\mathbb{R}+\mathrm{i}\epsilon$, we split $[G_\Psi - I_2](t) = C_\Psi \cdot t^{-1} + O(t^{-2})$. We compute directly the contour integral of the term in t^{-1}, and find the bound $\max_{a,b} \big|[C_\Psi]_{ab} \cdot \lambda^{-1}\big|$ if $\mathrm{Im}\, \lambda > \epsilon$, and 0 otherwise. Hence, it is bounded by $c_1/(1+|\lambda|)$ for some constant $c_1 > 0$. The contribution of the remainder $O(|t|^{-2})$ to the contour integral can be bounded thanks to the lower bound $\mathrm{dist}(\Sigma_\Psi, \lambda) \geq c_2/(1+|\lambda|)$ for some constant $c_2 > 0$. Collecting all these bounds justifies (4.2.15).

The Riemann–Hilbert problem for Ψ and Π have the same jump matrix G_Ψ, but Ψ must have a zero with prescribed leading coefficient at $\lambda = 0$, while Π has a finite value $\Pi(0)$. We then see that the formula:

$$\Psi(\lambda) \;=\; \Pi(\lambda) \cdot P_R(\lambda) \tag{4.2.19}$$

with:

$$P_R(\lambda) = I_2 + \frac{\theta_R}{\lambda} \cdot \Pi^{-1}(0)\sigma^{-}\Pi(0)\,, \quad \text{and} \quad \theta_R = \frac{1}{\upsilon^2(0)} \frac{\kappa_N}{1 + \kappa_N/(\omega_1 + \omega_2)} \tag{4.2.20}$$

yields the unique solution to the Riemann–Hilbert problem for Ψ. Tracking back the transformations $\Pi \rightsquigarrow \Psi \rightsquigarrow \chi$, gives the construction of the solution χ of the Riemann–Hilbert problem of Proposition 4.2.1, summarized in Figure 4.2. This concludes the proof of Proposition 4.2.1. ■

4.2.3 *Properties of the Solution* χ

Lemma 4.2.2 *The solution χ to the Riemann–Hilbert problem given in Proposition 4.2.1 admits the following symmetries*

$$\chi(-\lambda) = \begin{pmatrix} 1 & 0 \\ 0 & -1 \end{pmatrix} \cdot \chi(\lambda) \cdot \begin{pmatrix} 1 & -\lambda \\ 0 & 1 \end{pmatrix} \quad \text{and} \quad \big(\chi(\lambda^*)\big)^* = \begin{pmatrix} 1 & 0 \\ 0 & -1 \end{pmatrix} \cdot \chi(-\lambda) \cdot \begin{pmatrix} -1 & 0 \\ 0 & 1 \end{pmatrix} \tag{4.2.21}$$

where * *refers to the component-wise complex conjugation.*

Proof Since $G_\chi(-\lambda) = \mathrm{e}^{\frac{\mathrm{i}\pi\sigma_3}{2}} G_\chi^{-1}(\lambda) \mathrm{e}^{-\frac{\mathrm{i}\pi\sigma_3}{2}}$, the matrix:

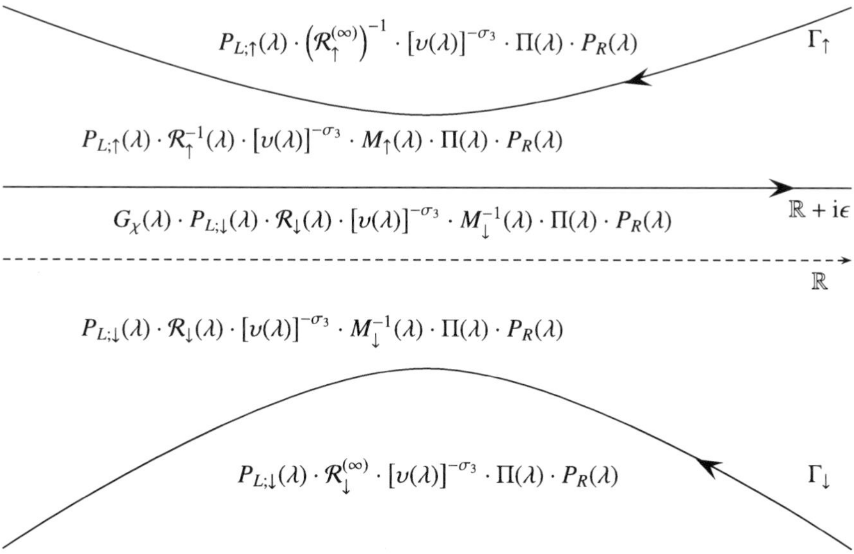

Fig. 4.2 Piecewise definition of the matrix χ. The curves $\Gamma_{\uparrow/\downarrow}$ separate all poles of $\lambda\mapsto R^{-1}(\lambda)$ from $\mathbb{R}$ and are such that $\mathrm{dist}(\Gamma_{\uparrow/\downarrow},\mathbb{R}) > \delta > \epsilon > 0$ for a sufficiently small δ. The matrix Π appearing here is defined through (4.2.14)

$$\Xi(\lambda) \;=\; \chi^{-1}(\lambda)\cdot \mathrm{e}^{-\frac{\mathrm{i}\pi\sigma_3}{2}}\cdot\chi(-\lambda) \tag{4.2.22}$$

is continuous across $\mathbb{R}$ and thus is an entire function. The asymptotic behaviour of $\Xi(\lambda)$ when $\lambda\to\infty$ is deduced from the growth conditions prescribed by the Riemann–Hilbert problem (*cf.* Proposition 4.2.1):

$$\Xi(\lambda) \;=\; \mathrm{i}\lambda\cdot\sigma^{+} \;-\mathrm{i}\big(\chi_1\cdot\sigma^{+} \;+\; \sigma^{+}\cdot\chi_1\big) \;+\; \mathrm{O}(\lambda^{-1})\,. \tag{4.2.23}$$

Since $\Xi(\lambda)$ is entire, by Liouville theorem this asymptotic expression is exact, namely

$$\Xi(\lambda) \;=\; \mathrm{i}\lambda\cdot\sigma^{+} \;-\mathrm{i}\big(\chi_1\cdot\sigma^{+} \;+\; \sigma^{+}\cdot\chi_1\big)\,. \tag{4.2.24}$$

Observe that

$$\chi_1\cdot\sigma^{+} \;+\; \sigma^{+}\cdot\chi_1 \;=\; \begin{pmatrix} [\chi_1]_{21} & \mathrm{tr}[\chi_1] \\ 0 & [\chi_1]_{21}\end{pmatrix}\,. \tag{4.2.25}$$

By expanding the relation $\det[\chi(\lambda)] = 1$ for $\lambda\in\mathbb{H}^{+}$ at large λ, we find that the matrix χ_1 is actually traceless. Finally, the jump condition at $\lambda=0$ takes the form

$$\chi_{-}(0) \;=\; \sigma_3\cdot\chi_{+}(0)\,. \tag{4.2.26}$$

Using this relation and the expression for Ξ given in (4.2.24), we get:

$$-\,\mathrm{i}\chi_{+}(0) \;=\; -\mathrm{i}\chi_{+}(0)\cdot\begin{pmatrix}[\chi_1]_{21} & 0\\ 0 & [\chi_1]_{21}\end{pmatrix} \qquad i.e. \qquad [\chi_1]_{21}=1 \tag{4.2.27}$$

since $\chi_{+}(0)$ is invertible. This proves the first relation in (4.2.21). In order to establish the second one, we consider:

$$\widetilde{\Xi}(\lambda) \;=\; \chi^{-1}(-\lambda)\cdot \mathrm{e}^{\frac{\mathrm{i}\pi\sigma_3}{2}}\cdot\big(\chi(\lambda^*)\big)^{*}\,. \tag{4.2.28}$$

With the relation $\big(G_{\chi}(\lambda^*)\big)^{*} \;=\; G_{\chi}^{-1}(\lambda)$ and the complex conjugate of the asymptotic behaviour for χ, one shows that $\widetilde{\Xi}$ is holomorphic on $\mathbb{C}\setminus\mathbb{R}$, continuous across $\mathbb{R}$ and hence entire. Furthermore, since it admits the asymptotic behaviour

$$\widetilde{\Xi}(\lambda) \;=\; \mathrm{e}^{-\frac{\mathrm{i}\pi\sigma_3}{2}}\cdot\Big(I_2 \,+\, \mathrm{O}(\lambda^{-1})\Big)\,, \tag{4.2.29}$$

by Liouville's theorem, $\widetilde{\Xi}(\lambda)=e^{-\frac{\mathrm{i}\pi\sigma_3}{2}}$. ■

Lemma 4.2.3 *The matrix χ admits the large-λ, $\lambda\in\mathbb{H}^{+}$ asymptotic expansion*

$$\chi(\lambda) \;\simeq\; \big(-\,\mathrm{i}\lambda\big)^{1/2}\cdot\sigma^{+} \,+\, \sum_{k\ge 0}\frac{K(\lambda)\cdot\chi_k-\mathrm{i}\sigma^{+}\cdot\chi_{k+1}}{\big(-\,\mathrm{i}\lambda\big)^{1/2}\lambda^{k}}\,, \tag{4.2.30}$$

where $(\chi_k)_k$ is a sequence of constant, 2×2 matrices, with $\chi_{-1}=0$, $\chi_0=I_2$, and:

$$K(\lambda) \;=\; \begin{pmatrix} -\mathrm{sgn}(\mathrm{Re}\,\lambda)\,\mathrm{e}^{\mathrm{i}\lambda\overline{x}_N} & 0\\ -\dfrac{\kappa_N}{\lambda}\cdot\mathrm{sgn}(\mathrm{Re}\,\lambda)\,\mathrm{e}^{\mathrm{i}\lambda\gamma\overline{x}_N}\;-\;1 & -\mathrm{i}\kappa_N\cdot\mathrm{sgn}(\mathrm{Re}\,\lambda)\,\mathrm{e}^{\mathrm{i}\lambda(\gamma-1)\overline{x}_N}\end{pmatrix}. \tag{4.2.31}$$

In particular, we have:

$$[\chi]_{11}(\lambda)\simeq\frac{1}{\big(-\,\mathrm{i}\lambda\big)^{1/2}}\sum_{k\ge 0}\frac{1}{\lambda^{k}}\Big[-\,\mathrm{sgn}(\mathrm{Re}\lambda)\mathrm{e}^{\mathrm{i}\lambda\overline{x}_N}[\chi_k]_{11}-\mathrm{i}\,[\chi_{k+1}]_{21}\Big], \tag{4.2.32}$$

$$\lambda^{-1}\cdot[\chi]_{12}(\lambda)\simeq\frac{1}{\big(-\,\mathrm{i}\lambda\big)^{1/2}}\sum_{k\ge 0}\frac{1}{\lambda^{k}}\Big[-\,\mathrm{sgn}(\mathrm{Re}\lambda)\mathrm{e}^{\mathrm{i}\lambda\overline{x}_N}[\chi_{k-1}]_{12}-\mathrm{i}[\chi_k]_{22}\Big]. \tag{4.2.33}$$

We remind that $[\chi_1]_{21}=1$. Also $\simeq$ means equality up to a $O(\lambda^{-\infty})$ remainder.

Proof It is enough to establish that Π admits, for any ℓ, the large-λ asymptotic expansion of the form:

$$\Pi(\lambda) = \sum_{\ell=0}^{k} \lambda^{-\ell}\,\Pi_\ell + \Delta_{[k]}\Pi(\lambda) \quad \text{with}$$

$$\Delta_{[k]}\Pi(\lambda) = \mathrm{O}\Big(\frac{1}{\lambda^{k+1-\delta}}\Big) \quad \text{for any } \delta > 0 \quad \text{and} \quad \Pi_0 = I_2\,. \tag{4.2.34}$$

Indeed, once this asymptotic expansion is established for Π, the results for χ follow from matrix multiplications prescribed on the top of Figure 4.2.

Equation (4.2.15) shows that the expansion (4.2.34) holds for $k = 0$ uniformly away from Σ_Ψ. This is actually valid everywhere, for the jump matrix $G_\Psi(\lambda)$ is analytic in a neighbourhood of Σ_Ψ and asymptotically close to I_2 at large λ in an open neighbourhood of Σ_Ψ, *c.f.* (4.2.17) and (4.2.18).

Now assume that the expansion holds up to some order k. Consider the integral representation (4.2.14) for Π. We recall that $(\Pi_- - I_2) \in L^2(\Sigma_\Psi)$ and $G_\Psi - I_2$ decays exponentially fast along $\Gamma_\uparrow \cup \Gamma_\downarrow$. Thus, standard manipulations give an asymptotic expansion of the form:

$$\int_{\Gamma_\uparrow\cup\Gamma_\downarrow} \frac{(G_\Psi - I_2)(t)\cdot\Pi_-(t)}{t-\lambda}\cdot\frac{\mathrm{d}t}{2\mathrm{i}\pi} \simeq \sum_{\ell\geq 1} T_\ell\,\lambda^{-\ell}\,. \tag{4.2.35}$$

It thus remains to focus on the integral on $\mathbb{R}+\mathrm{i}\epsilon$. We can first move the contour to $\mathbb{R}+\mathrm{i}\epsilon'$ for some $0 < \epsilon' < \epsilon$, and insert the assumed asymptotic expansion at order k:

$$\begin{aligned}\int_{\mathbb{R}+\mathrm{i}\epsilon} \frac{(G_\Psi - I_2)(t)\cdot\Pi_-(t)}{t-\lambda}\cdot\frac{\mathrm{d}t}{2\mathrm{i}\pi} &= \sum_{\ell=0}^{k} \int_{\mathbb{R}+\mathrm{i}\epsilon'} \frac{(G_\Psi - I_2)(t)\cdot\Pi_\ell}{t^\ell(t-\lambda)}\cdot\frac{\mathrm{d}t}{2\mathrm{i}\pi} \\ &\quad + \int_{\mathbb{R}+\mathrm{i}\epsilon'} \frac{(G_\Psi - I_2)(t)\cdot\Delta_{[k]}\Pi(t)}{t-\lambda}\cdot\frac{\mathrm{d}t}{2\mathrm{i}\pi}\,.\end{aligned} \tag{4.2.36}$$

It follows from (4.1.19) that we can decompose

$$r_N(\lambda) = r_N^{(+)}(\lambda)\mathrm{e}^{\mathrm{i}\lambda\gamma\bar{x}_N} + r_N^{(-)}(\lambda)\mathrm{e}^{-\mathrm{i}\lambda\gamma\bar{x}_N},$$

with $r_N^{(\pm)}(\lambda)$ bounded in λ away from its poles. This induces a decomposition $G_\Psi - I_2 = (G_\Psi - I_2)^{(+)} + (G_\Psi - I_2)^{(-)}$ on $\mathbb{R}+\mathrm{i}\epsilon'$. Inspecting the expression (4.2.5), one can convince oneself that there exist curves $\mathscr{C}_{G_\Psi}^{\pm} \subseteq \mathbb{H}^\pm$ going to ∞ when $\mathrm{Re}\,t \to \pm\infty$, $t \in \mathscr{C}_{G_\Psi}^{\pm}$ and such that:

- $t \mapsto \frac{(G_\Psi - I_2)^{(\pm)}(t)\cdot\Pi_\ell}{t^\ell\cdot(t-\lambda)}$ has no pole between $\mathbb{R}+\mathrm{i}\epsilon'$ and $\mathscr{C}_{G_\Psi}^{\pm}$,
- $(G_\Psi - I_2)^{(\pm)}(t)$ decays exponentially fast in t when $t \to \infty$ along $\mathscr{C}_{G_\Psi}^{\pm}$.

Therefore, we obtain:

$$\int_{\mathbb{R}+\mathrm{i}\epsilon'} \frac{(G_\Psi - I_2)(t)\cdot \Pi_\ell}{t^\ell (t-\lambda)}\cdot\frac{\mathrm{d}t}{2\mathrm{i}\pi} = \int_{\mathscr{C}^+_{G_\Psi}} \frac{(G_\Psi - I_2)^{(+)}(t)\cdot \Pi_\ell}{t^\ell (t-\lambda)}\cdot\frac{\mathrm{d}t}{2\mathrm{i}\pi} + \int_{\mathscr{C}^-_{G_\Psi}} \frac{(G_\Psi - I_2)^{(-)}(t)\cdot \Pi_\ell}{t^\ell (t-\lambda)}\cdot\frac{\mathrm{d}t}{2\mathrm{i}\pi} \qquad (4.2.37)$$

and the properties of this decomposition ensure the existence of an all order asymptotic expansion in λ^{-1} when $\lambda \to \infty$. It thus remains to focus on the last term present in (4.2.36). For $\delta > 0$ but small, we write:

$$\int_{\mathbb{R}+\mathrm{i}\epsilon'} \frac{(G_\Psi - I_2)(t)\cdot \Delta_{[k]}\Pi(t)}{t-\lambda}\cdot\frac{\mathrm{d}t}{2\mathrm{i}\pi} = -\sum_{\ell=0}^{k}\frac{1}{\lambda^{\ell+1}}\int_{\mathbb{R}+\mathrm{i}\epsilon'} t^\ell (G_\Psi - I_2)(t) \times \Delta_{[k]}\Pi(t)\cdot\frac{\mathrm{d}t}{2\mathrm{i}\pi} + \frac{\Delta_{[k]}T(\lambda)}{\lambda^{k+1}|\lambda|^{1-2\delta}}\,. \qquad (4.2.38)$$

The decay at ∞ of $\Delta_{[k]}\Pi$ and $(G_\Psi - I_2)$ guarantees the existence of an asymptotic expansion of the first term in the right-hand side, this up to a $\mathrm{O}(\lambda^{-k-2})$ remainder. Finally, we have:

$$|\Delta_{[k]}T(\lambda)| = \left|\int_{\mathbb{R}+\mathrm{i}\epsilon'} t^{k+1}\frac{(G_\Psi - I_2)(t)\cdot \Delta_{[k]}\Pi(t)}{|\lambda|^{2\delta-1}\cdot(t-\lambda)}\cdot\frac{\mathrm{d}t}{2\mathrm{i}\pi}\right| \le C\int_{\mathbb{R}+\mathrm{i}\epsilon'} \frac{|\lambda|^{1-2\delta}\,\mathrm{d}t}{|t|^{1-\delta}\,|t-\lambda|}\,, \qquad (4.2.39)$$

where we used the assumed bound given below of (4.2.34) for $\Delta_{[k]}\Pi(t)$ and the $O(1/t)$ decay (4.2.18) for $G_\Psi - I_2$. The growth of the right-hand side at large λ is then estimated by cutting the integral into pieces:

$$\frac{|\lambda|^{1-2\delta}}{|t|^{1-\delta}\,|t-\lambda|} \le \begin{cases} \widetilde{C}\,|\lambda|^{-2\delta}\,|t|^{-(1-\delta)} & \text{if } |\mathrm{Re}\,t| \le |\lambda|/2 \\ \widetilde{C}\,|\lambda|^{-\delta}\,|t-\lambda|^{-1} & \text{if } |\lambda|/2 \le |\mathrm{Re}\,t| \le 3|\lambda|/2 \\ \widetilde{C}\,|t|^{-(1+\delta)} & \text{if } |\mathrm{Re}\,t| \ge 3|\lambda|/2 \end{cases} \qquad (4.2.40)$$

for some $\widetilde{C} > 0$ independent of λ and t. The integral over t of the right-hand side on each of piece is finite, and collecting all the pieces, we get $\Delta_{[k]}T(\lambda) = \mathrm{o}(1)$ when $\lambda \to \infty$. ■

4.3 The Inverse of the Operator $\mathcal{S}_N$

4.3.1 Solving $\mathscr{S}_{N;\gamma}[\varphi] = h$ for $h \in H_s([0\,;\overline{x}_N])$, $-1 < s < 0$

With the 2×2 matrix χ in hand, we can come back to the inversion of the integral operator $\mathscr{S}_{N;\gamma}$ according to Lemma 4.1.1.

Proposition 4.3.1 *Assume* $-1 < s < 0$, *and* $h \in H_s([0\,;\overline{x}_N])$. *Any solution to* $\mathscr{S}_{N;\gamma}\big[\varphi\big](\xi) = h(\xi)$ *is of the form* $\varphi = \widetilde{\mathscr{W}}_{\vartheta;z_0}[h_\mathfrak{e}]$ *where*

$$\widetilde{\mathscr{W}}_{\vartheta;z_0}[h_\mathfrak{e}] = \mathcal{F}^{-1}\Big[(*-z_0)\,\chi_{11;+}\cdot\mathcal{C}_+\big[f_{1;z_0}\big] + \chi_{12;+}\cdot\mathcal{C}_+\big[f_{2;z_0}\big] + \vartheta\cdot\chi_{11;+}\Big]. \tag{4.3.1}$$

Above, $\vartheta \in \mathbb{C}$ *and* $z_0 \in \mathbb{C}\setminus\mathbb{R}$ *are arbitrary constants. We remind that* χ_+ *is the upper boundary value of* χ *on* $\mathbb{R}$, $\mathcal{C}$ *is the Cauchy transform (1.6.14),* $\mathcal{C}_\pm$ *its* $\pm$ *boundary values and* $h_\mathfrak{e}$ *is any extension of* h *to* $H_s(\mathbb{R})$.

$$\begin{pmatrix} f_{1;z_0}(\lambda) \\ f_{2;z_0}(\lambda)\end{pmatrix} = \mathrm{e}^{-\mathrm{i}\lambda\overline{x}_N}\mathcal{F}[h_\mathfrak{e}](\lambda)\cdot\begin{pmatrix}(\lambda-z_0)^{-1}\chi_{12;+}(\lambda) \\ -\chi_{11;+}(\lambda)\end{pmatrix}. \tag{4.3.2}$$

The transform $\widetilde{\mathscr{W}}_{\vartheta;z_0}$ *is continuous on* $H_s(\mathbb{R})$, $-1 < s < 0$:

$$||\widetilde{\mathscr{W}}_{0;z_0}[h_\mathfrak{e}]||_{H_s(\mathbb{R})} \le C_N\,||h_\mathfrak{e}||_{H_s(\mathbb{R})}\,, \tag{4.3.3}$$

the continuity constant C_N *being however dependent, a priori, on* N. *Finally, when* $h \in \mathcal{C}^1([0\,;\overline{x}_N])$ *the transform can be recast as*

$$\widetilde{\mathscr{W}}_{\vartheta;z_0}[h](\xi) = \int\limits_{\mathbb{R}+2\mathrm{i}\epsilon'}\frac{\mathrm{d}\lambda}{2\pi}\int\limits_{\mathbb{R}+\mathrm{i}\epsilon'}\frac{\mathrm{d}\mu}{2\mathrm{i}\pi}\frac{\mathrm{e}^{-\mathrm{i}\xi\lambda-\mathrm{i}\overline{x}_N\mu}}{\mu-\lambda}\Big\{\frac{\lambda-z_0}{\mu-z_0}\chi_{11}(\lambda)\chi_{12}(\mu)-\chi_{11}(\mu)\chi_{12}(\lambda)\Big\}\cdot\int\limits_0^{\overline{x}_N}\mathrm{e}^{\mathrm{i}\eta\mu}h(\eta)\cdot\mathrm{d}\eta$$
$$+\,\vartheta\int\limits_{\mathbb{R}+\mathrm{i}\epsilon'}\mathrm{e}^{-\mathrm{i}\lambda\xi}\chi_{11}(\lambda)\cdot\frac{\mathrm{d}\lambda}{2\pi}\,. \tag{4.3.4}$$

where $\epsilon' > \epsilon$ *is arbitrary but small enough and such that* $\operatorname{Im} z_0 > \epsilon'$ *in the case when* $z_0 \in \mathbb{H}^+$.

We stress that the integrals, as written in (4.3.4), are to be understood in the Riemann sense in that they only converge as oscillatory integrals.

Proof The proof is based on a Wiener–Hopf factorisation. For the moment, we only assume that $s < 0$. Let Φ be any solution to the vector Riemann–Hilbert problem for Φ outlined in Lemma 4.1.1. Then, define a piecewise holomorphic function Υ by

$$\Upsilon(\lambda) = \begin{cases} \chi^{-1}(\lambda)\Phi(\lambda) - \widehat{\boldsymbol{H}}(\lambda) & \lambda \in \mathbb{H}^+ \\ \chi^{-1}(\lambda)\Phi(\lambda) - \widehat{\boldsymbol{H}}(\lambda) & \lambda \in \mathbb{H}^- \end{cases} \tag{4.3.5}$$

where, for some $z_0 \in \mathbb{C}\setminus\mathbb{R}$

$$\widehat{\boldsymbol{H}}(\lambda) = \begin{pmatrix} (\lambda - z_0)^{\iota_s} \int\limits_{\mathbb{R}} \frac{g_{1;\iota_s}(t)\,\mathrm{d}t}{2\mathrm{i}\pi(t-\lambda)} \\ (\lambda - z_0)^{\iota_s - 1} \int\limits_{\mathbb{R}} \frac{g_{1;\iota_s-1}(t)\,\mathrm{d}t}{2\mathrm{i}\pi(t-\lambda)} \end{pmatrix} \quad \text{with} \quad \begin{pmatrix} g_1(\lambda) \\ g_2(\lambda) \end{pmatrix} = \chi_+^{-1}(\lambda) \cdot \boldsymbol{H}(\lambda) \,. \tag{4.3.6}$$

Above, taking into account that $s < 0$, we have set

$$g_{a;\iota_s}(t) = (t - z_0)^{-\iota_s} g_a(t) \quad \text{with} \quad \iota_s = k \quad \text{for} \ -k < s < -(k-1) \,. \tag{4.3.7}$$

It follows from the asymptotic behaviour for $\chi_+(\lambda)$ at large λ that $g_1 \in \mathcal{F}\big[H_{s-1/2}\big]$ and $g_2 \in \mathcal{F}\big[H_{s+1/2}\big]$. Recall that Theorem A.0.11 ensures that the $\pm$ boundary values $\mathcal{C}_\pm$ of the Cauchy transform on $\mathbb{R}$ are continuous operators on $H_\tau(\mathbb{R})$ for any $|\tau| < 1/2$. Thus, $\mathcal{C}_\pm[g_{1;\iota_s}] \in H_{s+k-1/2}(\mathbb{R})$ as well as $\mathcal{C}_\pm[g_{2;\iota_s-1}] \in H_{s+k-1/2}(\mathbb{R})$, which implies:

$$\widehat{\boldsymbol{H}}_{a;\pm} \in \mathcal{F}\big[H_{s_a}(\mathbb{R})\big] \qquad \text{with } s_1 = s - 1/2 \text{ and } s_2 = s + 1/2 \,. \tag{4.3.8}$$

Equation (4.1.6) ensures that, uniformly in $\mu > 0$,

$$\forall a \in \{1, 2\}, \qquad \int\limits_{\mathbb{R}} \big|\Upsilon_a(\lambda \pm \mathrm{i}\mu)\big|^2 \big(1 + |\lambda| + |\mu|\big)^{2s_a} \mathrm{d}\lambda < C \,. \tag{4.3.9}$$

The discontinuity equation satisfied by Φ along with $\widehat{\boldsymbol{H}}_{a;+} - \widehat{\boldsymbol{H}}_{a;-} = g_a$ guarantee that $\Upsilon_a \in \mathcal{O}(\mathbb{C}\setminus\mathbb{R})$ admits $\mathcal{F}\big[H_{s_a}(\mathbb{R})\big]$ $\pm$ boundary values that are equal. Then, straightforward manipulations show that, in fact, Υ is entire. Furthermore, for any $\ell \in \mathbb{N}$ such that $s_a + \ell > -1/2$ and for any $\mu > |\mathrm{Im}\, z|$, we have:

$$\partial_z^\ell \Upsilon_a(z) = \sum_{\epsilon=\pm} \epsilon \int\limits_{\mathbb{R}} \frac{\ell!\,\Upsilon_a(\lambda + \mathrm{i}\epsilon\mu)}{(\lambda + \mathrm{i}\epsilon\mu - z)^{\ell+1}} \frac{\mathrm{d}\lambda}{2\mathrm{i}\pi} \,. \tag{4.3.10}$$

Thus

$$\begin{aligned} \big|\partial_z^\ell \Upsilon_a(z)\big| \le & \ \frac{1}{\pi} \max_{\epsilon=\pm} \bigg(\int_{\mathbb{R}} \frac{\big(1 + |\lambda| + |\mu|\big)^{-2s_a}}{|\lambda + \mathrm{i}\epsilon\mu - z|^{2(\ell+1)}} \mathrm{d}\lambda \bigg)^{1/2} \\ & \times \bigg(\int_{\mathbb{R}} |\Upsilon_a(\lambda + \mathrm{i}\epsilon\mu)|^2 \big(1 + |\lambda| + |\mu|\big)^{2s_a} \mathrm{d}\lambda \bigg)^{1/2} \end{aligned} \tag{4.3.11}$$

where the last integral factor is bounded. So far, the parameter μ was arbitrary. We stress that the constant C in (4.3.9) is uniform in μ. Thus taking $\mu = 2|z|$ and assuming that $|z| > 1/2$, we find:

$$\left|\partial_z^\ell \Upsilon_a(z)\right| \leq C' \, |z|^{-(s_a+\ell+1/2)} \cdot \left(\int_{\mathbb{R}} \frac{\big(|\lambda|+2\big)^{-2s_a}}{\big[(\lambda-1)^2+1\big]^{\ell+1}} \, \mathrm{d}\lambda \right)^{\frac{1}{2}} . \tag{4.3.12}$$

In particular, reminding the values of s_a in (4.3.8), we find that $\partial_z^{k-1}\Upsilon_2(z)$ and $\partial_z^k \Upsilon_1(z)$ are entire and bounded, so they must be constant. These constants are zero due to (4.3.12). Hence, there exist polynomials $P_1 \in \mathbb{C}_{k-1}[X]$ and $P_2 \in \mathbb{C}_{k-2}[X]$ such that

$$\Upsilon(z) = \begin{pmatrix} P_1(z) \\ P_2(z) \end{pmatrix} . \tag{4.3.13}$$

Reciprocally, it is readily seen that the piecewise analytic vector

$$\Phi(\lambda) = \chi(\lambda) \cdot \widehat{\boldsymbol{H}}(\lambda) + \chi(\lambda) \cdot \begin{pmatrix} P_1(z) \\ P_2(z) \end{pmatrix} \quad \text{with} \quad P_a \in \mathbb{C}_{k-a}[X] \quad \text{for} \quad -k < s < -(k-1) \tag{4.3.14}$$

provides solutions to the Riemann–Hilbert problem for Φ.

From now on, we focus on the case $k = 1$, *i.e.* $h \in H_s([0\,;\overline{x}_N])$ for $-1 < s < 0$. Then, it follows from Lemma 4.1.1 that any solution to $\mathscr{S}_{N;\gamma}[\varphi] = h$ takes the form $\varphi = \widetilde{\mathscr{W}}_{\vartheta;z_0}[h_\mathfrak{e}]$, with:

$$\begin{aligned} \mathcal{F}\big[\widetilde{\mathscr{W}}_{\vartheta;z_0}[h_\mathfrak{e}]\big](\lambda) &= \Phi_{1;+}(\lambda) \\ &= \chi_{11;+}(\lambda)\cdot(\lambda - z_0)\mathcal{C}_+[f_{1;z_0}](\lambda) + \chi_{12;+}(\lambda)\cdot \mathcal{C}_+[f_{2;z_0}](\lambda) + \vartheta\cdot\chi_{11;+}(\lambda) \end{aligned} \tag{4.3.15}$$

with $f_{a;z_0}$'s given by (4.3.2).

It is then readily inferred from the asymptotic expansion for χ at $\lambda \to \infty$ given in Lemma 4.2.3, and from the jump conditions satisfied by χ, that indeed $\Phi_{1;+} \in \mathcal{F}\big[H_s([0\,;\overline{x}_N])\big]$. Also the continuity on $\mathcal{F}\big[H_\tau(\mathbb{R})\big]$ with $|\tau| < 1/2$ of the $\pm$ boundary values $\mathcal{C}_\pm$ of the Cauchy transform, *cf.* Theorem A.0.11, ensures that

$$||\Phi_{1;+}||_{\mathcal{F}[H_s(\mathbb{R})]} \leq C\,||h_\mathfrak{e}||_{H_s(\mathbb{R})} , \tag{4.3.16}$$

which in turn implies the bound (4.3.3).

It solely remains to prove the regularised expression (4.3.4). Given $h \in C^1([0\,;\overline{x}_N])$ it is clear that $h \in H_s([0\,;\overline{x}_N])$ for any $s < 1/2$. We chose the specific extension $h_\mathfrak{e} = h$. Then, it follows from the previous discussion that $\widetilde{\mathscr{W}}_{\vartheta;z_0}[h] \in H_s([0\,;\overline{x}_N])$. The integral in the right-hand side of (4.3.4), considered in the Riemann sense, defines a continuous function on $[0\,;\overline{x}_N]$, that we denote momentarily $\widetilde{\mathscr{V}}_{\vartheta;z_0}[h]$. Now, for any $f \in C^\infty([0\,;\overline{x}_N])$, starting with the expression (4.3.1) for $\widetilde{\mathscr{W}}_{\vartheta;z_0}[h]$, we have:

$$\begin{aligned}
\big(f,\widetilde{\mathscr{W}}_{\vartheta;z_0}[h]\big) = \big(\mathcal{F}[f],\Phi_{1;+}\big) &= \int_{\mathbb{R}} \mathcal{F}[f^*](-\lambda)\cdot\Phi_{1;+}(\lambda)\,\mathrm{d}\lambda \\
&= \int_{\mathbb{R}+2\mathrm{i}\epsilon'} \mathcal{F}[f^*](-\lambda)\cdot\Phi_1(\lambda)\,\mathrm{d}\lambda \\
&= \int_{\mathbb{R}} \Big(\mathcal{F}[\mathrm{e}^{2\epsilon'\bullet}f](\lambda)\Big)^*\cdot\mathcal{F}\big[\mathrm{e}^{-2\epsilon'\bullet}\widetilde{\mathscr{W}}_{\vartheta;z_0}[h]\big](\lambda)\mathrm{d}\lambda = \big(f,\widetilde{\mathscr{V}}_{\vartheta;z_0}[h]\big)\,.
\end{aligned} \tag{4.3.17}$$

in $\bullet$ represents the running variable with respect to which the Fourier transform is computed. There, we have equality $\widetilde{\mathscr{W}}_{\vartheta;z_0}[h]=\widetilde{\mathscr{V}}_{\vartheta;z_0}[h]$ for $h\in\mathcal{C}^1\cap H_s\big([0;\overline{x}_N]\big)$. ■

A priori, the solutions $\widetilde{\mathscr{W}}_{\vartheta;z_0}[h_\mathfrak{e}]$ given in (4.3.4) have two free parameters ϑ and z_0. This "double" freedom is, however, illusory.

Lemma 4.3.2 *Given $z_0,z_0'\in\mathbb{C}\setminus\mathbb{R}$ and $\vartheta\in\mathbb{C}$, there exists $\vartheta'\in\mathbb{C}$ such that $\widetilde{\mathscr{W}}_{\vartheta;z_0}=\widetilde{\mathscr{W}}_{\vartheta';z_0'}$.*

Proof By carrying out the decomposition $\lambda-z_0=\lambda-\mu+\mu-z_0$ in the first term present in the integrand of (4.3.4), we get that $\widetilde{\mathscr{W}}_{\vartheta;z_0}=\widetilde{\mathscr{W}}_{\vartheta(z_0);\infty}$

$$\vartheta(z_0) = \vartheta - \int_{\mathbb{R}+\mathrm{i}\epsilon'} \frac{\chi_{12}(\mu)\cdot\mathcal{F}\big[h_\mathfrak{e}\big](\mu)\cdot\mathrm{e}^{-\mathrm{i}\mu\overline{x}_N}}{\mu-z_0}\cdot\frac{\mathrm{d}\mu}{2\mathrm{i}\pi}\,, \tag{4.3.18}$$

and ∞ means that one should send $z_0\to\infty$ under the integral sign of (4.3.4). ■

Hence, with the above lemma in mind, we retrieve that the kernel of $\mathscr{S}_{N;\gamma}$ is one dimensional when considered as an operator on $H_s([0;\overline{x}_N])$, with $-1<s<0$. The above lemma of course implies that we can choose z_0 arbitrarily in (4.3.4). It is most suitable to consider the specific form of solutions obtained by taking $z_0\to 0$ with $\operatorname{Im} z_0<0$. For $h\in\mathcal{C}^1([0;\overline{x}_N])$, this yields a family of solution parametrized by $\vartheta\in\mathbb{C}$:

$$\begin{aligned}
\widetilde{\mathscr{W}}_\vartheta[h](\xi) = &\int_{\mathbb{R}+2\mathrm{i}\epsilon'}\frac{\mathrm{d}\lambda}{2\pi}\int_{\mathbb{R}+\mathrm{i}\epsilon'}\frac{\mathrm{d}\mu}{2\mathrm{i}\pi}\frac{\mathrm{e}^{-\mathrm{i}\lambda\xi-\mathrm{i}\mu\overline{x}_N}}{\mu-\lambda}\Big\{\frac{\lambda}{\mu}\cdot\chi_{11}(\lambda)\chi_{12}(\mu)-\chi_{11}(\mu)\chi_{12}(\lambda)\Big\}\mathcal{F}\big[h\big](\mu) \\
&+\vartheta\int_{\mathbb{R}+\mathrm{i}\epsilon'}\chi_{11}(\lambda)\mathrm{e}^{-\mathrm{i}\lambda\xi}\cdot\frac{\mathrm{d}\lambda}{2\pi}\,.
\end{aligned} \tag{4.3.19}$$

It is possible to find real-valued solutions to $\mathscr{S}_{N;\gamma}[\varphi]=h$ by taking h purely imaginary:

Lemma 4.3.3 *Let $\vartheta\in\mathrm{i}\mathbb{R}$ and let $h\in\mathcal{C}^1([0;\overline{x}_N])$ satisfy $h^*=-h$. Then, $\big(\widetilde{\mathscr{W}}_\vartheta[h_\mathfrak{e}]\big)^*=\widetilde{\mathscr{W}}_\vartheta[h_\mathfrak{e}]$.*

Proof From Lemma 4.2.2, we have $-\chi_{11}(-\lambda) = \big(\chi_{11}(\lambda^*)\big)^*$ and $\chi_{12}(-\lambda) = \big(\chi_{12}(\lambda^*)\big)^*$. Hence, under the assumptions of the present lemma

$$\begin{aligned}\big(\widetilde{\mathscr{W}}_{\vartheta}[h](\xi)\big)^* = \int_{\mathbb{R}} \frac{\mathrm{d}\lambda}{2\pi} \int_{\mathbb{R}} \frac{\mathrm{d}\mu}{-2\mathrm{i}\pi} \frac{\mathrm{e}^{\mathrm{i}\lambda\xi+\mathrm{i}\mu\overline{x}_N}\mathrm{e}^{2\epsilon'\xi+\epsilon'\overline{x}_N}}{\mu-\lambda+\mathrm{i}\epsilon'} \bigg\{ -\frac{\lambda-2\mathrm{i}\epsilon'}{\mu-\mathrm{i}\epsilon'} \cdot \chi_{11}(-\lambda+2\mathrm{i}\epsilon')\chi_{12}(-\mu+\mathrm{i}\epsilon') \\ + \chi_{11}(-\mu+\mathrm{i}\epsilon')\chi_{12}(-\lambda+2\mathrm{i}\epsilon')\bigg\} \underbrace{\mathcal{F}[h^*](-\mu+\mathrm{i}\epsilon')}_{-\mathcal{F}[h]} \\ - \underbrace{\vartheta^*}_{-\vartheta} \int_{\mathbb{R}} \mathrm{e}^{\mathrm{i}\lambda\xi}\mathrm{e}^{2\epsilon'\xi}\chi_{11}(-\lambda+\mathrm{i}\epsilon')\frac{\mathrm{d}\lambda}{2\pi} \,. \qquad (4.3.20)\end{aligned}$$

The change of variables $(\lambda, \mu) \mapsto (-\lambda, -\mu)$ in the first integral and $\lambda \mapsto -\lambda$ in the second integral entails the claim. ■

4.3.2 Local Behaviour of the Solution $\widetilde{\mathscr{W}}_{\vartheta}[h]$ at the Boundaries

In the present subsection, we shall establish the local behaviour of $\widetilde{\mathscr{W}}_{\vartheta}[h](\xi)$ at the boundaries of the segment $[0; \overline{x}_N]$, *viz.* when $\xi \to 0$ or $\xi \to \overline{x}_N$, this in the case where $h \in \mathcal{C}^1([0; \overline{x}_N])$. We shall demonstrate that there exist constants C_0, $C_{\overline{x}_N}$ affine in ϑ and depending on h, such that $\widetilde{\mathscr{W}}_{\vartheta}[h]$ exhibits the local behaviour

$$\begin{aligned}\widetilde{\mathscr{W}}_{\vartheta}[h](\xi) = \frac{C_0}{\sqrt{\xi}} + \mathrm{O}(1) \quad \text{for} \quad \xi \to 0^+ \\ \text{and} \quad \widetilde{\mathscr{W}}_{\vartheta}[h](\xi) = \frac{C_{\overline{x}_N}}{\sqrt{\overline{x}_N - \xi}} + \mathrm{O}(1) \quad \text{for} \quad \xi \to (\overline{x}_N)^- \,. \qquad (4.3.21)\end{aligned}$$

Let us recall that our motivation for studying $\widetilde{\mathscr{W}}_{\vartheta}$ takes its origin in the need to construct the density of equilibrium measure $\rho_{\mathrm{eq}}^{(N)}$ which solves $\mathcal{S}_N[\rho_{\mathrm{eq}}^{(N)}] = V'$, as well as to invert the master operator $\mathcal{U}_N$ arising in the Schwinger–Dyson equations described in Section 3.2. The density has a square root behaviour at the edges, what translates itself into a square root behaviour at $\xi = 0$ and $\xi = \overline{x}_N$ in the rescaled variables. Having this in mind, we would like to enforce $C_0 = C_{\overline{x}_N} = 0$. For this purpose, we can exploit the freedom of choosing ϑ. This is however not enough and, as it will be shown in the present section, in order to have a milder behaviour of $\widetilde{\mathscr{W}}_{\vartheta}[h]$ at the edges, one also needs to impose a linear constraint on h. In fact, we shall see later on that the latter solely translates the fact that $h \in \mathscr{S}_{N;\gamma}[H_s(\mathbb{R})]$ with $0 < s < 1/2$.

This informal discussion only serves as a guideline and motivation for the results of this subsection, in particular:

Proposition 4.3.4 *Let*

$$\mathscr{I}_{12}[h] = \int_{\mathbb{R}+\mathrm{i}\epsilon} \frac{\mathrm{e}^{-\mathrm{i}\mu\overline{x}_N}}{\mu}\chi_{12}(\mu)\cdot\mathcal{F}[h](\mu)\cdot\frac{\mathrm{d}\mu}{2\mathrm{i}\pi}\,. \tag{4.3.22}$$

Then, for any $h \in \mathcal{C}^1([0\,;\overline{x}_N])$ such that

$$\mathscr{I}_{11}[h] := \int_{\mathbb{R}+\mathrm{i}\epsilon} \mathrm{e}^{-\mathrm{i}\mu\overline{x}_N}\chi_{11}(\mu)\cdot\mathcal{F}[h](\mu)\cdot\frac{\mathrm{d}\mu}{2\mathrm{i}\pi} = 0 \tag{4.3.23}$$

we have $\widetilde{\mathscr{W}}_{\mathscr{I}_{12}[h]}[h] \in \big(L^1\cap L^\infty\big)([0\,;\overline{x}_N])$.

Prior to proving the above lemma, we shall first establish a lemma characterising the local behaviour at 0 and $\overline{x}_N$ of functions belonging to the kernel of $\mathscr{S}_{N;\gamma}$.

Lemma 4.3.5 *The function*

$$\psi(\xi) = \int_{\mathbb{R}+2\mathrm{i}\epsilon'} \mathrm{e}^{-\mathrm{i}\lambda\xi}\chi_{11}(\lambda)\frac{\mathrm{d}\lambda}{2\pi} \quad \textit{satisfies} \quad \mathscr{S}_{N;\gamma}[\psi](\xi)=0, \quad \xi\in]0\,;\overline{x}_N[\tag{4.3.24}$$

and admits the asymptotic behaviour

$$\psi(\xi) = \frac{1}{\mathrm{i}\sqrt{\pi\xi}} + \mathrm{O}(1) \quad \textit{when} \quad \xi\to 0^+$$
$$\textit{and} \quad \psi(\xi) = \frac{1}{\mathrm{i}\sqrt{\pi(\overline{x}_N-\xi)}} + \mathrm{O}(1) \quad \textit{when} \quad \xi \to \big(\overline{x}_N\big)^-\,. \tag{4.3.25}$$

Proof One has, for $\xi\in]0\,;\overline{x}_N[$ and in the distributional sense,

$$\begin{aligned}\mathscr{S}_{N;\gamma}[\psi](\xi) &= \int_{\mathbb{R}}\frac{\mathrm{d}\mu}{2\pi}\int_{\mathbb{R}}\frac{\mathrm{d}\lambda}{2\pi}\,\frac{\mathrm{e}^{-\mathrm{i}\mu\xi}\,\mathcal{F}[S_\gamma](\mu)}{2\mathrm{i}\pi\beta}\cdot\chi_{11;+}(\lambda)\cdot\frac{\mathrm{e}^{\mathrm{i}(\mu-\lambda)\overline{x}_N}-1}{\mathrm{i}(\lambda-\mu)}\\ &= \int_{\mathbb{R}}\frac{\mathrm{d}\mu}{2\pi}\int_{\mathbb{R}-\mathrm{i}\epsilon}\frac{\mathrm{d}\lambda}{2\pi}\,\frac{\mathrm{e}^{-\mathrm{i}\mu\xi}\,\mathcal{F}[S_\gamma](\mu)}{2\mathrm{i}\pi\beta}\cdot\frac{\chi_{11}(\lambda)\mathrm{e}^{\mathrm{i}\mu\overline{x}_N}-\mathrm{e}^{\mathrm{i}\lambda\overline{x}_N}\chi_{11}(\lambda)}{\mathrm{i}(\lambda-\mu)}\\ &= \int_{\mathbb{R}}\frac{\mathrm{d}\mu}{2\pi}\,\frac{\mathrm{e}^{-\mathrm{i}\mu\xi}\,\mathcal{F}[S_\gamma](\mu)}{2\mathrm{i}\pi\beta}\left\{-\chi_{11;+}(\mu) + \int_{\mathbb{R}-\mathrm{i}\epsilon}\frac{\mathrm{d}\lambda}{2\pi}\,\frac{\chi_{11}(\lambda)\,\mathrm{e}^{\mathrm{i}\mu\overline{x}_N}}{\mathrm{i}(\lambda-\mu)}\right\}\\ &= -\int_{\mathbb{R}}\frac{\mathrm{d}\mu}{2\pi}\left\{\chi_{21;-}(\mu)\mathrm{e}^{-\mathrm{i}\mu\xi} + \mathrm{e}^{\mathrm{i}\mu(\overline{x}_N-\xi)}\chi_{21;+}(\mu)\right\}.\end{aligned} \tag{4.3.26}$$

Note that, in the intermediate steps, we have used that $\chi_{11;+}(\lambda) = e^{i\lambda\overline{x}_N}\chi_{11;-}(\lambda)$, and deformed the integral over λ to the lower half-plane. Further, we have also used that

$$\chi_{21;-}(\lambda) + e^{i\lambda\overline{x}_N}\chi_{21;+}(\lambda) = \frac{\mathcal{F}[S_\gamma](\lambda)}{2i\pi\beta}\chi_{11;+}(\lambda) . \tag{4.3.27}$$

Observe that, when $0 < \xi < \overline{x}_N$, the function $\mu \mapsto \chi_{21;-}(\mu)e^{-i\mu\xi}$ (respectively, $\mu \mapsto \chi_{21;+}(\mu)e^{i\mu(\overline{x}_N-\xi)}$) admits an analytic continuation to the lower (respectively upper) half-plane that it is Riemann-integrable on $\mathbb{R}-i\tau$ (respectively $\mathbb{R}+i\tau$), this for any $\tau > 0$, and that it decays exponentially fast when $\tau \to +\infty$. As a consequence,

$$\forall \xi \in]0\,;\overline{x}_N[, \qquad \int_{\mathbb{R}} \frac{d\mu}{2\pi}\Big(e^{-i\mu\xi}\chi_{21;-}(\mu) + e^{i\mu(\overline{x}_N-\xi)}\chi_{21;+}(\mu)\Big) = 0 , \tag{4.3.28}$$

which is equivalent to $\mathscr{S}_{N;\gamma}[\psi](\xi) = 0$.

From the large-λ expansion of $\chi(\lambda)$ given in Lemma 4.2.3, we have for $\lambda \in \mathbb{R}+2i\epsilon'$,

$$W(\lambda) \equiv \chi_{11}(\lambda) + \frac{\operatorname{sgn}(\operatorname{Re}\lambda)\, e^{i\lambda\overline{x}_N} + i}{(-i\lambda)^{1/2}} = O\Big(|\lambda|^{-3/2}\Big) . \tag{4.3.29}$$

Hence,

$$\psi(\xi) = \int_{\mathbb{R}+2i\epsilon'} W(\lambda)\,e^{-i\lambda\xi}\cdot\frac{d\lambda}{2\pi} - \int_{\mathbb{R}+2i\epsilon'} \frac{\operatorname{sgn}(\operatorname{Re}\lambda)\, e^{i\lambda(\overline{x}_N-\xi)}}{(-i\lambda)^{1/2}}\cdot\frac{d\lambda}{2\pi} + \int_{\mathbb{R}+2i\epsilon'} \frac{e^{-i\lambda\xi}}{(-i\lambda)^{1/2}}\cdot\frac{d\lambda}{2i\pi} . \tag{4.3.30}$$

By dominated convergence, the first term is O(1) in the limit $\xi \to 0^+$. The second term is also a O(1). This is most easily seen by deforming the contour of integration into a loop in $\mathbb{H}_+$ around $i\mathbb{R}^+ + 2i\epsilon'$, hence making the integral strongly convergent, and then applying dominated convergence. Finally, the third term (4.3.30) can be explicitly computed by deforming the integration contour to $-i\mathbb{R}^+$:

$$\int_{\mathbb{R}+2i\epsilon'} \frac{e^{-i\lambda\xi}}{(-i\lambda)^{1/2}}\cdot\frac{d\lambda}{2i\pi} = \frac{-1}{\sqrt{\xi}}\int_0^{+\infty}\bigg\{\frac{1}{(-e^{i0^+}t)^{1/2}} - \frac{1}{(-e^{-i0^+}t)^{1/2}}\bigg\}\frac{e^{-t}\,dt}{2\pi} = \frac{\Gamma(1/2)}{i\pi\sqrt{\xi}} = \frac{1}{i\sqrt{\pi\xi}} . \tag{4.3.31}$$

Similar arguments ensure that the first and last term in (4.3.30) are a O(1) in the $\xi \to (\overline{x}_N)^-$ limit. The middle term can be estimated as

$$\int_{\mathbb{R}+2i\epsilon'} \frac{\operatorname{sgn}(\operatorname{Re}\lambda)}{(-i\lambda)^{\frac{1}{2}}} e^{i\lambda(\overline{x}_N-\xi)} \cdot \frac{d\lambda}{2\pi} = \frac{e^{-2(\overline{x}_N-\xi)\epsilon'}}{2\pi\sqrt{\overline{x}_N-\xi}} \int_{\mathbb{R}} \frac{\operatorname{sgn}(\lambda)\, e^{i\lambda}\, d\lambda}{\big(-i\lambda+2\epsilon'(\overline{x}_N-\xi)\big)^{1/2}}$$
$$= i\frac{e^{-2(\overline{x}_N-\xi)\epsilon'}}{\pi\sqrt{\overline{x}_N-\xi}} \int_0^{+\infty} \frac{e^{-t}\, dt}{\big(t+2\epsilon'(\overline{x}_N-\xi)\big)^{1/2}}$$
$$= \frac{i}{\sqrt{\pi(\overline{x}_N-\xi)}} + O\big(\sqrt{\overline{x}_N-\xi}\big)\,. \tag{4.3.32}$$

Putting together all of the terms entails the claim. ■

Before carrying on with the proof of Proposition 4.3.4 we still need to prove a technical lemma relative to the large-λ behaviour of certain building blocks of $\widetilde{\mathscr{W}_\vartheta}[h]$.

Lemma 4.3.6 *Let $h \in \mathcal{C}^{p+1}\big([0\,;\overline{x}_N]\big)$. Then, the integrals*

$$\mathscr{J}_{1a}[h](\lambda) = \int_{\mathbb{R}+i\epsilon'} \frac{\chi_{1a}(\mu)\cdot\mathcal{F}[h](\mu)\cdot e^{-i\mu\overline{x}_N}}{\mu^{\delta_{2a}}(\mu-\lambda)} \cdot \frac{d\mu}{2i\pi} \quad \textit{with} \quad \delta_{2a} = \begin{cases} 1 \text{ if } a=2 \\ 0 \text{ if } a=1 \end{cases} \tag{4.3.33}$$

admit the $|\lambda| \to \infty$, $\operatorname{Im}\lambda > 2\epsilon' > 0$, *asymptotic behaviour:*

$$\mathscr{J}_{1a}[h](\lambda) = -\lambda^{-1}\mathscr{I}_{1a}[h] + \sum_{k=1}^{p} \frac{w_{k;a}^{(1/2)}(\lambda)}{\big(-i\lambda\big)^{1/2}\lambda^k} + \sum_{k=1}^{p} \frac{w_{k;a}^{(1)}}{\lambda^{k+1}} + O(\lambda^{-(p+3/2)}) \tag{4.3.34}$$

where

$$w_{k;a}^{(1/2)}(\lambda) = \sum_{\ell=0}^{k-1} i^{k-\ell} h^{(k-\ell-1)}(\overline{x}_N)\Big\{\operatorname{sgn}(\operatorname{Re}\lambda)\, e^{i\lambda\overline{x}_N}\,[\chi_{\ell-\delta_{2a}}]_{1a} + i\,[\chi_{\ell+1-\delta_{2a}}]_{2a}\Big\}, \tag{4.3.35}$$

and $w_{k;a}^{(1)}$ are constants whose explicit expression is given in the core of the proof.

Proof The regularity of h implies the following decomposition for its Fourier transform:

$$\mathcal{F}[h](\mu) = -\sum_{k=0}^{p} \frac{h^{(k)}(\overline{x}_N)\, e^{i\mu\overline{x}_N} - h^{(k)}(0)}{(-i\mu)^{k+1}} + \frac{(-1)^{p+1}}{(i\mu)^{p+1}} \int_0^{\overline{x}_N} h^{(p+1)}(t)\, e^{i\xi\mu}\, d\xi\,. \tag{4.3.36}$$

It gives directly access to the large-μ expansion:

$$\mu^{-\delta_{2a}}\chi_{1a}(\mu)\cdot\mathcal{F}[h](\mu) = \sum_{k=1}^{p} \frac{T_a^{(k)}(\mu)}{(-i\mu)^{1/2}\mu^k} + R_{1a}^{(p)}(\mu)\,. \tag{4.3.37}$$

The remainder is $R^{(p)}_{1a}(\mu) = \mathrm{O}(\mu^{-p-3/2})$ when μ is large, whereas $T^{(k)}_a(\mu)$ remains bounded as long as $\operatorname{Im}\mu$ is bounded. Explicitly, these functions read:

$$T^{(k)}_a(\mu) = \sum_{\ell=0}^{k-1} \mathrm{i}^{k-\ell}\Big(h^{(k-1-\ell)}(0) - \mathrm{e}^{\mathrm{i}\mu\bar{x}_N} h^{(k-1-\ell)}(\bar{x}_N)\Big)\Big\{ - \operatorname{sgn}(\operatorname{Re}\mu)\,\mathrm{e}^{\mathrm{i}\mu\bar{x}_N}\,[\chi_{\ell-\delta_{2a}}]_{a1} - \mathrm{i}\,[\chi_{\ell+1-\delta_{2a}}]_{a2}\Big\} \tag{4.3.38}$$

where χ_m are the matrices appearing in the asymptotic expansion of χ, see (4.2.30). The integral of interest can be recast as

$$\begin{aligned}\mathscr{I}_{1a}[h](\lambda) = & \sum_{k=1}^{p} \int_{\mathbb{R}+\mathrm{i}\epsilon'} \frac{T^{(k)}_a(\mu)\,\mathrm{e}^{-\mathrm{i}\mu\bar{x}_N}}{(-\mathrm{i}\mu)^{1/2}\mu^k(\mu-\lambda)} \cdot \frac{\mathrm{d}\mu}{2\mathrm{i}\pi} \\ & - \sum_{\ell=0}^{p} \frac{1}{\lambda^{\ell+1}} \int_{\mathbb{R}+\mathrm{i}\epsilon'} \mu^{\ell} R^{(p)}_{1a}(\mu)\,\mathrm{e}^{-\mathrm{i}\mu\bar{x}_N} \cdot \frac{\mathrm{d}\mu}{2\mathrm{i}\pi} \\ & + \int_{\mathbb{R}+\mathrm{i}\epsilon'} \frac{\mu^{p+1} R^{(p)}_{1a}(\mu)}{\lambda^{p+1}(\mu-\lambda)}\,\mathrm{e}^{-\mathrm{i}\mu\bar{x}_N} \cdot \frac{\mathrm{d}\mu}{2\mathrm{i}\pi} .\end{aligned} \tag{4.3.39}$$

In virtue of the bound on $R^{(p)}_{1a}$, the last term is a $\mathrm{O}\big(\lambda^{-p-\frac{3}{2}}\big)$. In order to obtain the asymptotic expansion of the first term we study the model integral

$$J_k(\lambda) = \int_{\mathbb{R}+\mathrm{i}\epsilon'} \frac{\big(c_1 \operatorname{sgn}(\operatorname{Re}\mu) - c_2\mathrm{e}^{-\mathrm{i}\mu\bar{x}_N}\big)\big(\kappa_1\mathrm{e}^{\mathrm{i}\mu\bar{x}_N} - \kappa_2\big)}{(-\mathrm{i}\mu)^{1/2}\mu^k(\mu-\lambda)} \cdot \frac{\mathrm{d}\mu}{2\mathrm{i}\pi} , \tag{4.3.40}$$

where $\operatorname{Im}\lambda > \epsilon'$ while c_1, c_2 and κ_1, κ_2 are free parameters. By deforming appropriately the contours, we get that:

$$\begin{aligned} J_k(\lambda) = & \kappa_1 \frac{c_1 \operatorname{sgn}(\operatorname{Re}\lambda)\mathrm{e}^{\mathrm{i}\lambda\bar{x}_N} - c_2}{(-\mathrm{i}\lambda)^{1/2}\lambda^k} - c_1\kappa_2 \oint_{-\Gamma([0\,;\mathrm{i}\epsilon'])} \frac{\operatorname{sgn}\big(\operatorname{Re}\mu\big)}{\big(-\mathrm{i}\mu\big)^{1/2}\mu^k(\mu-\lambda)} \cdot \frac{\mathrm{d}\mu}{2\mathrm{i}\pi} \\ & + c_1\kappa_1(-\mathrm{i})^k \int_{\epsilon'}^{+\infty} \frac{t^{-k-1/2}\mathrm{e}^{-t\bar{x}_N}}{\mathrm{i}t-\lambda} \cdot \frac{\mathrm{d}t}{\pi} + c_2\kappa_2 \oint_{-\Gamma(\mathrm{i}\mathbb{R}^-)} \frac{\mathrm{e}^{-\mathrm{i}\mu\bar{x}_N}\mu^{-k}}{\big(-\mathrm{i}\mu\big)^{1/2}(\mu-\lambda)} \cdot \frac{\mathrm{d}\mu}{2\mathrm{i}\pi} \\ = & \kappa_1 \frac{c_1 \operatorname{sgn}(\operatorname{Re}\lambda)\,\mathrm{e}^{\mathrm{i}\lambda\bar{x}_N} - c_2}{(-\mathrm{i}\lambda)^{1/2}\lambda^k} - \sum_{q=0}^{p} \lambda^{-(q+1)} L^{(q)}_k + \lambda^{-(p+2)}\Delta_{[p]} M_k(\lambda) . \end{aligned} \tag{4.3.41}$$

The constant $L_k^{(q)}$ occurring above is expressed in terms of integrals

$$L_k^{(q)} = -c_1\kappa_2 \oint\limits_{-\Gamma([0;\mathrm{i}\epsilon'])} \frac{\mathrm{sgn}(\mathrm{Re}\mu)\cdot\mu^{q-k}}{\big(-\mathrm{i}\mu\big)^{1/2}}\cdot\frac{\mathrm{d}\mu}{2\mathrm{i}\pi}$$
$$+ c_1\kappa_1(-\mathrm{i})^{k-q}\int\limits_{\epsilon'}^{+\infty} t^{q-k-1/2}\cdot\mathrm{e}^{-\overline{x}_N t}\cdot\frac{\mathrm{d}t}{\pi} + c_2\kappa_2\oint\limits_{-\Gamma(\mathrm{i}\mathbb{R}^-)}\frac{\mathrm{e}^{-\mathrm{i}\mu\overline{x}_N}\cdot\mu^{q-k}}{\big(-\mathrm{i}\mu\big)^{1/2}}\cdot\frac{\mathrm{d}\mu}{2\mathrm{i}\pi} \qquad (4.3.42)$$

and the remainder function reads:

$$\Delta_{[p]}M_k(\lambda) = c_1\kappa_2\oint\limits_{-\Gamma([0;\mathrm{i}\epsilon'])}\frac{\lambda\cdot\mathrm{sgn}(\mathrm{Re}\,\mu)\cdot\mu^{p+1-k}}{(\mu-\lambda)\cdot(-\mathrm{i}\mu)^{1/2}}\cdot\frac{\mathrm{d}\mu}{2\mathrm{i}\pi}$$
$$- c_2\kappa_2\oint\limits_{\Gamma(\mathrm{i}\mathbb{R}^-)}\frac{\lambda\cdot\mathrm{e}^{-\mathrm{i}\mu\overline{x}_N}\cdot\mu^{p+1-k}}{\big(-\mathrm{i}\mu\big)^{1/2}(\mu-\lambda)}\cdot\frac{\mathrm{d}\mu}{2\mathrm{i}\pi}$$
$$+ c_1\kappa_1(-\mathrm{i})^{k-p}\int\limits_{\epsilon'}^{+\infty}\frac{\lambda\cdot t^{p-k+1/2}\cdot\mathrm{e}^{-t\overline{x}_N}}{(t+\mathrm{i}\lambda)}\cdot\frac{\mathrm{d}t}{\pi}\,. \qquad (4.3.43)$$

If we define:

$$\widetilde{w}_{k;a}^{(1)} = -\sum_{k=1}^{p}\sum_{\ell=0}^{k-1}\left\{L_k^{(q)}\;\middle|\;\begin{matrix} c_1\to-[\chi_{\ell-\delta_{2a}}]_{1a} & \kappa_1\to-\mathrm{i}^{k-\ell}h^{(k-\ell-1)}(\overline{x}_N) \\ c_2\to\mathrm{i}\,[\chi_{\ell+1-\delta_{2a}}]_{2a} & \kappa_2\to\mathrm{i}^{k-\ell}h^{(k-\ell-1)}(0)\end{matrix}\right\}. \qquad (4.3.44)$$

we obtain:

$$\sum_{k=1}^{p}\int\limits_{\mathbb{R}+\mathrm{i}\epsilon'}\frac{T_a^{(k)}(\mu)\mathrm{e}^{-\mathrm{i}\mu\overline{x}_N}}{(-\mathrm{i}\mu)^{1/2}\mu^k(\mu-\lambda)}\cdot\frac{\mathrm{d}\mu}{2\mathrm{i}\pi} = \sum_{k=1}^{p}\frac{w_{k;a}^{(1/2)}(\lambda)}{(-\mathrm{i}\lambda)^{1/2}\lambda^k} + \sum_{q=0}^{p}\frac{\widetilde{w}_{k;a}^{(1)}}{\lambda^{q+1}} + \mathrm{O}(\lambda^{-(p+2)}). \qquad (4.3.45)$$

Furthermore, the above relation and Equations (4.3.33) and (4.3.39), ensure that

$$\int\limits_{\mathbb{R}+\mathrm{i}\epsilon'} R_{1a}^{(p)}(\mu)\mathrm{e}^{-\mathrm{i}\mu\overline{x}_N}\cdot\frac{\mathrm{d}\mu}{2\mathrm{i}\pi} = \mathscr{I}_{1a}[h] + \widetilde{w}_{0;a}^{(1)}. \qquad (4.3.46)$$

Hence, putting all the terms together, we arrive to the expansion (4.3.34) with the constants $w_{k;a}^{(1)}$ given by

$$w_{k;a}^{(1)} = \widetilde{w}_{k;a}^{(1)} - \int\limits_{\mathbb{R}+i\epsilon'} \mu^k R_{1a}^{(p)}(\mu) \mathrm{e}^{-\mathrm{i}\mu \overline{x}_N} \cdot \frac{\mathrm{d}\mu}{2\mathrm{i}\pi} \qquad k \geq 1. \tag{4.3.47}$$

■

Proof (*of Proposition* 4.3.4). Given $h \in \mathcal{C}^1([0\,;\overline{x}_N])$ and for $\xi \in]0\,;\overline{x}_N[$, we can represent $\widetilde{\mathscr{W}_0}$ as an integral taken in the Riemann sense[1]

$$\widetilde{\mathscr{W}_0}[h](\xi) = \int\limits_{\mathbb{R}+2\mathrm{i}\epsilon'} \mathrm{e}^{-\mathrm{i}\lambda\xi} \Big[\lambda\cdot\chi_{11}(\lambda)\mathscr{J}_{12}[h](\lambda) - \chi_{12}(\lambda)\mathscr{J}_{11}[h](\lambda)\Big] \frac{\mathrm{d}\lambda}{2\pi}\,, \tag{4.3.48}$$

where we remind that $\mathscr{J}_{1a}[h](\lambda)$ have been defined in (4.3.33). Using the asymptotic expansions of Lemma 4.2.3 for χ and those of Lemma 4.3.6 for $\mathscr{J}_{1a}[h]$, we can decompose:

$$\lambda\cdot\chi_{11}(\lambda)\mathscr{J}_{12}[h](\lambda) - \chi_{12}(\lambda)\mathscr{J}_{11}[h](\lambda) = \mathscr{I}_{12}[h]\cdot\frac{\operatorname{sgn}(\operatorname{Re}\lambda)e^{\mathrm{i}\lambda\overline{x}_N}+\mathrm{i}}{(-\mathrm{i}\lambda)^{1/2}} - \frac{\mathrm{i}\mathscr{I}_{11}[h]}{(-\mathrm{i}\lambda)^{1/2}}$$
$$+\frac{w_{1;2}^{(1/2)}(\lambda)\big\{\operatorname{sgn}(\operatorname{Re}\lambda)e^{\mathrm{i}\lambda\overline{x}_N}+\mathrm{i}\big\} - \mathrm{i}\,w_{1;1}^{(1/2)}(\lambda)}{\mathrm{i}\lambda} + \mathrm{O}(\lambda^{-3/2})\,. \tag{4.3.49}$$

As a matter of fact, the coefficient of $1/(-\mathrm{i}\lambda)$ in this formula vanishes, as can be seen from the expressions (4.3.35) for $w_{1;a}^{(1/2)}$. Besides, integrating the $O(\lambda^{-3/2})$ in (4.3.48) yields a contribution remaining finite at $\xi = 0$ and $\xi = \overline{x}_N$, that we denote $\widetilde{\mathscr{W}_0^{\mathrm{c}}}[h] \in \mathcal{C}^0\big([0\,;\overline{x}_N]\big)$. Eventually, the effect of the first line of (4.3.49) once inserted in (4.3.48) is already described in (4.3.31) and (4.3.32). All in all, we find:

$$\widetilde{\mathscr{W}_0}[h](\xi) = \mathscr{I}_{12}[h]\bigg\{\frac{\mathrm{i}}{\sqrt{\pi\xi}} + \frac{\mathrm{i}}{\sqrt{\pi(\overline{x}_N-\xi)}} + \mathrm{O}\big(\sqrt{\overline{x}_N-\xi}\big)\bigg\} - \frac{\mathrm{i}\mathscr{I}_{11}[h]}{\sqrt{\pi\xi}} + \widetilde{\mathscr{W}_0^{\mathrm{c}}}[h](\xi)\,. \tag{4.3.50}$$

Since we have $\widetilde{\mathscr{W}_\vartheta}[h](\xi) = \widetilde{\mathscr{W}_0}[h](\xi) + \vartheta\psi(\xi)$ in terms of the function ψ of Lemma 4.3.5, we deduce that:

$$\widetilde{\mathscr{W}}_{\mathscr{I}_{12}[h]}[h](\xi) = -\frac{\mathrm{i}\mathscr{I}_{11}[h]}{\sqrt{\pi\xi}} + O\big(\sqrt{\overline{x}_N-\xi}\big) + \widetilde{\mathscr{W}_0^{\mathrm{c}}}[h](\xi) \tag{4.3.51}$$

and this function is continuous on $[0\,;\overline{x}_N]$ if and only if $\mathscr{I}_{11}[h] = 0$. ■

[1] The fact that the integral (4.3.48) is well-defined in the Riemann sense will follow from the analysis carried out in this proof.

4.3.3 A Well-Behaved Inverse Operator of $\mathcal{S}_{N;\gamma}$

Since, *in fine*, we are solely interested in solutions belonging to $\big(L^1 \cap L^\infty\big)([0\,;\overline{x}_N])$ we shall henceforth only focus on $\widetilde{\mathscr{W}}_{\mathscr{I}_{12}[h]}[h]$ and denote this specific solution as $\mathscr{W}_{N;\gamma}[h]$. Furthermore, we shall restrict our reasoning to a class of functions such that $\mathscr{I}_{11}[h]=0$. We now establish:

Proposition 4.3.7 *Let $0<s<1/2$. The subspace*

$$\mathscr{X}_s\big([-\gamma\overline{x}_N\,;(\gamma+1)\overline{x}_N]\big) \;=\; \Big\{h\in H_s\big([-\gamma\overline{x}_N\,;(\gamma+1)\overline{x}_N]\big)\;:\;\mathscr{I}_{11}[h]\;=\;0\Big\} \tag{4.3.52}$$

is closed in $H_s\big([-\gamma\overline{x}_N\,;(\gamma+1)\overline{x}_N]\big)$, and the operator:

$$\mathscr{S}_{N;\gamma}\;:\;H_s\big([0\,;\overline{x}_N]\big)\longrightarrow\mathscr{S}_{N;\gamma}\big[H_s\big([0\,;\overline{x}_N]\big)\big]\;=\;\mathscr{X}_s\big([-\gamma\overline{x}_N\,;(\gamma+1)\overline{x}_N]\big) \tag{4.3.53}$$

is continuously invertible. Its inverse is the operator

$$\mathscr{W}_{N;\gamma}\;:\;\mathscr{X}_s\big([-\gamma\overline{x}_N\,;(\gamma+1)\overline{x}_N]\big)\longrightarrow H_s\big([0\,;\overline{x}_N]\big)\,. \tag{4.3.54}$$

On functions $h\in\mathcal{C}^1([0\,;\overline{x}_N])$, it is defined as:

$$\mathscr{W}_{N;\gamma}[h](\xi)=\int\limits_{\mathbb{R}+2\mathrm{i}\epsilon'}\frac{\mathrm{d}\lambda}{2\pi}\int\limits_{\mathbb{R}+\mathrm{i}\epsilon'}\frac{\mathrm{d}\mu}{2\mathrm{i}\pi}\,\frac{\mathrm{e}^{-\mathrm{i}\lambda\xi-\mathrm{i}\mu\overline{x}_N}}{\mu-\lambda}\Big\{\chi_{11}(\lambda)\chi_{12}(\mu)-\frac{\mu}{\lambda}\cdot\chi_{11}(\mu)\chi_{12}(\lambda)\Big\}\mathcal{F}[h](\mu)\,. \tag{4.3.55}$$

For $h\in\mathcal{C}^1([0\,;\overline{x}_N])$, $\mathscr{W}_{N;\gamma}[h](\xi)$ is a continuous function on $[0\,;\overline{x}_N]$, which vanishes at least like a square root at 0 and $\overline{x}_N$. The operator $\mathscr{W}_{N;\gamma}$ extends continuously to $H_s([0\,;\overline{x}_N])$, $0<s<1/2$ although the constant of continuity of $\mathscr{W}_{N;\gamma}$ depends, a priori, on N.

Comparing (4.3.55) with the double integral defining $\widetilde{\mathscr{W}}_{\vartheta}$ in (4.3.19), one observes that λ/μ in front of $\chi_{11}(\lambda)$ is absent and that there is an additional pre-factor μ/λ in front of $\chi_{12}(\lambda)$.

Proof Continuity of $\mathscr{W}_{N;\gamma}$.

Take $h\in\mathcal{C}^1\big([-\gamma\overline{x}_N\,;(\gamma+1)\overline{x}_N]\big)$. We first establish that $\mathscr{W}_{N;\gamma}[h]$, as defined by (4.3.55), extends as a continuous operator from $H_s\big([-\gamma\overline{x}_N\,;(\gamma+1)\overline{x}_N]\big)$ to $H_s(\mathbb{R})$. We observe that:

$$\begin{aligned}\mathcal{F}\big[\mathrm{e}^{-2\epsilon'*}\mathscr{W}_{N;\gamma}[h]\big](\lambda)\;&=\;\chi_{11}(\lambda+2\mathrm{i}\epsilon')\cdot\mathcal{C}\big[\widehat{\chi}_{12}\mathcal{F}[h_{\epsilon'}]\big](\lambda+\mathrm{i}\epsilon')\\&\quad-\;\frac{\chi_{12}(\lambda+2\mathrm{i}\epsilon')}{\lambda+2\mathrm{i}\epsilon'}\cdot\mathcal{C}\big[\widehat{\chi}_{11}\mathcal{F}[h_{\epsilon'}]\big](\lambda+\mathrm{i}\epsilon')\end{aligned} \tag{4.3.56}$$

with $h_{\epsilon'}(\xi) = \mathrm{e}^{-\epsilon'\xi} h(\xi)$,

$$\widehat{\chi}_{11}(\mu) = (\mu + \mathrm{i}\epsilon')\chi_{11}(\mu + \mathrm{i}\epsilon')\mathrm{e}^{-\mathrm{i}(\mu+\mathrm{i}\epsilon')\overline{x}_N} \quad \text{and} \quad \widehat{\chi}_{12}(\mu) = \chi_{12}(\mu + \mathrm{i}\epsilon')\mathrm{e}^{-\mathrm{i}(\mu+\mathrm{i}\epsilon')\overline{x}_N} .$$

It thus follows from the growth at infinity of χ_{11} and χ_{12} and the continuity on $H_\tau(\mathbb{R})$, $|\tau| \leq 1/2$, of the transforms $\mathcal{C}_\epsilon$, where $\mathcal{C}_\epsilon[f](\lambda) = C[f](\lambda + \mathrm{i}\epsilon')$, *cf.* Proposition A.0.12, that

$$\begin{aligned} ||\mathscr{W}_{N;\gamma}[h]||_{H_s(\mathbb{R})} &\leq C\Big\{ \big|\big|\mathcal{C}_{\epsilon'}\big[\widehat{\chi}_{12}\cdot\mathcal{F}[h_{\epsilon'}]\big]\big|\big|_{\mathcal{F}[H_{s-1/2}(\mathbb{R})]} + \big|\big|\mathcal{C}_{\epsilon'}\big[\widehat{\chi}_{11}\cdot\mathcal{F}[h_{\epsilon'}]\big]\big|\big|_{\mathcal{F}[H_{s-1/2}(\mathbb{R})]} \Big\} \\ &\leq C'\Big\{ ||\widehat{\chi}_{12}\cdot\mathcal{F}[h_{\epsilon'}]||_{\mathcal{F}[H_{s-1/2}(\mathbb{R})]} + ||\widehat{\chi}_{11}\cdot\mathcal{F}[h_{\epsilon'}]||_{\mathcal{F}[H_{s-1/2}(\mathbb{R})]} \Big\} \\ &\leq C''\, ||h_{\epsilon'}||_{H_s(\mathbb{R})} \leq C'''\, ||h||_{H_s([-\gamma\overline{x}_N\, ;\,(\gamma+1)\overline{x}_N])} \, . \end{aligned} \tag{4.3.57}$$

■

Proof The space $\mathscr{X}_s\big([-\gamma\overline{x}_N\,;\,(\gamma+1)\overline{x}_N]\big)$.

Given $h \in H_s\big([-\gamma\overline{x}_N\,;\,(\gamma+1)\overline{x}_N]\big)$, we have:

$$\big|\mathscr{I}_{11}[h]\big| \leq \bigg(\int_{\mathbb{R}} (1+|\mu|)^{-2s}|\chi_{11}(\mu)|^2\,\mathrm{d}\mu\bigg)^{1/2} \cdot ||h||_{H_s([-\gamma\overline{x}_N\,;\,(\gamma+1)\overline{x}_N])} \, . \tag{4.3.58}$$

As a consequence, $\mathscr{I}_{11}$ is a continuous linear form on $H_s\big([-\gamma\overline{x}_N\,;\,(\gamma+1)\overline{x}_N]\big)$. In particular, its kernel is closed, what ensures that $\mathscr{X}_s\big([-\gamma\overline{x}_N\,;\,(\gamma+1)\overline{x}_N]\big)$ is a closed subspace of $H_s\big([-\gamma\overline{x}_N\,;\,(\gamma+1)\overline{x}_N]\big)$. We now establish that:

$$\mathscr{S}_{N;\gamma}\big[H_s\big([0\,;\,\overline{x}_N]\big)\big] \subseteq \mathscr{X}_s\big([-\gamma\overline{x}_N\,;\,(\gamma+1)\overline{x}_N]\big) \, . \tag{4.3.59}$$

Let $\varphi \in C^1\big([0\,;\,\overline{x}_N]\big)$ and define $h = \mathscr{S}_{N;\gamma}[\varphi]$. Then, using the jump condition (4.3.27):

$$\begin{aligned} \mathscr{I}_{11}[h] &= \int_0^{\overline{x}_N} \mathrm{d}\eta\,\varphi(\eta) \int_{\mathbb{R}} \mathrm{e}^{-\mathrm{i}\mu\overline{x}_N}\,\chi_{11;+}(\mu)\,\frac{\mathcal{F}[S_\gamma](\mu)}{2\mathrm{i}\pi\beta}\,\mathrm{e}^{\mathrm{i}\mu\eta} \\ &= \int_0^{\overline{x}_N} \mathrm{d}\eta\,\varphi(\eta) \int_{\mathbb{R}} \Big(\chi_{21;-}(\mu)\mathrm{e}^{\mathrm{i}\mu(\eta-\overline{x}_N)} + \chi_{21;+}(\mu)\mathrm{e}^{\mathrm{i}\mu\eta}\Big)\,\mathrm{d}\mu \end{aligned} \tag{4.3.60}$$

and this quantity vanishes according to (4.3.28). The equality can then be extended to the whole of $H_s\big([0\,;\,\overline{x}_N]\big)$, $0 < s < 1/2$ since $\mathscr{I}_{11}$ and $\mathscr{S}_{N;\gamma}$ are continuous on this space and $C^1\big([0\,;\,\overline{x}_N]\big)$ is dense in $H_s\big([0\,;\,\overline{x}_N]\big)$. ■

Proof Relation to the inverse.

By definition, for any $h \in \big(H_s \cap \mathcal{C}^1\big)([-\gamma\overline{x}_N\,;\,(1+\gamma)\overline{x}_N])$, we have:

$$
\begin{aligned}
&\widetilde{\mathscr{W}}_{\mathscr{I}_{12}}[h][h](\xi) \\
&= \int\limits_{\mathbb{R}+2\mathrm{i}\epsilon'} \frac{\mathrm{d}\lambda}{2\pi} \int\limits_{\mathbb{R}+\mathrm{i}\epsilon'} \frac{\mathrm{d}\mu}{2\mathrm{i}\pi} \frac{\mathrm{e}^{-\mathrm{i}\lambda\xi-\mathrm{i}\mu\overline{x}_N}}{\mu-\lambda} \bigg\{ \frac{\lambda}{\mu}\cdot \chi_{11}(\lambda)\chi_{12}(\mu) - \chi_{11}(\mu)\chi_{12}(\lambda) \bigg\} \cdot \mathcal{F}[h](\mu) \\
&\quad + \Bigg(\int\limits_{\mathbb{R}+2\mathrm{i}\epsilon'} \frac{\mathrm{d}\lambda}{2\pi}\mathrm{e}^{-\mathrm{i}\lambda\xi}\chi_{11}(\lambda) \Bigg) \cdot \Bigg(\int\limits_{\mathbb{R}+\mathrm{i}\epsilon'} \frac{\mathrm{d}\mu}{2\mathrm{i}\pi} \frac{\mathrm{e}^{-\mathrm{i}\mu\overline{x}_N}}{\mu} \chi_{12}(\mu)\cdot\mathcal{F}[h](\mu) \Bigg) \\
&= \int\limits_{\mathbb{R}+2\mathrm{i}\epsilon'} \frac{\mathrm{d}\lambda}{2\pi} \int\limits_{\mathbb{R}+\mathrm{i}\epsilon'} \frac{\mathrm{d}\mu}{2\mathrm{i}\pi} \frac{\mathrm{e}^{-\mathrm{i}\lambda\xi-\mathrm{i}\mu\overline{x}_N}}{\mu-\lambda} \bigg\{ \chi_{11}(\lambda)\chi_{12}(\mu) - \chi_{11}(\mu)\chi_{12}(\lambda) \bigg\} \cdot \mathcal{F}[h](\mu) \\
&\quad - \Bigg(\int\limits_{\mathbb{R}+2\mathrm{i}\epsilon'} \frac{\mathrm{d}\lambda}{2\pi}\mathrm{e}^{-\mathrm{i}\lambda\xi} \frac{\chi_{12}(\lambda)}{\lambda} \Bigg) \cdot \underbrace{\Bigg(\int\limits_{\mathbb{R}+\mathrm{i}\epsilon'} \frac{\mathrm{d}\mu}{2\mathrm{i}\pi}\mathrm{e}^{-\mathrm{i}\mu\overline{x}_N} \chi_{11}(\mu)\cdot\mathcal{F}[h](\mu) \Bigg)}_{=0} \\
&= \mathscr{W}_{N;\gamma}[h](\xi)\,. \qquad (4.3.61)
\end{aligned}
$$

In the last line, we used the freedom to add a term proportional to $\mathscr{I}_{11}[h] = 0$, so that the combination retrieves the announced expression (4.3.55). The continuity of the linear functional $\mathscr{I}_{12}$ on $H_s([0\,;\,\overline{x}_N])$ is proven analogously to (4.3.58), hence ensuring the continuity of the operator $\widetilde{\mathscr{W}}_{\mathscr{I}_{12}[h]}$. Since both operators $\mathscr{W}_{N;\gamma}$ and $\widetilde{\mathscr{W}}_{\mathscr{I}_{12}[h]}$ are continuous on $H_s([0\,;\,\overline{x}_N])$ and coincide on $\mathcal{C}^1$ functions which form a dense subspace, they coincide on the whole $H_s([0\,;\,\overline{x}_N])$. From there we deduce two facts:

- we indeed have $\mathscr{S}_{N;\gamma}\big[\mathscr{W}_{N;\gamma}[h]\big] = h$, as a consequence of $\mathscr{S}_{N;\gamma}\big[\widetilde{\mathscr{W}}_{\mathscr{I}_{12}[h]}[h]\big] = h$. This shows that the reverse inclusion to (4.3.59) holds as well.
- The function $\mathscr{W}_{N;\gamma}[h]$ is supported on $[0\,;\,\overline{x}_N]$ (and thus belongs to $H_s([0\,;\,\overline{x}_N])$) since Lemma 4.1.1 ensures that $\widetilde{\mathscr{W}}_{\mathscr{I}_{12}[h]}[h]$ is supported on $[0\,;\,\overline{x}_N]$ this for any $h \in H_s\big([-\gamma\overline{x}_N\,;\,(\gamma+1)\overline{x}_N]\big) \subseteq H_\tau\big([-\gamma\overline{x}_N\,;\,(\gamma+1)\overline{x}_N]\big)$ with $0 < s < 1/2$ and $-1 < \tau < 0$. ■

Proof Local behaviour for $\mathcal{C}^1([0\,;\,\overline{x}_N])$ functions.

It follows from a slight improvement of the local estimates carried out in the proof of Proposition 4.3.4 that, given $h \in \mathcal{C}^1([0\,;\,\overline{x}_N])$, we have:

$$
\begin{aligned}
\mathscr{W}_{N;\gamma}[h](\xi) &= C_L^{(0)} + C_L^{(1/2)}\sqrt{\xi} + \mathrm{O}(\xi) \\
\mathscr{W}_{N;\gamma}[h](\xi) &= C_R^{(0)} + C_R^{(1/2)}\sqrt{\overline{x}_N-\xi} + \mathrm{O}\big(\overline{x}_N-\xi\big)\,, \qquad (4.3.62)
\end{aligned}
$$

form some constants $C_{L/R}^{(a)}$ with $a \in \{0, 1/2\}$. It thus remains to check that $C_L^{(0)} = C_R^{(0)} = 0$. It follows also from the proof of Proposition 4.3.4 that $\mathscr{W}_{N;\gamma}[h]$ is, in fact, continuous on $\mathbb{R}$. Since $\operatorname{supp}\big[\mathscr{W}_{N;\gamma}[h]\big] = [0\,;\,\overline{x}_N]$, the function has to vanish at 0 and $\overline{x}_N$ so as to ensure its continuity. Thence, $C_L^{(0)} = C_R^{(0)} = 0$. ■

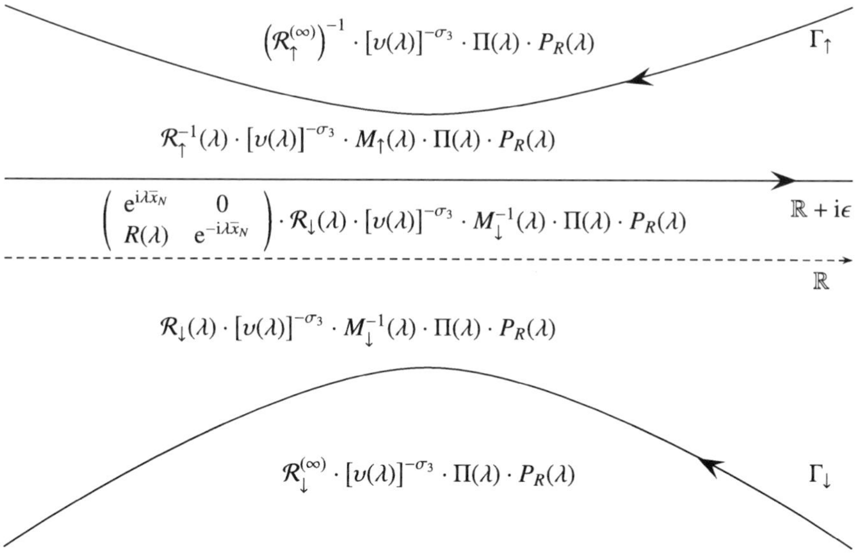

Fig. 4.3 Piecewise definition of the matrix χ (at $\gamma \to +\infty$). The curves $\Gamma_{\uparrow/\downarrow}$ separate all poles of $\lambda \mapsto \lambda R(\lambda)$ from $\mathbb{R}$ and are such that $\mathrm{dist}(\Gamma_{\uparrow/\downarrow}, \mathbb{R}) > \delta$ for some $\delta > 0$ but sufficiently small

4.3.4 $\mathcal{W}_N$: The Inverse Operator of $\mathcal{S}_N$

In order to construct the inverse to $\mathscr{S}_N$, we should take the limit $\gamma \to +\infty$ in the previous formulae. It so happens that this limit is already well-defined at the level of the solution to the Riemann–Hilbert problem for χ as defined through Figure 4.2. More precisely, from now on, let χ be as defined in Figure 4.3 where the matrix Π is as defined through (4.2.12)–(4.2.14) with the exception that one should send $\gamma \to +\infty$ in the jump matrices for Ψ (4.2.4) and (4.2.5). Note that, in this limit, $G_\Psi = I_2$ on $\mathbb{R} + i\epsilon$, *viz.* Ψ is continuous across $\mathbb{R} + i\epsilon$. Then, we can come back to the inversion of the initial operator $\mathcal{S}_N$ in unrescaled variables—compare (4.0.1), (4.1.1) and (4.1.2).

Proposition 4.3.8 *Let* $0 < s < 1/2$. *The operator* $\mathcal{S}_N \; : \; H_s\big([a_N \,; b_N]\big) \longrightarrow H_s\big(\mathbb{R}\big)$ *is continuous and invertible on its image:*

$$\mathfrak{X}_s\big(\mathbb{R}\big) = \left\{ H \in H_s(\mathbb{R}) \quad : \quad \int\limits_{\mathbb{R}+i\epsilon} \chi_{11}(\mu)\mathcal{F}[H](N^{\alpha}\mu)\mathrm{e}^{-iN^{\alpha}\mu b_N} \cdot \frac{\mathrm{d}\mu}{2i\pi} \; = \; 0 \right\}. \tag{4.3.63}$$

The inverse is then given by the operator $\mathcal{W}_N \; : \; \mathfrak{X}_s(\mathbb{R}) \longrightarrow H_s([a_N \,; b_N])$ *defined in (2.4.17):*

$$\mathcal{W}_N[H](\xi) = \frac{N^{2\alpha}}{2\pi\beta} \int\limits_{\mathbb{R}+2\mathrm{i}\epsilon} \frac{\mathrm{d}\lambda}{2\mathrm{i}\pi} \int\limits_{\mathbb{R}+\mathrm{i}\epsilon} \frac{\mathrm{d}\mu}{2\mathrm{i}\pi} \frac{\mathrm{e}^{-\mathrm{i}N^{\alpha}\lambda(\xi-a_N)}}{\mu-\lambda} \Big\{ \chi_{11}(\lambda)\chi_{12}(\mu) - \frac{\mu}{\lambda}\cdot\chi_{11}(\mu)\chi_{12}(\lambda) \Big\} \mathrm{e}^{-\mathrm{i}N^{\alpha}\mu b_N} \mathcal{F}\big[H\big](N^{\alpha}\mu) \tag{4.3.64}$$

with χ being understood as defined in Figure 4.3.

Proof Starting from the expression for the inverse operator to $\mathscr{S}_{N;\gamma}$ and carrying out the change of variables, one obtains an operator $\mathcal{W}_{N;\gamma}$ which corresponds to the inverse of the operator $\mathcal{S}_{N;\gamma}$. Then, in this expression we replace χ at finite γ by the solution χ at $\gamma \to +\infty$, as it is given in Figure 4.3. This corresponds to the operator $\mathcal{W}_N$, as defined in (2.4.17). One can then verify explicitly on the integral representation for $\mathcal{W}_N$ by using certain elements of the Riemann–Hilbert problem satisfied by χ that the equation $\mathcal{S}_N\big[\mathcal{W}_N[H]\big] = H$ does hold on $[a_N\,;\,b_N]$. All the other conclusions of the theorem can be proved similarly to Proposition 4.3.7. ■

We describe a symmetry of the integral transform $\mathcal{W}_N$ that will appear handy in the remaining of the text.

Lemma 4.3.9 *The operator $\mathcal{W}_N$ has the reflection symmetry:*

$$\mathcal{W}_N[H](a_N + b_N - \xi) = -\mathcal{W}_N[H^{\wedge}](\xi) \tag{4.3.65}$$

where we agree upon $H^{\wedge}(\xi) = H(a_N + b_N - \xi)$.

Proof It follows from the jump conditions satisfies by χ and from Lemma 4.2.2 that, for $\lambda \in \mathbb{R}$,

$$\chi_{11;+}(-\lambda) = \mathrm{e}^{-\mathrm{i}\lambda\overline{x}_N}\cdot\chi_{11;+}(\lambda)$$
$$\text{and} \quad \chi_{12;+}(-\lambda) = \mathrm{e}^{-\mathrm{i}\lambda\overline{x}_N}\cdot\Big(\chi_{12;+}(\lambda) - \lambda\,\chi_{11;+}(\lambda)\Big)\,. \tag{4.3.66}$$

Upon squeezing the contours of integration in the integral representation for $\mathcal{W}_N$ to $\mathbb{R}$ we get, in particular, the + boundary values of χ_{1a}. It is then enough to implement the change of variables $(\lambda, \mu, \eta) \mapsto (-\lambda, -\mu, b_N + a_N - \eta)$ and observe that, all in all, the unwanted terms cancel out. ■

In the case of a constant argument (which clearly does *not* belong to $\mathfrak{X}_s(\mathbb{R})$) the expression for $\mathcal{W}_N$ simplifies:

Lemma 4.3.10 *The function* $\mathcal{W}_N[1](\xi)$ *admits the one-fold integral representation*

$$\mathcal{W}_N[1](\xi) = -\frac{N^{\alpha}\chi_{12;+}(0)}{2i\pi\beta}\int\limits_{\mathbb{R}+i\epsilon'}\frac{\chi_{11}(\lambda)}{\lambda}e^{-iN^{\alpha}\lambda(\xi-a_N)}\cdot\frac{d\lambda}{2i\pi}\,. \tag{4.3.67}$$

Proof Starting from the representation (2.4.17) we get, for any $\xi\in]a_N\,;b_N[$,

$$\mathcal{W}_N[1](\xi) = \frac{N^{\alpha}}{2\pi i\beta}\int\limits_{\mathbb{R}+2i\epsilon}\frac{d\lambda}{2i\pi}$$
$$\int\limits_{\mathbb{R}+i\epsilon}\frac{d\mu}{2i\pi}\frac{e^{-iN^{\alpha}(\xi-a_N)\lambda}}{\mu-\lambda}\left\{\frac{1}{\mu}\cdot\chi_{11}(\lambda)\chi_{12}(\mu)-\frac{1}{\lambda}\cdot\chi_{11}(\mu)\chi_{12}(\lambda)\right\}\cdot\left(1-e^{-i\mu\overline{x}_N}\right). \tag{4.3.68}$$

One should then treat the terms involving the function 1 and $e^{-i\mu\overline{x}_N}$ arising in the right-hand side differently. The part involving 1 is zero as can be seen by deforming the μ-integral up to $+i\infty$. In what concerns the part involving $e^{-i\mu\overline{x}_N}$, we deform the μ-integral up to $-i\infty$ by using the jump conditions $e^{-i\lambda\overline{x}_N}\chi_{1a;+}(\lambda) = \chi_{1a;-}(\lambda)$. Solely the pole at $\mu=0$ contributes, hence leading to (4.3.67). ■

References

1. Novokshenov, V.Yu.: Convolution equations on a finite segment and factorization of elliptic matrices. Mat. Zam. **27**, 449–455 (1980)
2. Beals, R., Coifman, R.R.: Scattering and inverse scattering for first order systems. Commun. Pure Appl. Math. **37**, 39–90 (1984)
3. Deift, P.A.: Orthogonal polynomials and random matrices: a Riemann-Hilbert approach. In: Courant Lecture Notes, vol. 3. New-York University (1999)

Chapter 5
The Operators $\mathcal{W}_N$ and $\mathcal{U}_N^{-1}$

Abstract In this chapter we derive a local (in ξ), uniform (in N), behaviour of the inverse $\mathcal{W}_N[H](\xi)$. This will allow an effective simplification, in the large-N limit, of the various integrals involving $\mathcal{W}_N[H]$ arising from the Schwinger–Dyson equations of Proposition 3.2.3. Furthermore, these local asymptotics will provide a base for estimating the W_p^∞ norms of the inverse of the master operator $\mathcal{U}_N$, *cf.* (3.2.1). In fact, such estimates demand to have a control on the leading and sub-leading contributions issuing from $\mathcal{W}_N$ with respect to W_p^∞ norms. We shall demonstrate in Section 5.1.1 that the operator $\mathcal{W}_N$ can be decomposed as

$$\mathcal{W}_N \;=\; \mathcal{W}_R + \mathcal{W}_{\text{bk}} + \mathcal{W}_L \; + \; \mathcal{W}_{\exp} \, . \tag{5.0.1}$$

The operator $\mathcal{W}_{\exp}$ represents an exponentially small remainder in W_p^∞ norm, while the three other operators contribute to the leading order asymptotics when $N \to \infty$. Their expression is constructed solely out of the leading asymptotics in N of the solution χ to the Riemann–Hilbert problem given in Proposition 4.2.1. In Section 5.1.2 we shall build on this decomposition so as to show that there arise two regimes for the large-N asymptotic behaviour of $\mathcal{W}_N[H]$ namely when

- ξ is in the "bulk" of $[a_N ; b_N]$, *i.e.* uniformly in N away from the endpoints a_N and b_N.
- ξ is close to the boundaries, *viz.* in the vicinity of the endpoints a_N (resp b_N).

In addition to providing the associated asymptotic expansions, we shall also establish certain properties of the remainders which will turn out to be crucial for our further purposes.

5.1 Local Behaviour of $\mathcal{W}_N[H](\xi)$ in ξ, uniformly in N

5.1.1 An Appropriate Decomposition of $\mathcal{W}_N$

We remind that any function $H \in \mathcal{C}^k\big([a_N ; b_N]\big)$ admits a continuation into a function $\mathcal{C}^k_{\mathrm{c}}\big(]a_N - \eta ; b_N + \eta[\big)$ for some $\eta > 0$. We denote any such extension by H_{e}, as

G. Borot et al., *Asymptotic Expansion of a Partition Function Related to the Sinh-model*, Mathematical Physics Studies,
DOI 10.1007/978-3-319-33379-3_5

it was already specified in the notation and basic definition section. In the present subsection we establish a decomposition that is adapted for deriving the local and uniform in N asymptotic expansion for $\mathcal{W}_N$.

In this section and the next ones, we will use extensively the following notations:

Definition 5.1.1 To a variable ξ on the real line, we associate $x_R = N^{\alpha}(b_N - \xi)$ and $x_L = N^{\alpha}(\xi - a_N)$ the corresponding rescaled and centred around the right (respectively left) boundary variables. Similarly, for a variable η, we denote y_R and y_L its rescaled and centred variable.

Definition 5.1.2 If H is a function of a variable ξ, we denote $H^{\wedge}(\xi) = H(a_N + b_N - \xi)$ its reflection around the centre of $[a_N\,;\,b_N]$ (as already met in Lemma 4.3.9). This exchanges the role of the left and right boundaries. If H is a function of many variables, by $H^{\wedge}$ we mean that all variables are simultaneously reflected. If $\mathcal{O}$ is an operator, we define the reflected operator by:

$$\mathcal{O}^{\wedge}[H] = \big(\mathcal{O}[H^{\wedge}]\big)^{\wedge} \tag{5.1.1}$$

Definition 5.1.3 Let $\mathscr{C}_{\text{reg}}^{(+)}$ (respectively $\mathscr{C}_{\text{reg}}^{(-)}$) be a contour such that:

- it passes between $\mathbb{R}$ and $\Gamma_{\uparrow}$ (respectively $\Gamma_{\downarrow}$).
- it comes from infinity in the direction of angle $\mathrm{e}^{\pm 3\mathrm{i}\pi/4}$ and goes to infinity in the direction of angle $\mathrm{e}^{\pm \mathrm{i}\pi/4}$.

These contours are depicted in Figure 5.1, and we denote $\varsigma/2 = \mathrm{dist}(\mathscr{C}_{\text{reg}}^{(+)}, \mathbb{R}) > 0$. We also introduce an odd function J by setting, for $x > 0$:

$$J(x) = \int_{\mathscr{C}_{\text{reg}}^{(+)}} \frac{\mathrm{e}^{\mathrm{i}\lambda x}}{R(\lambda)} \frac{\mathrm{d}\lambda}{2\mathrm{i}\pi} . \tag{5.1.2}$$

Proposition 5.1.4 *Given any function $H \in \mathcal{C}^k\big([a_N\,;\,b_N]\big)$ with $k \geq 1$ belonging $\mathfrak{X}_s(\mathbb{R})$ (the image of $\mathcal{S}_N$, see (4.3.68)), the function $\mathcal{W}_N[H]$ is $\mathcal{C}^{k-1}\big(]a_N\,;\,b_N[\big)$ and admits the representation*

$$\mathcal{W}_N[H](\xi) = \mathcal{W}_R[H_{\mathrm{e}}](x_R, \xi) + \mathcal{W}_{\text{bk}}[H_{\mathrm{e}}](\xi) + \mathcal{W}_L[H_{\mathrm{e}}](x_L, \xi) + \mathcal{W}_{\text{exp}}[H](\xi) \tag{5.1.3}$$

with:

$$\mathcal{W}_{\text{bk}}[H_{\mathrm{e}}](\xi) = \frac{N^{\alpha}}{2\pi\beta} \int_{\mathbb{R}} \big[H_{\mathrm{e}}(\xi + N^{-\alpha}y) - H_{\mathrm{e}}(\xi)\big]\, J(y) \cdot \mathrm{d}y\, , \tag{5.1.4}$$

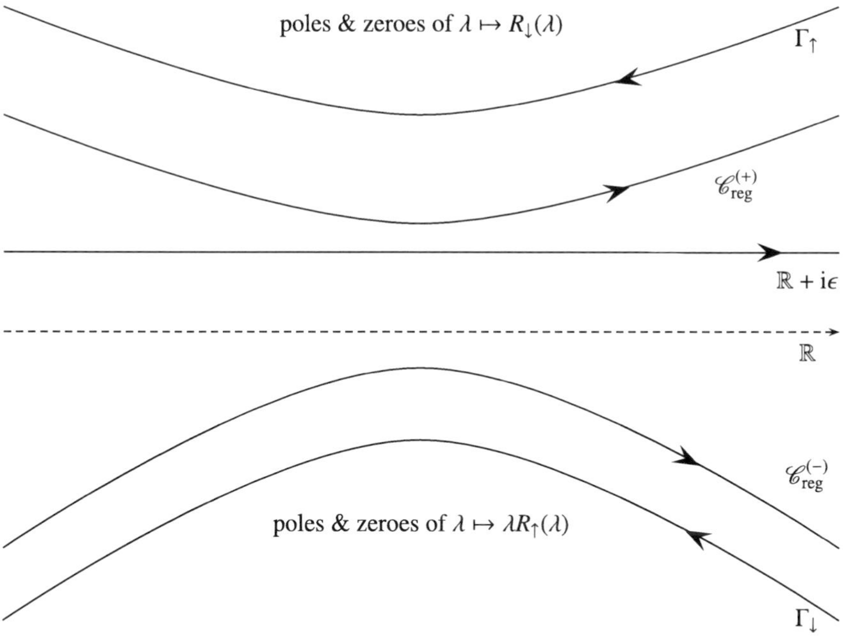

Fig. 5.1 The curves $\mathscr{C}_{\mathrm{reg}}^{(\pm)}$

$$\mathcal{W}_R[H_{\mathfrak{e}}](x,\xi) = -\frac{N^{\alpha}}{2\pi\beta}\int_x +\infty\big[H_{\mathfrak{e}}\big(\xi + N^{-\alpha}y\big) - H_{\mathfrak{e}}(\xi)\big]\, J(y)\cdot \mathrm{d}y$$

$$-\frac{N^{2\alpha}}{2\pi\beta}\int\limits_{\mathscr{C}_{\mathrm{reg}}^{(+)}}\frac{\mathrm{d}\lambda}{2\mathrm{i}\pi}\int\limits_{\mathscr{C}_{\mathrm{reg}}^{(-)}}\frac{\mathrm{d}\mu}{2\mathrm{i}\pi}\,\frac{\mathrm{e}^{\mathrm{i}\lambda x}}{(\mu-\lambda)R_{\downarrow}(\lambda)R_{\uparrow}(\mu)}\cdot\left\{\int\limits_{a_N}^{b_N} H_{\mathfrak{e}}(\eta)\mathrm{e}^{\mathrm{i}\mu N^{\alpha}(\eta-b_N)}\mathrm{d}\eta - \frac{H_{\mathfrak{e}}(\xi)}{\mathrm{i}\mu N^{\alpha}}\right\}, \tag{5.1.5}$$

$$\mathcal{W}_L[H_{\mathfrak{e}}](x,\xi) = -\mathcal{W}_R[H_{\mathfrak{e}}^{\wedge}](x, a_N + b_N - \xi)\,. \tag{5.1.6}$$

The remainder operator $\mathcal{W}_{\exp}[H_{\mathfrak{e}}]$ reads:

$$\mathcal{W}_{\exp} = \mathcal{W}_N^{(++)} - \big(\mathcal{W}_N^{(++)}\big)^{\wedge} + \mathcal{W}_{\mathrm{res}} - \big(\mathcal{W}_{\mathrm{res}}\big)^{\wedge} + \Delta\mathcal{W}_N^{(+-)} - \big(\Delta\mathcal{W}_N^{(+-)}\big)^{\wedge}\,, \tag{5.1.7}$$

where the operators $\mathcal{W}_N^{(++)}$ and $\Delta\mathcal{W}_N^{(+-)}$ are given by

$$\mathcal{W}_N^{(++)}[H](\xi) = \frac{N^{2\alpha}}{2\pi\beta}\int\limits_{\mathscr{C}_{\mathrm{reg}}^{(+)}}\frac{\mathrm{d}\lambda}{2\mathrm{i}\pi}\int\limits_{\mathscr{C}_{\mathrm{reg}}^{(+)}}\frac{\mathrm{d}\mu}{2\mathrm{i}\pi}\int\limits_{a_N}^{b_N}\mathrm{d}\eta\,\frac{\mathrm{e}^{-\mathrm{i}N^{\alpha}\lambda(\xi-b_N)+\mathrm{i}N^{\alpha}\mu(\eta-a_N)}}{(\mu-\lambda)R_{\downarrow}(\lambda)R_{\downarrow}(\mu)}$$

$$\times\left\{\Psi_{11}(\lambda)\Psi_{12}(\mu) - \frac{\mu}{\lambda}\cdot\Psi_{11}(\mu)\Psi_{12}(\lambda)\right\}H(\eta)\,,$$

$$\Delta\mathcal{W}_N^{(+-)}[H](\xi) = \frac{N^{2\alpha}}{2\pi\beta}\int\limits_{\mathscr{C}_{\text{reg}}^{(+)}}\frac{\mathrm{d}\lambda}{2\mathrm{i}\pi}\int\limits_{\mathscr{C}_{\text{reg}}^{(-)}}\frac{\mathrm{d}\mu}{2\mathrm{i}\pi}\int\limits_{a_N}^{b_N}\mathrm{d}\eta\frac{\mathrm{e}^{-\mathrm{i}N^\alpha\lambda(\xi-b_N)+\mathrm{i}N^\alpha\mu(\eta-b_N)}}{(\mu-\lambda)R_\downarrow(\lambda)R_\uparrow(\mu)}$$
$$\times\left\{1+\frac{\mu}{\lambda}\cdot\Psi_{21}(\mu)\Psi_{12}(\lambda)-\Psi_{11}(\lambda)\Psi_{22}(\mu)\right\}H(\eta) \quad (5.1.8)$$

while $\mathcal{W}_{\text{res}}$ is the one-form:

$$\mathcal{W}_{\text{res}}[H] = -\frac{N^{2\alpha}}{2\pi\beta}\frac{\Pi_{12}(0)\theta_R}{R_\downarrow(0)}\int\limits_{\mathscr{C}_{\text{reg}}^{(+)}}\frac{\mathrm{d}\mu}{2\mathrm{i}\pi}\frac{\Psi_{11}(\mu)}{R_\downarrow(\mu)}\int\limits_{a_N}^{b_N}\mathrm{d}\eta\, H(\eta)\,\mathrm{e}^{\mathrm{i}N^\alpha\mu(\eta-a_N)}\,. \quad (5.1.9)$$

The piecewise holomorphic matrix $\Psi(\mu)$ corresponds to the solution to the Riemann–Hilbert problem for Ψ described in Section 4.2.2 in which we have taken the limit $\gamma\to+\infty$.

In the expressions above, we have used an extension $H_{\mathfrak{e}}$ of H whenever it was necessary to integrate H over the whole real line, but we can keep H when only the integrals over $[a_N\,;b_N]$ are involved—*e.g.* in (5.1.8). The decomposition given in Proposition 5.1.4 splits $\mathcal{W}_N$ into a sum of four operators. The operator $\mathcal{W}_{\text{bk}}$ takes into account the purely bulk-type contribution of the inverse, namely those which do not feel the presence of the boundaries a_N, b_N of the support of the equilibrium measure. This operator does not localise at a specific point but rather takes values which are of the same order of magnitude throughout the whole of the interval $[a_N\,;b_N]$. In their turn the operators $\mathcal{W}_{R/L}$ represent the contributions of the right/left boundaries of the support of the equilibrium measure. These operators localise, with exponential precision, on their respective left or right boundary. Namely, they decay exponentially fast in $x_{R/L}$ when $x_{R/L}\to+\infty$. This fact is a consequence of the exponential decay at $\pm\infty$ of $J(x)$ in what concerns the first integral in (5.1.5) and an immediate bound of the second one which follows from $\inf\left\{\operatorname{Im}\lambda,\ \lambda\in\mathscr{C}_{\text{reg}}^{(+)}\right\}>0$.

Proof We remind that since we are considering the $\gamma\to+\infty$ limit, the matrix Ψ has no jump across $\mathbb{R}+\mathrm{i}\epsilon$. A straightforward calculation based on the identity:

$$\chi(\lambda) = \begin{pmatrix}-R_\downarrow^{-1}(\lambda)\,\mathrm{e}^{\mathrm{i}\lambda\overline{x}_N} & R_\uparrow^{-1}(\lambda)\\ -R_\uparrow(\lambda) & 0\end{pmatrix}\cdot\Psi(\lambda)\quad\text{valid for }\lambda\text{ between }\mathbb{R}\text{ and }\Gamma_\uparrow \quad (5.1.10)$$

shows that, for such λ's and μ's,

$$\frac{N^{2\alpha}}{2\pi\beta}\cdot\mathrm{e}^{-\mathrm{i}N^\alpha\lambda(\xi-a_N)}\left\{\chi_{11}(\lambda)\chi_{12}(\mu)-\frac{\mu}{\lambda}\cdot\chi_{11}(\mu)\chi_{12}(\lambda)\right\}\mathrm{e}^{\mathrm{i}N^\alpha\mu(\eta-b_N)}$$
$$=\sum_{\epsilon_1,\epsilon_2\in\{\pm\}}K_{\epsilon_1,\epsilon_2}\big(\lambda,\mu\mid\xi,\eta\big)\,. \quad (5.1.11)$$

The above decomposition contains four kernels

$$K_{--}(\lambda, \mu \mid \xi, \eta) = \frac{N^{2\alpha}}{2\pi\beta} \frac{\mathrm{e}^{-\mathrm{i}N^{\alpha}\lambda(\xi-a_N)+\mathrm{i}N^{\alpha}\mu(\eta-b_N)}}{R_{\uparrow}(\lambda)R_{\uparrow}(\mu)} \left\{ \Psi_{21}(\lambda)\Psi_{22}(\mu) - \frac{\mu}{\lambda}\Psi_{21}(\mu)\Psi_{22}(\lambda) \right\}, \tag{5.1.12}$$

$$K_{++}(\lambda, \mu \mid \xi, \eta) = \frac{N^{2\alpha}}{2\pi\beta} \frac{\mathrm{e}^{-\mathrm{i}N^{\alpha}\lambda(\xi-b_N)+\mathrm{i}N^{\alpha}\mu(\eta-a_N)}}{R_{\downarrow}(\lambda)R_{\downarrow}(\mu)} \left\{ \Psi_{11}(\lambda)\Psi_{12}(\mu) - \frac{\mu}{\lambda}\Psi_{11}(\mu)\Psi_{12}(\lambda) \right\}, \tag{5.1.13}$$

$$K_{+-}(\lambda, \mu \mid \xi, \eta) = -\frac{N^{2\alpha}}{2\pi\beta} \frac{\mathrm{e}^{-\mathrm{i}N^{\alpha}\lambda(\xi-b_N)+\mathrm{i}N^{\alpha}\mu(\eta-b_N)}}{R_{\downarrow}(\lambda)R_{\uparrow}(\mu)} \left\{ \Psi_{11}(\lambda)\Psi_{22}(\mu) - \frac{\mu}{\lambda}\Psi_{21}(\mu)\Psi_{12}(\lambda) \right\}, \tag{5.1.14}$$

$$K_{-+}(\lambda, \mu \mid \xi, \eta) = -\frac{N^{2\alpha}}{2\pi\beta} \frac{\mathrm{e}^{-\mathrm{i}N^{\alpha}\lambda(\xi-a_N)+\mathrm{i}N^{\alpha}\mu(\eta-a_N)}}{R_{\uparrow}(\lambda)R_{\downarrow}(\mu)} \left\{ \Psi_{21}(\lambda)\Psi_{12}(\mu) - \frac{\mu}{\lambda}\Psi_{11}(\mu)\Psi_{22}(\lambda) \right\}. \tag{5.1.15}$$

The labeling of the kernels $K_{\epsilon_1,\epsilon_2}(\lambda, \mu \mid \xi, \eta)$ by the subscripts ϵ_1, ϵ_2 refers to the half-planes $\mathbb{H}^{\epsilon_1} \times \mathbb{H}^{\epsilon_2}$ in which they are exponentially small when $N \to \infty$, provided that the variables $\xi, \eta \in [a_N ; b_N]$ are uniformly away from the boundaries a_N or b_N.

One should note that the above kernels $K_{\epsilon_1,\epsilon_2}$ have a simple pole at $\lambda = 0$. In particular,

$$\mathrm{Res}\Big(K_{-+}(\lambda, \mu \mid \xi, \eta)\,\mathrm{d}\lambda, \lambda = 0\Big) = \frac{\mu\Psi_{11}(\mu)}{R_{\downarrow}(\mu)} \cdot \mathrm{e}^{\mathrm{i}N^{\alpha}\mu(\eta-a_N)} \frac{N^{2\alpha}\cdot\theta_R\cdot\Pi_{12}(0)}{2\pi\beta\cdot\big(\lambda R_{\uparrow}(\lambda)\big)_{|\lambda=0}}, \tag{5.1.16}$$

$$\mathrm{Res}\Big(K_{--}(\lambda, \mu \mid \xi, \eta)\,\mathrm{d}\lambda, \lambda = 0\Big) = -\frac{\mu\Psi_{21}(\mu)}{R_{\uparrow}(\mu)} \cdot \mathrm{e}^{\mathrm{i}N^{\alpha}\mu(\eta-b_N)} \frac{N^{2\alpha}\cdot\theta_R\cdot\Pi_{12}(0)}{2\pi\beta\cdot\big(\lambda R_{\uparrow}(\lambda)\big)_{|\lambda=0}}. \tag{5.1.17}$$

Furthermore, the kernels are related. Indeed, according to the definition of Ψ in terms of χ in Figure 4.1, we have for λ between $\Gamma_{\downarrow}$ and $\mathbb{R}$:

$$\chi(\lambda) = \begin{pmatrix} -R_{\downarrow}^{-1}(\lambda) & R_{\uparrow}^{-1}(\lambda)\,\mathrm{e}^{-\mathrm{i}\lambda\overline{x}_N} \\ 0 & R_{\downarrow}(\lambda) \end{pmatrix} \cdot \Psi(\lambda) \tag{5.1.18}$$

and by invoking the reflection relation for χ obtained in Lemma 4.2.2, we can show that:

$$\Psi(-\lambda) = \begin{pmatrix} 0 & \lambda \\ -\lambda^{-1} & 0 \end{pmatrix} \cdot \Psi(\lambda) \cdot \begin{pmatrix} 1 & -\lambda \\ 0 & 1 \end{pmatrix}. \tag{5.1.19}$$

The above equation ensures that

$$K_{-+}\big(-\lambda, -\mu \mid a_N + b_N - \xi, a_N + b_N - \eta\big) = K_{+-}\big(\lambda, \mu \mid \xi, \eta\big), \tag{5.1.20}$$

$$K_{--}\big(-\lambda, -\mu \mid a_N + b_N - \xi, a_N + b_N - \eta\big) = K_{++}\big(\lambda, \mu \mid \xi, \eta\big). \tag{5.1.21}$$

The decomposition (5.1.11) of the integral kernel allows one recasting the operator $\mathcal{W}_N$ as:

$$\mathcal{W}_N[H](\xi) = \sum_{\epsilon_1,\epsilon_2\in\{\pm 1\}} \widetilde{\mathcal{W}}_N^{(\epsilon_1\epsilon_2)}[H](\xi) \tag{5.1.22}$$

where

$$\widetilde{\mathcal{W}}_N^{(\epsilon_1\epsilon_2)}[H](\xi) = \int_{\mathbb{R}+2\mathrm{i}\epsilon} \frac{\mathrm{d}\lambda}{2\mathrm{i}\pi} \int_{\mathbb{R}+\mathrm{i}\epsilon} \frac{\mathrm{d}\mu}{2\mathrm{i}\pi} \int_{a_N}^{b_N} \mathrm{d}\eta \, \frac{K_{\epsilon_1\epsilon_2}\big(\lambda,\mu \mid \xi,\eta\big)}{\mu-\lambda} H(\eta)\,. \tag{5.1.23}$$

The next step consists in deforming the contours arising in the definition of $\widetilde{\mathcal{W}}_N^{(\epsilon_1\epsilon_2)}[H]$. We shall discuss these handlings on the example of $\widetilde{\mathcal{W}}_N^{(-+)}[H]$. In this case, one should deform the λ-integration to $\mathbb{R}-2\mathrm{i}\epsilon$. In doing so, we pick the residues at the poles at $\lambda=0$ and $\lambda=\mu$ leading to

$$\widetilde{\mathcal{W}}_N^{(-+)}[H](\xi) = \mathcal{W}_{\mathrm{res}}[H] + \int_{\mathbb{R}} \frac{\mathrm{d}\lambda}{2\mathrm{i}\pi} \int_{a_N}^{b_N} \mathrm{d}\eta \, K_{-+}\big(\lambda,\lambda \mid \xi,\eta\big) H(\eta)$$
$$+ \int_{\mathbb{R}-2\mathrm{i}\epsilon} \frac{\mathrm{d}\lambda}{2\mathrm{i}\pi} \int_{\mathbb{R}-\mathrm{i}\epsilon} \frac{\mathrm{d}\mu}{2\mathrm{i}\pi} \int_{a_N}^{b_N} \mathrm{d}\eta \, \frac{K_{-+}\big(\lambda,\mu \mid \xi,\eta\big)}{\mu-\lambda} H(\eta)\,.$$

It remains to implement the change of variables $(\lambda,\mu)\mapsto(-\lambda,-\mu)$ in the last integral and observe that

$$K_{-+}\big(\lambda,\lambda \mid \xi,\eta\big) = \frac{N^{2\alpha}}{2\pi\beta} \frac{\mathrm{e}^{\mathrm{i}N^{\alpha}\mu(\eta-\xi)}}{R(\lambda)} \quad \text{since} \quad \det\Psi(\lambda)=1\,, \tag{5.1.24}$$

so as to obtain

$$\widetilde{\mathcal{W}}_N^{(-+)}[H](\xi) = \mathcal{W}_{\mathrm{res}}[H] + \mathcal{W}_{\mathrm{bk}}^{(0)}[H](\xi) - \widetilde{\mathcal{W}}_N^{(+-)}\big[H^{\wedge}\big](a_N+b_N-\xi)\,, \tag{5.1.25}$$

with $\mathcal{W}_{\mathrm{res}}$ being given by (5.1.9) and

$$\mathcal{W}_{\mathrm{bk}}^{(0)}[H](\xi) = \frac{N^{2\alpha}}{2\pi\beta} \fint_{a_N}^{b_N} J\big(N^{\alpha}(\eta-\xi)\big)\, H(\eta)\,\mathrm{d}\eta \tag{5.1.26}$$

with the function J given in (5.1.2). A similar reasoning applied to the case of $\widetilde{\mathcal{W}}_N^{(--)}[H](\xi)$ yields

$$\widetilde{\mathcal{W}}_N^{(--)}[H](\xi) = -\widetilde{\mathcal{W}}_N^{(++)}\big[H^{\wedge}\big](a_N+b_N-\xi) - \mathcal{W}_{\mathrm{res}}[H^{\wedge}]\,. \tag{5.1.27}$$

Hence, eventually, upon deforming the contours to $\mathscr{C}^{(+)}_{\mathrm{reg}}$ or $\mathscr{C}^{(-)}_{\mathrm{reg}}$ in the $\mathcal{W}_N^{(\epsilon,\epsilon')}$ operators,

$$\mathcal{W}_N[H](\xi) = \mathcal{W}_N^{(++)}[H](\xi) - \mathcal{W}_N^{(++)}[H^{\wedge}](a_N + b_N - \xi) + \mathcal{W}_N^{(+-)}[H](\xi) \\ - \mathcal{W}_N^{(+-)}[H^{\wedge}](a_N + b_N - \xi) + \mathcal{W}_{\mathrm{res}}[H] - \mathcal{W}_{\mathrm{res}}[H^{\wedge}] + \mathcal{W}_{\mathrm{bk}}^{(0)}[H](\xi) . \tag{5.1.28}$$

The operator $\mathcal{W}_N^{(++)}$ appearing above has been defined in (5.1.8) whereas

$$\mathcal{W}_N^{(+-)}[H](\xi) = \int_{\mathscr{C}^{(+)}_{\mathrm{reg}}} \frac{\mathrm{d}\lambda}{2\mathrm{i}\pi} \int_{\mathscr{C}^{(-)}_{\mathrm{reg}}} \frac{\mathrm{d}\mu}{2\mathrm{i}\pi} \int_{a_N}^{b_N} \mathrm{d}\eta \, \frac{K_{+-}(\lambda, \mu \mid \xi, \eta)}{\mu - \lambda} H(\eta). \tag{5.1.29}$$

At this stage, it remains to observe that

$$\mathcal{W}_N^{(+-)}[H](\xi) = \mathcal{W}_R^{(0)}[H](x_R) + \Delta\mathcal{W}_N^{(+-)}[H](\xi) , \tag{5.1.30}$$

where $\Delta\mathcal{W}_N^{(+-)}$ is as defined in (5.1.8), while

$$\mathcal{W}_R^{(0)}[H](x) = -\frac{N^{2\alpha}}{2\pi\beta} \int_{\mathscr{C}^{(+)}_{\mathrm{reg}}} \frac{\mathrm{d}\lambda}{2\mathrm{i}\pi} \int_{\mathscr{C}^{(-)}_{\mathrm{reg}}} \frac{\mathrm{d}\mu}{2\mathrm{i}\pi} \int_{a_N}^{b_N} \mathrm{d}\eta \, \frac{H(\eta)\, \mathrm{e}^{\mathrm{i}\lambda x + \mathrm{i}N^{\alpha}\mu(\eta - b_N)}}{(\mu - \lambda)\, R_{\downarrow}(\lambda) R_{\uparrow}(\mu)} . \tag{5.1.31}$$

As a consequence, we obtain the decomposition:

$$\mathcal{W}_N[H](\xi) = \mathcal{W}_L^{(0)}[H](x_L) + \mathcal{W}_{\mathrm{bk}}^{(0)}[H](\xi) + \mathcal{W}_R^{(0)}[H](x_R) + \mathcal{W}_{\mathrm{exp}}[H](\xi) , \tag{5.1.32}$$

where we have set $\mathcal{W}_L^{(0)}[H](x) = -\mathcal{W}_R^{(0)}[H^{\wedge}](x)$. In order to obtain the representation (5.1.3) it is enough to incorporate certain terms present in $\mathcal{W}_{\mathrm{bk}}^{(0)}[H](\xi)$ into the R and L-type operators. Namely, we can recast $\mathcal{W}_{\mathrm{bk}}^{(0)}[H](\xi)$ as

$$\begin{aligned}\mathcal{W}_{\mathrm{bk}}^{(0)}[H](\xi) &= \frac{N^{2\alpha}}{2\pi\beta} \int_{a_N}^{b_N} J\big(N^{\alpha}(\eta - \xi)\big) \big[H(\eta) - H(\xi)\big] \mathrm{d}\eta - N^{\alpha} H(\xi) \big[\varrho_0(x_R) - \varrho_0(x_L)\big] \\ &= \mathcal{W}_{\mathrm{bk}}[H_{\mathrm{e}}](\xi) - N^{\alpha}\big[\varrho_0(x_R) - \varrho_0(x_L)\big] H_{\mathrm{e}}(\xi) \\ &\quad - \frac{N^{\alpha}}{2\pi\beta} \bigg\{ \int_{x_R}^{+\infty} + \int_{-\infty}^{-x_L} \bigg\} \cdot \big[H_{\mathrm{e}}(\xi + N^{-\alpha} y) - H_{\mathrm{e}}(\xi)\big] J(y)\, \mathrm{d}y.\end{aligned}$$

There, we have introduced

$$\varrho_0(x) \;=\; \frac{-1}{2\mathrm{i}\pi\beta}\int\limits_{\mathscr{C}_{\mathrm{reg}}^{(+)}} \frac{\mathrm{e}^{\mathrm{i}\lambda x}}{\lambda R(\lambda)}\frac{\mathrm{d}\lambda}{2\mathrm{i}\pi} \quad i.e. \quad \varrho_0'(x) = -\frac{J(x)}{2\pi\beta}\,. \tag{5.1.33}$$

The representation (5.1.3) for $\mathcal{W}_N[H]$ follows by redistributing the terms. This decomposition also ensures that $\mathcal{W}_N[H] \in \mathcal{C}^{k-1}(]a_N\,;b_N[)$. Indeed, this regularity follows from the exponential decay of the integrands in Fourier space when $\xi \in]a_N\,;b_N[$ and derivation under the integral theorems. ■

Note that the integral defining ϱ_0 in (5.1.33) can be evaluated explicitly leading to:

$$\varrho_0(x) \;=\; \frac{1}{2\pi^2\beta}\cdot\ln\left|\frac{1-\mathrm{e}^{-\frac{2\pi|x|\omega_1\omega_2}{\omega_1+\omega_2}}}{1+\mathrm{e}^{-\frac{2\pi|x|\omega_1\omega_2}{\omega_1+\omega_2}+\mathrm{i}\pi\frac{\omega_2-\omega_1}{\omega_1+\omega_2}}}\right|\,. \tag{5.1.34}$$

In particular, it exhibits a logarithmic singularity at the origin meaning that $J(x)$ has a $1/x$ behaviour around 0.

5.1.2 *Local Approximants for $\mathcal{W}_N$*

In this subsection, we obtain uniform—in the running variable—asymptotic expansions for the operators $\mathcal{W}_{\mathrm{bk}}$, $\mathcal{W}_R$ and $\mathcal{W}_{\mathrm{exp}}$. In particular, we shall establish that if ξ is uniformly away from b_N (respectively a_N), $\mathcal{W}_R$ (respectively $\mathcal{W}_L$) will only generate exponentially small (in N) corrections. Finally, this exponentially small bound will hold uniformly after a finite number of ξ-differentiations. Prior to discussing these matters we need to introduce two families of auxiliary functions on $\mathbb{R}^+$ and constants that come into play during the description of these behaviours.

Definition 5.1.5 For any integer $\ell \geq 0$:

$$\varpi_\ell(x) = \frac{1}{2\pi\beta}\int\limits_x^{+\infty} y^\ell\, J(y)\,\mathrm{d}y\,, \tag{5.1.35}$$

$$\varrho_\ell(x) = \frac{\mathrm{i}^{\ell+1}}{2\pi\beta}\int\limits_{\mathscr{C}_{\mathrm{reg}}^{(+)}}\frac{\mathrm{d}\lambda}{2\mathrm{i}\pi}\int\limits_{\mathbb{R}-\mathrm{i}\epsilon'}\frac{\mathrm{d}\mu}{2\mathrm{i}\pi}\,\frac{\mathrm{e}^{\mathrm{i}\lambda x}}{\mu^{\ell+1}R_\uparrow(\mu)(\mu-\lambda)R_\downarrow(\lambda)}\,, \tag{5.1.36}$$

$$u_\ell = \frac{\mathrm{i}^\ell}{2\mathrm{i}\pi\beta\,\ell!}\frac{\partial^\ell}{\partial\lambda^\ell}\left(\frac{1}{R(\lambda)}\right)_{|\lambda=0}\,. \tag{5.1.37}$$

Note that $u_{2p} = 0$ since R is an odd function—given in (4.1.18).

For $\ell = 0$, this definition of ϱ_0 coincides with (5.1.33), whose explicit expression is (5.1.34). Indeed, we remember from Section 4.1.2 that $\uparrow$ means that we can move the contour of integration over μ up to $+\mathrm{i}\infty$ without hitting a pole of $R_{\uparrow}^{-1}(\mu)$. According to (4.1.26), $\mu\, R_{\uparrow}(\mu)$ has a non-zero limit when $\mu \to 0$, so we just pick up the residue at $\mu = \lambda$, which leads to the expression (5.1.33). For $\ell \geq 1$, the function ϖ_ℓ is continuous at $x = 0$. Furthermore, for any $\ell \geq 0$, $\varrho_\ell(x)$ and $\varpi_p(x)$ decay exponentially fast in x when $x \to +\infty$. Indeed, it is readily seen on the basis of their explicit integral representations that there exists $C_\ell > 0$ such that:

$$|\,\varrho_\ell(x)| + |\,\varpi_\ell(x)| \;\leq\; C_\ell\, \mathrm{e}^{-C'_\ell x} \qquad \text{for } \ell \geq 1\,. \tag{5.1.38}$$

Proposition 5.1.6 *Let $k \geq 0$ be an integer, $H \in \mathcal{C}^{2k+1}\big([a_N\,;\,b_N]\big)$, and define:*

$$\mathcal{W}_{R;k}[H](x,\xi) = \frac{H(\xi) - H(b_N)}{\xi - b_N}\cdot x\varrho_0(x) \;-\; \sum_{\ell=1}^{k} \frac{H^{(\ell)}(\xi)}{N^{(\ell-1)\alpha}\cdot \ell!}\cdot \varpi_\ell(x)$$
$$+ \sum_{\ell=1}^{k} \frac{H^{(\ell)}(b_N)}{N^{(\ell-1)\alpha}}\cdot \varrho_\ell(x)\,, \tag{5.1.39}$$

$$\mathcal{W}_{\mathrm{bk};k}[H](\xi) = \sum_{\ell=1}^{k} \frac{H^{(\ell)}(\xi)}{N^{\alpha(\ell-1)}}\cdot u_\ell\,. \tag{5.1.40}$$

These operators provide the asymptotic expansions, uniform for $\xi \in [a_N\,;\,b_N]$:

$$\mathcal{W}_R[H_\mathfrak{e}](x_R,\xi) = \mathcal{W}_{R;k}[H](x_R,\xi) \;+\; \Delta_{[k]}\mathcal{W}_R[H_\mathfrak{e}](x_R,\xi)\,, \tag{5.1.41}$$
$$\mathcal{W}_{\mathrm{bk}}[H_\mathfrak{e}](\xi) = \mathcal{W}_{\mathrm{bk};k}[H](\xi) \;+\; \Delta_{[k]}\mathcal{W}_{\mathrm{bk}}[H_\mathfrak{e}](\xi)\,. \tag{5.1.42}$$

The remainder in (5.1.41) *takes the form:*

$$\Delta_{[k]}\mathcal{W}_R[H_\mathfrak{e}](x,\xi) \;=\; \mathcal{R}^{(0)}_{R;[k]}[H_\mathfrak{e}](x,\xi) \;+\; \sum_{\ell=0}^{k} x^{\ell+1/2}\,\mathcal{R}^{(1/2)}_{R;[k];\ell}[H_\mathfrak{e}](x)\,, \tag{5.1.43}$$

with $\mathcal{R}^{(0)}_{R;[k]}[H_\mathfrak{e}] \in W_k^{(\infty)}\big(\mathbb{R}^+ \times [a_N\,;\,b_N]\big)$ and $\mathcal{R}^{(1/2)}_{R;[k];\ell}[H_\mathfrak{e}] \in W_k^{(\infty)}\big(\mathbb{R}^+\big)$. One has the more precise bound: $\forall m \in [\![\,0\,;\,k\,]\!]$,

$$\max_{\substack{p\in[\![\,0\,;\,m\,]\!]\\ \ell\in[\![\,0\,;\,k\,]\!]}} \Big\{ \big|\partial_\xi^p \mathcal{R}^{(0)}_{R;[k]}[H_\mathfrak{e}](x_R,\xi)\big| \;+\; \big|\partial_\xi^p \mathcal{R}^{(1/2)}_{R;[k];\ell}[H_\mathfrak{e}](x_R)\big| \Big\}$$
$$\leq\; \frac{C\mathrm{e}^{-C' x_R}}{N^{(k-m)\alpha}}\, ||H_\mathfrak{e}^{(k+1)}||_{W_m^\infty(\mathbb{R})} \tag{5.1.44}$$

for some $C, C' > 0$ independent of N and H. The remainder in (5.1.42) is bounded by:

$$\big|\big|\Delta_{[k]}\mathcal{W}_{\mathrm{bk}}[H_{\mathrm{e}}]\big|\big|_{W_m^{\infty}([a_N\,;b_N])} \;\leq\; C\,N^{-k\alpha}\,||H_{\mathrm{e}}^{(k+1)}||_{W_m^{\infty}(\mathbb{R})}\;. \tag{5.1.45}$$

Proposition 5.1.7 *Let $k \geq 0$ be an integer, and $H \in \mathcal{C}^{2k+1}\big([a_N\,;\,b_N]\big)$. The operator $\mathcal{W}_{\mathrm{exp}}$ takes the form:*

$$\mathcal{W}_{\mathrm{exp}}[H](\xi) = \mathcal{R}^{(0)}_{\mathrm{exp};\,R}[H](x_R,\xi) \;+\; \sum_{\ell=0}^{k} x_R^{\ell+1/2}\,\mathcal{R}^{(1/2)}_{\mathrm{exp};\,R;\ell}[H](x_R)$$
$$+\,\mathcal{R}^{(0)}_{\mathrm{exp};\,L}[H](x_L,\xi) \;+\; \sum_{\ell=0}^{k} x_L^{\ell+1/2}\,\mathcal{R}^{(1/2)}_{\mathrm{exp};\,L;\ell}[H](x_L) \tag{5.1.46}$$

with $\mathcal{R}^{(0)}_{\mathrm{exp};\,R/L}[H_{\mathrm{e}}] \in W_k^{(\infty)}\big(\mathbb{R}^+ \times [a_N\,;\,b_N]\big)$ and $\mathcal{R}^{(1/2)}_{\mathrm{exp};\,R/L;\ell}[H_{\mathrm{e}}] \in W_k^{(\infty)}\big(\mathbb{R}^+\big)$. One has the more precise bound: $\forall m \in [\![\,0\,;\,k\,]\!]$,

$$\max_{\substack{p\in[\![\,0\,;\,m\,]\!] \\ \ell\in[\![\,0\,;\,k\,]\!]}}\Big\{\big|\partial_\xi^p\mathcal{R}^{(0)}_{\mathrm{exp};\,R/L}[H_{\mathrm{e}}](x_{R/L},\xi)\big| \;+\; \big|\partial_\xi^p\mathcal{R}^{(1/2)}_{\mathrm{exp};\,R/L;\ell}[H_{\mathrm{e}}](x_{R/L})\big|\Big\}$$
$$\leq\; C N^{m\alpha}\mathrm{e}^{-C'N^{\alpha}}\;||H_{\mathrm{e}}^{(k+1)}||_{W_m^{\infty}(\mathbb{R})} \tag{5.1.47}$$

for some $C, C' > 0$ independent of N and H.

The idea for obtaining the above form of the asymptotic expansions is to represent H in terms of its Taylor-integral expansion of order k. We can then compute explicitly the contributions issuing from the polynomial part of the Taylor series expansion for H and obtain sharp bounds on the remainder by exploiting the structure of the integral remainder in the Taylor integral series. In particular, the analysis of this integral remainder allows uniform bounds for the remainder as given in (5.1.44), (5.1.45) and (5.1.47). The reason for such handlings instead of more direct bounds issues from the fact that the integrals we manipulate are only weakly convergent. One thus has first to build on the analytic structure of the integrand so as to obtain the desired bounds and expressions and, in particular, carry out some contour deformations. Clearly, such handlings cannot be done anymore upon inserting the absolute value under the integral sign, as then the integrand is no more analytic.

Proof We carry out the analysis, individually, for each operator.

The operator $\mathcal{W}_{\mathrm{bk}}$

The Taylor integral expansion of H up to order k yields the representation

$$\mathcal{W}_{\mathrm{bk}}[H_{\mathrm{e}}](\xi) \;=\; \sum_{p=1}^{k}\frac{1}{2\pi\beta\,N^{(p-1)\alpha}}\,\frac{H^{(p)}(\xi)}{p!}\int_{\mathbb{R}} y^p J(y)\,\mathrm{d}y \;+\; \Delta_{[k]}\mathcal{W}_{\mathrm{bk}}[H_{\mathrm{e}}](\xi)\,, \tag{5.1.48}$$

where

$$\Delta_{[k]}\mathcal{W}_{\mathrm{bk}}[H_{\mathrm{e}}](\xi) \;=\; \frac{1}{2\pi\beta N^{k\alpha}}\int_0^1 \mathrm{d}t\,\frac{(1-t)^k}{k!}\int_{\mathbb{R}} \mathrm{d}y\, y^{k+1} J(y)\, H_{\mathrm{e}}^{(k+1)}(\xi+N^{-\alpha}ty)\,. \tag{5.1.49}$$

In the first terms of (5.1.48) we identify:

$$\int_{\mathbb{R}} y^{\ell} J(y)\mathrm{d}y \;=\; \mathrm{i}^{\ell-1}\frac{\partial^{\ell}}{\partial\lambda^{\ell}}\Big(\frac{1}{R(\lambda)}\Big)_{|\lambda=0} \;=\; 2\pi\beta\,\ell!\,u_{\ell}\,, \tag{5.1.50}$$

and we remind that this is zero when ℓ is even. Finally, we get that the remainder is a $\mathcal{C}^k$ function of ξ, and:

$$\forall m\in[\![\,0\,;\,k\,]\!],\qquad \|\Delta_{[k]}\mathcal{W}_{\mathrm{bk}}[H_{\mathrm{e}}]\|_{W_m^{\infty}([a_N\,;b_N])} \;\le\; \frac{\|H_{\mathrm{e}}^{(k+1)}\|_{W_m^{\infty}(\mathbb{R})}}{N^{k\alpha}}\int_{\mathbb{R}} |y|^{k+1}|J(y)|\,\frac{\mathrm{d}y}{2\pi\beta}\,. \tag{5.1.51}$$

Since J decays exponentially at ∞ (see (5.1.33) and (5.1.34)), the last integral gives a finite, k-dependent constant.

The operator $\mathcal{W}_R$

The contribution arising in the first line of (5.1.5) can be treated analogously to $\mathcal{W}_{\mathrm{bk}}$, what leads to

$$-\frac{N^{\alpha}}{2\pi\beta}\int_x^{+\infty} J(y)\big[H_{\mathrm{e}}(\xi+N^{-\alpha}y)-H_{\mathrm{e}}(\xi)\big]\mathrm{d}y \;=\; -\sum_{\ell=1}^{k}\frac{H_{\mathrm{e}}^{(\ell)}(\xi)}{N^{(\ell-1)\alpha}\,\ell!}\,\varpi_{\ell}(x)\;+\;\Delta_{[k]}\mathcal{W}_R^{(1)}[H_{\mathrm{e}}](x,\xi) \tag{5.1.52}$$

with

$$\Delta_{[k]}\mathcal{W}_R^{(1)}[H_{\mathrm{e}}](x,\xi) = \frac{-1}{2\pi\beta\,N^{k\alpha}}\int_x^{+\infty}\mathrm{d}y\,y^{k+1}J(y)\int_0^1\mathrm{d}t\,\frac{(1-t)^k}{k!}\,H_{\mathrm{e}}^{(k+1)}(\xi+N^{-\alpha}ty)\,. \tag{5.1.53}$$

Since J decays exponentially at infinity, we clearly have:

$$\max_{p\in[\![\,0\,;\,m\,]\!]}\big|\partial_{\xi}^{p}\cdot\Delta_{[k]}\mathcal{W}_R^{(1)}[H_{\mathrm{e}}](x_R,\xi)\big| \;\le\; C\,\mathrm{e}^{-C'x_R}\cdot\frac{\|H_{\mathrm{e}}^{(k+1)}\|_{W_m^{\infty}(\mathbb{R})}}{N^{(k-m)\alpha}} \tag{5.1.54}$$

for some constants C, C' independent of H and N. We remind that the ξ-derivative can act on both variables ξ *and* $x_R = N^{\alpha}(b_N-\xi)$.

We now focus on the contributions issuing from the second line of (5.1.5). For this purpose, observe that the Taylor-integral series representation for H yields the following representation for the Fourier transform of H:

$$\int_{a_N}^{b_N} H(\eta)\mathrm{e}^{\mathrm{i}\mu N^{\alpha}(\eta-b_N)}\,\mathrm{d}\eta \;=\; \mathcal{F}_{1;k}[H](\mu) \,+\, \mathcal{F}_{2;k}[H_{\mathfrak{e}}](\mu) \,+\, \mathcal{F}_3[H_{\mathfrak{e}}](\mu)\,, \quad (5.1.55)$$

where we complete the integral over $[a_N\,;\,b_N]$ to $]-\infty\,;\,b_N]$ in the first term, while the two last terms come from subtracting the right and left contributions:

$$\mathcal{F}_{1;k}[H](\mu) = -\sum_{p=0}^{k}\Big(\frac{\mathrm{i}}{N^{\alpha}\mu}\Big)^{p+1}\cdot H^{(p)}(b_N),$$
$$\mathcal{F}_3[H_{\mathfrak{e}}](\mu) = \int_{-\infty}^{a_N} H_{\mathfrak{e}}(\eta)\,\mathrm{e}^{\mathrm{i}\mu N^{\alpha}(\eta-b_N)}\,\mathrm{d}\eta \quad (5.1.56)$$

and

$$\mathcal{F}_{2;k}[H_{\mathfrak{e}}](\mu) \;=\; \int_{-\infty}^{b_N}\mathrm{d}\eta\int_0^1\mathrm{d}t\,\frac{(1-t)^k}{k!}\mathrm{e}^{\mathrm{i}\mu N^{\alpha}(\eta-b_N)}(\eta-b_N)^{k+1}\,H_{\mathfrak{e}}^{(k+1)}\big(b_N+t(\eta-b_N)\big)\,. \quad (5.1.57)$$

Thus,

$$-\frac{N^{2\alpha}}{2\pi\beta}\int_{\mathscr{C}_{\mathrm{reg}}^{(+)}}\frac{\mathrm{d}\lambda}{2\mathrm{i}\pi}\int_{\mathscr{C}_{\mathrm{reg}}^{(-)}}\frac{\mathrm{d}\mu}{2\mathrm{i}\pi}\frac{\mathrm{e}^{\mathrm{i}\lambda x}}{(\mu-\lambda)R_{\downarrow}(\lambda)R_{\uparrow}(\mu)}\int_{a_N}^{b_N}H(\eta)\mathrm{e}^{-\mathrm{i}\mu y_R}\,\mathrm{d}\eta$$
$$=\sum_{\ell=0}^{k}\frac{H^{(\ell)}(b_N)}{N^{(\ell-1)\alpha}}\,\varrho_{\ell}(x)+\mathcal{L}_{\Lambda_0}\big[\mathcal{F}_{2;k}[H_{\mathfrak{e}}]\,+\,\mathcal{F}_3[H_{\mathfrak{e}}]\big](x)\,. \quad (5.1.58)$$

$\mathcal{L}_{\Lambda_0}$ is an operator with integral kernel—see later Equation (5.1.65):

$$\Lambda_0(\lambda,\mu) \;=\; \frac{-1}{R_{\uparrow}(\mu)R_{\downarrow}(\lambda)} \quad (5.1.59)$$

which satisfies the assumptions of Lemma 5.1.8 appearing below. Thence, Lemma 5.1.8 entails the decomposition:

$$\mathcal{L}_{\Lambda_0}\big[\mathcal{F}_{2;k}[H_{\mathfrak{e}}]\,+\,\mathcal{F}_3[H_{\mathfrak{e}}]\big](x) \;=\; \sum_{\ell=0}^{k}\Big\{x^{\ell+1/2}\mathrm{e}^{-\varsigma x}\mathcal{L}_{\Lambda_0;\ell}\big[\mathcal{F}_{2;k}[H_{\mathfrak{e}}]\,+\,\mathcal{F}_3[H_{\mathfrak{e}}]\big](x)\Big\}$$
$$+\,(\Delta_{[k]}\mathcal{L}_{\Lambda_0})\big[\mathcal{F}_{2;k}[H_{\mathfrak{e}}]\,+\,\mathcal{F}_3[H_{\mathfrak{e}}]\big](x) \quad (5.1.60)$$

in which both $\mathcal{L}_{\Lambda_0;\ell}\big[\mathcal{F}_{2;k}[H_{\mathfrak{e}}]+\mathcal{F}_3[H_{\mathfrak{e}}]\big](x)$ and $(\Delta_{[k]}\mathcal{L}_{\Lambda_0})\big[\mathcal{F}_{2;k}[H_{\mathfrak{e}}]+\mathcal{F}_3[H_{\mathfrak{e}}]\big](x)$ belong to $W_k^{\infty}(\mathbb{R}^+)$ and are as given in (5.1.67) and (5.1.68)

By using the bounds:

$$\left|\mathcal{F}_{2;k}[H_{\mathfrak{e}}](\mu)\right| \;\leq\; \frac{c_k||H_{\mathfrak{e}}^{(k+1)}||_{L^{\infty}(\mathbb{R})}}{(N^{\alpha}|\mu|)^{k+2}} \quad \text{since} \quad \frac{1}{|\mathrm{Im}\,\mu|} \;\leq\; \frac{c'}{|\mu|} \quad \text{for } \mu\in\mathscr{C}_{\mathrm{reg}}^{(-)}\,, \tag{5.1.61}$$

and

$$\left|\mathcal{F}_{3}[H_{\mathfrak{e}}](\mu)\right| \;\leq\; c\frac{||H_{\mathfrak{e}}||_{L^{\infty}(\mathbb{R})}}{|\mu|N^{\alpha}}\cdot \mathrm{e}^{-\overline{x}_N|\mathrm{Im}\,\mu|}\,. \tag{5.1.62}$$

we get that there exists N-independent constants C, C' such that

$$\max_{\substack{p\in[\![\,0\,;\,m\,]\!]\\ \ell\in[\![\,0\,;\,k\,]\!]}}\left|\partial_{\xi}^{p}\cdot\left\{\left(\mathrm{e}^{-\varsigma x_R}\mathcal{L}_{\Lambda_0;\ell} \;+\; \Delta_{[k]}\mathcal{L}_{\Lambda_0}\right)\left[\mathcal{F}_{2;k}[H_{\mathfrak{e}}] \;+\; \mathcal{F}_3[H_{\mathfrak{e}}]\right](x_R)\right\}\right|$$
$$\leq\; C\,\mathrm{e}^{-C'x_R}\,\frac{||H_{\mathfrak{e}}^{(k+1)}||_{L^{\infty}(\mathbb{R})}}{N^{(k-m)\alpha}}\,. \tag{5.1.63}$$

We have relied on:

Lemma 5.1.8 *Let $\Lambda(\lambda,\mu)$ be a holomorphic function of λ and μ belonging to the region of the complex plane delimited by $\mathscr{C}_{\mathrm{reg}}^{(+)}$ and $\mathscr{C}_{\mathrm{reg}}^{(-)}$ and such that it admits an asymptotic expansion*

$$\Lambda(\lambda,\mu)=\sum_{\ell=0}^{k}\frac{\Lambda_{\ell}(\mu)}{\left[\mathrm{i}(\lambda-\mathrm{i}\varsigma)\right]^{\ell+1/2}}+\Delta_{[k]}\Lambda(\lambda,\mu)\ \text{with}\ \begin{cases}|\Lambda_{\ell}(\mu)| & =\mathrm{O}\big(|\mu|^{1/2}\big)\\ |\Delta_{[k]}\Lambda(\lambda,\mu)| & =\mathrm{O}\big(|\lambda|^{-(k+3/2)}\cdot|\mu|^{1/2}\big)\end{cases}. \tag{5.1.64}$$

Then, the integral operator on $\mu\cdot L^{\infty}\big(\mathscr{C}_{\mathrm{reg}}^{(-)}\big)\;\equiv\;\left\{f\;:\;\mu\mapsto\mu f(\mu)\in L^{\infty}\big(\mathscr{C}_{\mathrm{reg}}^{(-)}\big)\right\}$

$$\mathcal{L}_{\Lambda}\big[f\big](x)\;=\;\frac{N^{2\alpha}}{2\pi\beta}\int_{\mathscr{C}_{\mathrm{reg}}^{(+)}}\frac{\mathrm{d}\lambda}{2\mathrm{i}\pi}\int_{\mathscr{C}_{\mathrm{reg}}^{(-)}}\frac{\mathrm{d}\mu}{2\mathrm{i}\pi}\,\frac{\Lambda(\lambda,\mu)}{\mu-\lambda}\mathrm{e}^{\mathrm{i}\lambda x}f(\mu) \tag{5.1.65}$$

can be recast as

$$\mathcal{L}_{\Lambda}\big[f\big](x)\;=\;\sum_{\ell=0}^{k}x^{\ell+1/2}\mathrm{e}^{-\varsigma x}\mathcal{L}_{\Lambda;\ell}[f](x)+\Delta_{[k]}\mathcal{L}_{\Lambda}\big[f\big](x)\,, \tag{5.1.66}$$

where the operators

$$\mathcal{L}_{\Lambda;k}[f](x)=\frac{N^{2\alpha}}{2\pi\beta}\int_{\Gamma(\mathrm{i}\mathbb{R}^{+})}\frac{\mathrm{d}\lambda}{2\mathrm{i}\pi}\int_{\mathscr{C}_{\mathrm{reg}}^{(-)}}\frac{\mathrm{d}\mu}{2\mathrm{i}\pi}\,\frac{\Lambda_{k}(\mu)\mathrm{e}^{\mathrm{i}\lambda}f(\mu)}{\left[x\big(\mu-\mathrm{i}\varsigma\big)-\lambda\right](\mathrm{i}\lambda)^{\ell+\frac{1}{2}}}\,, \tag{5.1.67}$$

$$\Delta_{[k]}\mathcal{L}_\Lambda[f](x) = \frac{N^{2\alpha}}{2\pi\beta}\int_{\mathscr{C}_{\mathrm{reg}}^{(+)}}\frac{\mathrm{d}\lambda}{2\mathrm{i}\pi}\int_{\mathscr{C}_{\mathrm{reg}}^{(-)}}\frac{\mathrm{d}\mu}{2\mathrm{i}\pi}\frac{\Delta_{[k]}\Lambda(\lambda,\mu)}{\mu-\lambda}\mathrm{e}^{\mathrm{i}\lambda x}f(\mu)\,, \tag{5.1.68}$$

are continuous as operators $\mu\cdot L^\infty\big(\mathscr{C}_{\mathrm{reg}}^{(-)}\big)\to W_k^\infty\big(\mathbb{R}^+\big)$. *Note that, above,* $\Gamma\big(\mathrm{i}\mathbb{R}^+\big)$ *corresponds to a small counterclockwise loop around* $\mathrm{i}\mathbb{R}^+$.

Proof It is enough to insert the large-μ expansion of Λ and then, in the part subordinate to the inverse power-law expansion, deform the λ-integrals to $\Gamma\big(\mathrm{i}\mathbb{R}+\mathrm{i}\varsigma\big)$, translate by $+\mathrm{i}\varsigma$ and, finally, rescale by x. The statements about continuity are evident. ■

The operator $\mathcal{W}_{\mathrm{exp}}$ (Proposition 5.1.7)

The analysis relative to the structure of $\mathcal{W}_{\mathrm{exp}}[H_\mathfrak{e}]$ follows basically the same steps as above so we shall not detail them here again. The main point, though, is the presence of an exponential prefactor e^{-cN^α} which issues from the bound (4.2.15) on $\Pi - I_2$. ■

5.1.3 Large N Asymptotics of the Approximants of $\mathcal{W}_N$

The results of Propositions 5.1.6 and 5.1.7 induce the representation

$$\mathcal{W}_N[H](\xi) = \mathcal{W}_{R;k}[H](x_R,\xi) + \mathcal{W}_{\mathrm{bk};k}[H](\xi) - \mathcal{W}_{R;k}[H](x_L,a_N+b_N-\xi) + \Delta_{[k]}\mathcal{W}_N[H_\mathfrak{e}](\xi)\,, \tag{5.1.69}$$

with all remainders at order k being collected in the last term. In this subsection, we shall derive asymptotic expansion (in N) of the approximants $\mathcal{W}_{\mathrm{bk};k}$ and $\mathcal{W}_{R;k}$ in the case when their unrescaled variable ξ scales towards b_N as $\xi = b_N - N^{-\alpha}x$ with x being independent of N. We, however, first need to establish properties of certain auxiliary functions that appear in this analysis.

Definition 5.1.9 Let $\ell \geq 0$ be an integer. As a supplement to Definition 5.1.5, we introduce, for any integer $\ell \geq 0$:

$$\mathfrak{b}_\ell(x) = \varrho_{\ell+1}(x) - \frac{(-x)^{\ell+1}}{(\ell+1)!}\varrho_0(x) - \sum_{\substack{s+p=\ell\\ s,p\geq 0}}\frac{(-x)^p\varpi_{s+1}(x)}{p!(s+1)!}$$

$$\textit{and}\quad \mathfrak{u}_\ell(x) = \sum_{\substack{s+p=\ell\\ s,p\geq 0}}\frac{(-x)^p u_{s+1}}{p!} \tag{5.1.70}$$

and:

$$\mathfrak{a}_0(x) = \mathfrak{b}_0(x)+\mathfrak{u}_0(x)\,,\qquad \mathfrak{a}_\ell(x) = \frac{\mathfrak{b}_\ell(x)+\mathfrak{u}_\ell(x)}{\mathfrak{a}_0(x)}\quad \text{for } \ell\geq 1\,. \tag{5.1.71}$$

It will be important for the estimates of Section 5.2.2 to remark that $x^{-1/2}\mathfrak{a}_0(x)$ is a smooth and positive function:

Lemma 5.1.10 *Let $\ell, n, m \geq 0$ be three integers such that $n > m$. There exist polynomials $p_{\ell;m,n}$ of degree at most $n+\ell$ and functions $f_{\ell;m,n} \in W^{\infty}_{n-m}(\mathbb{R}^+)$ such that, for any $x > 0$:*

$$\begin{aligned} \mathfrak{a}_0(x) &= \sqrt{x}\, p_{0;m,n}(x)\mathrm{e}^{-\varsigma x} \;+\; x^m\, f_{0;m,n}(x) \\ \text{and} \quad \mathfrak{a}_0(x)\cdot\mathfrak{a}_\ell(x) &= \sqrt{x}\, p_{\ell;m,n}(x)\mathrm{e}^{-\varsigma x} \;+\; x^m\, f_{\ell;m,n}(x)\,. \end{aligned} \tag{5.1.72}$$

The function $\mathfrak{a}_0(x)$ is positive for $x > 0$ and satisfies

$$\mathfrak{a}_0(x) \underset{x\to 0}{=} \frac{1}{\pi\beta}\sqrt{\frac{x}{\pi(\omega_1+\omega_2)}} \;+\; \mathrm{O}(x) \tag{5.1.73}$$

Finally, one has, in the $x \to +\infty$ regime,

$$\mathfrak{a}_0(x) \;=\; u_1 \;+\; \mathrm{O}(\mathrm{e}^{-\varsigma x}), \quad \mathfrak{a}_0(x)\cdot\mathfrak{a}_\ell(x) \;=\; \mathfrak{u}_\ell(x) \;+\; \mathrm{O}(\mathrm{e}^{-\varsigma x}) \tag{5.1.74}$$

and the bound on the remainder is stable with respect to finite-order differentiations.

Proof By using the integral representation (5.1.2) for the function J, we can readily recast $\varpi_\ell(x)$, for $x > 0$ as:

$$\varpi_\ell(x) \;=\; \frac{\mathrm{i}^{\ell+1}}{2\pi\beta}\int\limits_{\mathscr{C}^{(+)}_{\mathrm{reg}}} \frac{\mathrm{e}^{\mathrm{i}\lambda x}}{\lambda}\frac{\partial^\ell}{\partial\lambda^\ell}\Big(\frac{1}{R(\lambda)}\Big)\frac{\mathrm{d}\lambda}{2\mathrm{i}\pi}\,. \tag{5.1.75}$$

The μ-integral arising in the definition (5.1.36) of ϱ_ℓ can be computed by moving the contour of integration over μ up to $+\mathrm{i}\infty$, and picking the residues at $\mu = \lambda$ and $\mu = 0$:

$$\begin{aligned} \varrho_\ell(x) &= \frac{\mathrm{i}^{\ell+1}}{2\pi\beta}\int\limits_{\mathscr{C}^{(+)}_{\mathrm{reg}}} \frac{\mathrm{e}^{\mathrm{i}\lambda x}}{\lambda^{\ell+1}R(\lambda)}\frac{\mathrm{d}\lambda}{2\mathrm{i}\pi} \;+\; \tau_\ell(x) \\ \text{with} \quad \tau_\ell(x) &= \frac{\mathrm{i}^{\ell+1}}{2\pi\beta}\int\limits_{\mathscr{C}^{(+)}_{\mathrm{reg}}} \frac{\mathrm{e}^{\mathrm{i}\lambda x}}{\ell!R_\downarrow(\lambda)}\cdot\frac{\partial^\ell}{\partial\mu^\ell}\Big(\frac{1}{(\mu-\lambda)R_\uparrow(\mu)}\Big)_{|\mu=0}\frac{\mathrm{d}\lambda}{2\mathrm{i}\pi}\,. \end{aligned}$$

The first term can be related to the functions ϱ_0 and ϖ_s of Definition 5.1.5 by an ℓ-fold integration by parts based on the identities:

$$\frac{1}{\lambda^{\ell+1}} \;=\; \frac{\partial^\ell}{\partial\lambda^\ell}\Big\{\frac{(-1)^\ell}{\lambda\,\ell!}\Big\} \quad \text{and} \quad \frac{\partial^\ell}{\partial\lambda^\ell}\Big\{\frac{\mathrm{e}^{\mathrm{i}\lambda x}}{R(\lambda)}\Big\} \;=\; \sum_{\substack{s+p=\ell \\ s,p\geq 0}}\frac{\ell!}{s!p!}(\mathrm{i}x)^p\mathrm{e}^{\mathrm{i}\lambda x}\cdot\frac{\partial^s}{\partial\lambda^s}\Big\{\frac{1}{R(\lambda)}\Big\}\,. \tag{5.1.76}$$

Namely, we obtain—writing the identity for $\ell + 1$ instead of ℓ—that:

$$\varrho_{\ell+1}(x) \; - \; \tau_{\ell+1}(x) \; = \; \frac{(-x)^{\ell+1}}{(\ell+1)!}\, \varrho_0(x) \; + \sum_{\substack{s+p=\ell \\ s,p\geq 0}} \frac{(-x)^p \varpi_{s+1}(x)}{p!(s+1)!} \, . \tag{5.1.77}$$

According to Definition 5.1.9, we can thus identify $\tau_{\ell+1}(x) = \mathfrak{b}_\ell(x)$—in this proof, we will nevertheless keep the notation τ_ℓ. Hence, it remains to focus on $\tau_\ell(x)$. Computing the ℓth-order μ-derivative appearing in its integrand and then repeating the same integration by parts trick, we obtain that:

$$\tau_\ell(x) \; = \; -\frac{\mathrm{i}^{\ell+1}}{2\pi\beta} \sum_{s+r+p=\ell} \frac{(\mathrm{i}x)^r}{s!p!r!} \frac{\partial^s}{\partial \mu^s}\bigg(\frac{1}{R_\uparrow(\mu)} \bigg)_{|\mu=0} \int_{\mathscr{C}^{(+)}_{\text{reg}}} \frac{\mathrm{e}^{\mathrm{i}\lambda x}}{\lambda} \cdot \frac{\partial^p}{\partial \lambda^p}\bigg\{ \frac{1}{R_\downarrow(\lambda)} \bigg\} \frac{\mathrm{d}\lambda}{2\mathrm{i}\pi} \, . \tag{5.1.78}$$

In the second integral, let us move a bit the contour $\mathscr{C}^{(+)}_{\text{reg}}$ to a contour $\mathscr{C}^{(+)}_{\text{reg},0}$ which passes below 0 while keeping the same asymptotic directions as $\mathscr{C}^{(+)}_{\text{reg}}$. Doing so, we pick up the residue at $\lambda = 0$:

$$\int_{\mathscr{C}^{(+)}_{\text{reg}}} \frac{\mathrm{e}^{\mathrm{i}\lambda x}}{\lambda} \cdot \frac{\partial^p}{\partial \lambda^p}\bigg\{ \frac{1}{R_\downarrow(\lambda)} \bigg\} \frac{\mathrm{d}\lambda}{2\mathrm{i}\pi} = -\frac{\partial^p}{\partial \lambda^p}\bigg\{ \frac{1}{R_\downarrow(\lambda)} \bigg\}_{|\lambda=0} + \int_{\mathscr{C}^{(+)}_{\text{reg},0}} \frac{\mathrm{e}^{\mathrm{i}\lambda x}}{\lambda} \cdot \frac{\partial^p}{\partial \lambda^p}\bigg\{ \frac{1}{R_\downarrow(\lambda)} \bigg\} \frac{\mathrm{d}\lambda}{2\mathrm{i}\pi} \tag{5.1.79}$$

We observe that there exist constants $c_{p;q}$ such that:

$$\frac{1}{\lambda} \cdot \frac{\partial^p}{\partial \lambda^p}\bigg\{ \frac{1}{R_\downarrow(\lambda)} \bigg\} \; = \; \sum_{q=p+1}^{n} \frac{c_{p;q}}{\big[\mathrm{i}(\lambda - \mathrm{i}\varsigma)\big]^{q+1/2}} \; + \; \Delta^{(p)}_{[n]}\big[R_\downarrow^{-1}\big](\lambda) \, . \tag{5.1.80}$$

This decomposition ensures that $\Delta^{(p)}_{[n]}\big[R_\downarrow^{-1}\big](\lambda)$ is holomorphic in $\mathbb{H}^-$, has a simple pole at $\lambda = 0$ and satisfies $\Delta^{(p)}_{[n]}\big[R_\downarrow^{-1}\big](\lambda) = \mathrm{O}\Big(\lambda^{-(n+3/2)}\Big)$.

Since $\varsigma/2$ is the distance between $\mathscr{C}^{(+)}_{\text{reg}}$ and $\mathbb{R}$, we can choose this contour—for a fixed ς—such that the branch cut of the denominators in (5.1.80) is located on a vertical half-line above $\mathscr{C}^{(+)}_{\text{reg},0}$. This implies that the remainder in (5.1.80) is holomorphic below $\mathscr{C}^{(+)}_{\text{reg},0}$. So, in the second integral of (5.1.79), we obtain with the first sum contributions involving:

$$\int_{\mathscr{C}^{(+)}_{\text{reg}}} \frac{e^{\mathrm{i}\lambda x}}{\big[\mathrm{i}(\lambda - \mathrm{i}\varsigma)\big]^{q+1/2}} \frac{\mathrm{d}\lambda}{2\mathrm{i}\pi} = \frac{e^{-\varsigma x}\, x^{q-1/2}}{\mathrm{i}\Gamma(q+1/2)} \tag{5.1.81}$$

in which (after the change of variable $t = -\mathrm{i}x(\lambda - \mathrm{i}\varsigma)$) we have recognised the Hankel contour integral representation of $\{\Gamma(q+1/2)\}^{-1}$. In its turn, the contribution of the remainder in (5.1.80) can be written:

$$\int_{\mathscr{C}^{(+)}_{\mathrm{reg},0}} \mathrm{e}^{\mathrm{i}\lambda x} \Delta^{(p)}_{[n]}\big[R_{\downarrow}^{-1}\big](\lambda) \,\frac{\mathrm{d}\lambda}{2\mathrm{i}\pi} = \int_{\mathscr{C}^{(+)}_{\mathrm{reg}}} \Big(\mathrm{e}^{\mathrm{i}\lambda x} - \sum_{r=0}^{m-1} \frac{(\mathrm{i}\lambda)^r x^r}{r!}\Big) \cdot \Delta^{(p)}_{[n]}\big[R_{\downarrow}^{-1}\big](\lambda) \cdot \frac{\mathrm{d}\lambda}{2\mathrm{i}\pi}$$
$$+ \sum_{r=0}^{m-1} \underbrace{\int_{\mathscr{C}^{(+)}_{\mathrm{reg},0}} \frac{(\mathrm{i}\lambda)^r x^r}{r!} \cdot \Delta^{(p)}_{[n]}\big[R_{\downarrow}^{-1}\big](\lambda) \cdot \frac{\mathrm{d}\lambda}{2\mathrm{i}\pi}}_{=0} . \tag{5.1.82}$$

Note that the last sum vanishes since we can deform the contour of integration to $-\mathrm{i}\infty$ provided $m \leq n$. Also, we could deform $\mathscr{C}^{(+)}_{\mathrm{reg},0}$ back to $\mathscr{C}^{(+)}_{\mathrm{reg}}$ in the first term since the integrand has no pole at $\lambda = 0$. All-in-all, we get

$$\int_{\mathscr{C}^{(+)}_{\mathrm{reg}}} \frac{\mathrm{e}^{\mathrm{i}\lambda x}}{\lambda} \frac{\partial^p}{\partial \lambda^p}\Big\{\frac{1}{R_{\downarrow}(\lambda)}\Big\} \frac{\mathrm{d}\lambda}{2\mathrm{i}\pi} = -\frac{\partial^p}{\partial \lambda^p}\Big\{\frac{1}{R_{\downarrow}(\lambda)}\Big\}_{|\lambda=0} + \sum_{q=p+1}^{n} \frac{c_{p;q}\, e^{-\varsigma x}\, x^{q-1/2}}{\mathrm{i}\Gamma(q+1/2)}$$
$$+ \int_{\mathscr{C}^{(+)}_{\mathrm{reg}}} \Delta^{(p)}_{[n]}\big[R_{\downarrow}^{-1}\big](\lambda) \Big(\mathrm{e}^{\mathrm{i}\lambda x} - \sum_{r=0}^{m-1} \frac{(\mathrm{i}x)^r \lambda^r}{r!}\Big) \frac{\mathrm{d}\lambda}{2\mathrm{i}\pi} . \tag{5.1.83}$$

With the bound

$$\Big|\mathrm{e}^{\mathrm{i}\lambda x} - \sum_{r=0}^{m-1} \frac{(\mathrm{i}x)^r \lambda^r}{r!}\Big| \leq x^m |\lambda|^m \tag{5.1.84}$$

and theorems of derivation under the integral, we can conclude that the last term in (5.1.83) is at least $n-m$ times differentiable and that it has, at least, an m-fold zero at $x = 0$. With the decomposition (5.1.83), we can come back to τ_ℓ given by (5.1.78). The second term in (5.1.83)—which contain derivatives of $1/R_{\downarrow}$—can be recombined with its prefactor—containing derivatives of $1/R_{\uparrow}$—by using the Leibniz rule backwards for the representation of the derivative at 0 of $1/R = 1/(R_{\uparrow} R_{\downarrow})$. Subsequently, we find there exist a polynomial $p_{\ell;m,n}$ of degree at most $n+\ell$ and a function $f_{\ell;m,n} \in W^{\infty}_{n-m}(\mathbb{R}^+)$ such that

$$\tau_{\ell+1}(x) = \sqrt{x}\, p_{\ell;m,n}(x) \mathrm{e}^{-\varsigma x} + x^m f_{\ell;m,n}(x) - \frac{\mathrm{i}^\ell}{2\pi\beta} \sum_{s+p=\ell} \frac{(\mathrm{i}x)^p}{(s+1)!\,p!} \frac{\partial^{s+1}}{\partial\lambda^{s+1}}\Big\{\frac{1}{R(\lambda)}\Big\}_{|\lambda=0} . \tag{5.1.85}$$

The claim then follows upon adding up all of the terms. Finally, the estimates at $x \to +\infty$ of $\mathfrak{a}_\ell$ follow readily from the exponential decay at $x \to +\infty$ of the functions ϱ and ϖ.

To compute the behaviour at $x \to 0$, we remind that:

$$\mathfrak{a}_0(x) \;=\; \mathfrak{b}_0(x) + \mathfrak{u}_1 \;=\; \tau_1(x) + \mathfrak{u}_1 \,. \tag{5.1.86}$$

We already know from (5.1.85) that $\mathfrak{a}_0(0) = 0$, and we just have to look in (5.1.78)–(5.1.83) for the coefficient of $\sqrt{x}$ in the case $\ell = 1$. For this purpose, it is enough to write (5.1.78) with $n = 1$. Then, the squareroot behaviour occur for $p = r = 0$ and $s = 1$ in the sum, and gives:

$$\mathfrak{a}_0(x) = \frac{c_{0;1}\, x^{1/2}\, \mathrm{e}^{-\varsigma x}}{2\mathrm{i}\pi\beta \cdot \Gamma(3/2)}\, \partial_\mu R_\uparrow^{-1}(\mu)|_{\mu=0} \;+\; \mathrm{O}(x)\,. \tag{5.1.87}$$

The coefficient $c_{0;1}$ is given by the large λ asymptotics in (5.1.80), coming from that of $R_\downarrow(\lambda)$ given by (4.1.29):

$$c_{0;1} = -1. \tag{5.1.88}$$

On the other hand, we know from (4.1.26) that:

$$\partial_\mu R_\uparrow^{-1}(\mu)|_{\mu=0} = \frac{1}{\mathrm{i}\sqrt{\omega_1+\omega_2}}\,. \tag{5.1.89}$$

Therefore:

$$\mathfrak{a}_0(x) = \frac{1}{\pi\beta}\sqrt{\frac{x}{\pi(\omega_1+\omega_2)}} \;+\; \mathrm{O}(x)\,. \tag{5.1.90}$$

We finally turn to proving that $\mathfrak{a}_0 > 0$ on $\mathbb{R}^+$. It follows from the previous calculations that

$$\mathfrak{a}_0(x) \;=\; \frac{1}{2\pi\beta}\bigg(\frac{1}{\mu\cdot R_\uparrow(\mu)}\bigg)_{|\mu=0} \int\limits_{\mathscr{C}_{\mathrm{reg}}^{(+)}} \frac{\mathrm{e}^{\mathrm{i}\lambda x}-1}{\lambda R_\downarrow(\lambda)}\, \frac{\mathrm{d}\lambda}{2\mathrm{i}\pi}\,. \tag{5.1.91}$$

The integral can be computed by deforming the contour up to $+\mathrm{i}\infty$ and, in doing so, we pick up the residues of the poles located at

$$\lambda = \frac{2\mathrm{i}\pi n\, \omega_1\omega_2}{\omega_1+\omega_2}\,, \qquad n \geq 1\,. \tag{5.1.92}$$

All-in-all this yields

$$\begin{aligned} \mathfrak{a}_0(x) \;&=\; \sum_{n\geq 1} \mathfrak{a}_{0;n}\Big(1 - \mathrm{e}^{-\frac{2\pi\omega_1\omega_2}{\omega_1+\omega_2}nx}\Big) \\ \text{with}\quad \mathfrak{a}_{0;n} \;&=\; \frac{(\omega_1+\omega_2)\cdot(-1)^{n-1}\cdot\kappa^{-n\kappa}\cdot(1-\kappa)^{-n(1-\kappa)}}{2\pi\beta\omega_1\omega_2\cdot n^2\cdot n!\cdot\Gamma\big(-\kappa n\big)\cdot\Gamma\big(-(1-\kappa)n\big)} \end{aligned} \tag{5.1.93}$$

and $\kappa = \omega_2/(\omega_1 + \omega_2) < 1$. By using the Euler reflection formula, we can recast $\mathfrak{a}_{0;n}$ into a manifestly strictly positive form

$$\mathfrak{a}_{0;n} = \frac{(\omega_1 + \omega_2)}{2\pi\beta\omega_1\omega_2} \cdot \left(\frac{\sin[\pi\kappa n]}{\pi}\right)^2 \cdot \frac{\Gamma\big(1+\kappa n\big)\cdot\Gamma\big(1+(1-\kappa)n\big)}{n^2\cdot n!\cdot\kappa^{n\kappa}\cdot(1-\kappa)^{n(1-\kappa)}} . \tag{5.1.94}$$

The asymptotics of $\mathfrak{a}_{0;n}$ then takes the form

$$\mathfrak{a}_{0;n} \underset{n\to+\infty}{\sim} \frac{(\omega_1+\omega_2)}{2\beta\omega_1\omega_2} \cdot \sqrt{\frac{2\kappa(1-\kappa)}{\pi n^3}} \cdot \left(\frac{\sin[\pi\kappa n]}{\pi}\right)^2 . \tag{5.1.95}$$

Thus the series (5.1.93) defining $\mathfrak{a}_0(x)$ converges uniformly for $x \in \mathbb{R}^+$. Since the series only contains positive summands, $\mathfrak{a}_0(x)$ is positive for $x > 0$. ■

The main reason for investigating the properties of the functions $\mathfrak{a}_\ell(x)$ lies in the fact that they describe the large-N asymptotics of the function $\mathcal{W}_{R;k}[H](x, b_N - N^{-\alpha}x) + \mathcal{W}_{\mathrm{bk};k}[H](b_N - N^{-\alpha}x)$. In particular, $\mathfrak{a}_0(x)$ arises as the first term and plays a particularly important role in the analysis that will follow. Let us remind Definition 3.3.1 for the weighted norm:

$$\mathcal{N}_N^{(\ell)}[H] = \sum_{k=0}^{\ell} \frac{||H||_{W_k^\infty(\mathbb{R})}}{N^{k\alpha}} . \tag{5.1.96}$$

Lemma 5.1.11 *Let $k \geq 0$ be an integer, $H \in \mathcal{C}^{2k+1}\big([a_N\,;\,b_N]\big)$. Define the functions:*

$$\mathcal{W}^{(\mathrm{as})}_{R;k}[H](x) = H'(b_N)\,\mathfrak{b}_0(x) + \sum_{\ell=1}^{k-1} \frac{H^{(\ell+1)}(b_N)\,\mathfrak{b}_\ell(x)}{N^{\ell\alpha}} , \tag{5.1.97}$$

$$\mathcal{W}^{(\mathrm{as})}_{\mathrm{bk};k}[H](x) = H'(b_N)\,u_1 + \sum_{\ell=1}^{k-1} \frac{H^{(\ell+1)}(b_N)\,\mathfrak{u}_\ell(x)}{N^{\ell\alpha}} . \tag{5.1.98}$$

The approximants at order k, $\mathcal{W}_{R;k}[H](x, b_N - N^{-\alpha}x)$ and $\mathcal{W}_{\mathrm{bk};k}[H](b_N - N^{-\alpha}x)$, admit the large-$N$ asymptotic expansions:

$$\mathcal{W}_{R;k}[H](x, b_N - N^{-\alpha}x) = \mathcal{W}^{(\mathrm{as})}_{R;k}[H](x) + \Delta_{[k]}\mathcal{W}^{(\mathrm{as})}_{R}[H](x) , \tag{5.1.99}$$

$$\mathcal{W}_{\mathrm{bk};k}[H](b_N - N^{-\alpha}x) = \mathcal{W}^{(\mathrm{as})}_{\mathrm{bk};k}[H](x) + \Delta_{[k]}\mathcal{W}^{(\mathrm{as})}_{\mathrm{bk}}[H](x) . \tag{5.1.100}$$

The remainders have the following structure:

$$\Delta_{[k]}\mathcal{W}^{(\mathrm{as})}_{R}[H](x) = N^{-k\alpha}\cdot \mathrm{e}^{-\varsigma x}\left\{(\ln x)\,\mathcal{R}^{(1)}_{\mathrm{as};k}[H](x) + \mathcal{R}^{(2)}_{\mathrm{as};k}[H](x)\right\} , \tag{5.1.101}$$

$$\Delta_{[k]}\mathcal{W}^{(\mathrm{as})}_{\mathrm{bk}}[H](x) = N^{-k\alpha}\cdot\mathcal{R}^{(3)}_{\mathrm{as};k}[H](x) , \tag{5.1.102}$$

where $\mathcal{R}^{(a)}_{\mathrm{as};k}[H] \in W^{\infty}_{\ell}(\mathbb{R}^+)$ *for* $a = 1, 2, 3$. *For* $a = 1$, *we have:*

$$\left|\mathcal{R}^{(1)}_{\mathrm{as};k}[H](x)\right| = \mathrm{O}(x^{k+1}) \tag{5.1.103}$$

uniformly in N. *Moreover, we have uniform bounds for* $x \in [0\,; \epsilon N^{\alpha}]$, *namely for* $\ell \in [\![\, 0\,;\, k\,]\!]$:

$$\left|\partial_{\xi}^{\ell}\mathcal{R}^{(1)}_{\mathrm{as};k}[H](x_R)\right| \le C_{k,\ell} \cdot x_R^{k-\ell+1} \cdot N^{\ell\alpha} \cdot \mathcal{N}_N^{(\ell)}\big[H_{\mathfrak{e}}^{(k+1)}\big], \tag{5.1.104}$$

$$a = 2, 3, \qquad \left|\partial_{\xi}^{\ell}\mathcal{R}^{(a)}_{\mathrm{as};k}[H](x_R)\right| \le C_{k,\ell} \cdot N^{\ell\alpha} \cdot \mathcal{N}_N^{(\ell)}\big[H_{\mathfrak{e}}^{(k+1)}\big], \tag{5.1.105}$$

where we remind $x_R = N^{\alpha}(b_N - \xi)$.

Note that we can combine the operators into the asymptotic expansion

$$\mathcal{W}^{(\mathrm{as})}_{R;k}[H](x) \;+\; \mathcal{W}^{(\mathrm{as})}_{\mathrm{bk};k}[H](x) \;=\; H'(b_N)\,\mathfrak{a}_0(x)\left\{1 + \sum_{\ell=1}^{k} \frac{H^{(\ell+1)}(b_N)\,\mathfrak{a}_{\ell}(x)}{H'(b_N)\,N^{\alpha\ell}}\right\}. \tag{5.1.106}$$

Proof The form of the large-N asymptotic expansion follows from straightforward manipulations on the Taylor integral representation for $H^{(\ell)}(\xi)$ around $\xi = b_N$ for $\ell \in [\![\, 0\,;\, k\,]\!]$. The control on the remainder arising in (5.1.99), (5.1.100) and (5.1.106) follows from the explicit integral representation for the remainder in the Taylor-integral series:

$$\begin{aligned}\Delta_{[k]}\mathcal{W}^{(\mathrm{as})}_{R}[H](x) &= N^{-k\alpha}\int_0^1 \mathrm{d}t\, H^{(k+1)}(b_N - N^{-\alpha}\,tx)\left\{-\frac{(1-t)^k\,(-x)^{k+1}\varrho_0(x)}{k!}\right.\\ &\qquad\left. - \sum_{\ell=1}^{k}\frac{(1-t)^{k-\ell}(-x)^{1+k-\ell}\,\varpi_{\ell}(x)}{\ell!(k-\ell)!}\right\},\\ \Delta_{[k]}\mathcal{W}^{(\mathrm{as})}_{\mathrm{bk}}[H](x) &= N^{-k\alpha}\sum_{\ell=1}^{k} u_{\ell}\,\frac{(-x)^{k+1-\ell}}{(k-\ell)!}\int_0^1 \mathrm{d}t\,(1-t)^{k-\ell}\,H^{(k+1)}(b_N - N^{-\alpha}\,tx)\,.\end{aligned} \tag{5.1.107}$$

and we remark that $\varrho_0(x)$—given by (5.1.34)—has a logarithmic singularity when $x \to 0$. The details to arrive to (5.1.104) and (5.1.105) are left to the reader. ■

Collecting the bounds we have obtained in sup norms, we find in particular $\mathcal{W}_N[H]$ is bounded when H is $\mathcal{C}^1$:

Corollary 5.1.12 *There exists* $C > 0$ *independent of* N *such that, for any* $H \in \mathcal{C}^1([a_N\,; b_N])$,

$$||\mathcal{W}_N[H]||_{W_0^{\infty}([a_N\,;b_N])} \le C\,||H_{\mathfrak{e}}||_{W_1^{\infty}(\mathbb{R})}\,. \tag{5.1.108}$$

5.2 The Operator $\mathcal{U}_N^{-1}$

Let us remind the definition of the operators $\mathcal{U}_N$ and $\mathcal{S}_N$:

$$\mathcal{U}_N[\phi](\xi) = \phi(\xi) \cdot \Big\{ V'(\xi) \; - \; \mathcal{S}_N[\rho_{\mathrm{eq}}^{(N)}](\xi) \Big\} + \mathcal{S}_N[\phi \cdot \rho_{\mathrm{eq}}^{(N)}](\xi) \quad (5.2.1)$$

$$\mathcal{S}_N\big[\phi\big](\xi) = \fint_{a_N}^{b_N} S\big[N^{\alpha}(\xi - \eta)\big]\phi(\eta)\,\mathrm{d}\eta$$

$$\text{and} \quad S(\xi) = \sum_{p=1}^{2} \beta\pi\omega_p \operatorname{cotanh}\big[\pi\omega_p\xi\big] \quad (5.2.2)$$

and the fact that $\mathcal{W}_N$ defined in Section 4.3.4 is the inverse operator to $\mathcal{S}_N$. We also remind that the density $\rho_{\mathrm{eq}}^{(N)}$ of the N-dependent equilibrium measure satisfies the integral equation:

$$\forall \xi \in [a_N\,;b_N], \qquad \mathcal{S}_N[\rho_{\mathrm{eq}}^{(N)}](\xi) = V'(\xi)\,. \quad (5.2.3)$$

This makes the first term of (5.2.1) vanish for $\xi \in [a_N\,;b_N]$, but it can be non-zero outside of this segment.

In this section we obtain an integral representation for the inverse of $\mathcal{U}_N$, which shows that $\mathcal{U}_N^{-1}[H]$ is smooth as long as H is. Then, in Section 5.2.2, we shall provide explicit, N-dependent, bounds on the $W_{\ell}^{\infty}(\mathbb{R})$ norms of $\mathcal{U}_N^{-1}[H]$. This technical result is crucial in the analysis of the Schwinger–Dyson equation performed in Section 3.3.

5.2.1 An Integral Representation for $\mathcal{U}_N^{-1}$

Proposition 5.2.1 *The operator $\mathcal{U}_N$ is invertible on $\big(\mathfrak{X}_s \cap \mathcal{C}^1\big)(\mathbb{R})$, $0 < s < 1/2$, and its inverse admits the representation*

$$\mathcal{U}_N^{-1}[H](\xi) \;=\; \frac{\mathcal{V}_N[H](\xi)}{\mathcal{V}_N[V'](\xi)}\,, \quad (5.2.4)$$

where $\mathcal{V}_N \;=\; \mathcal{V}_N^{[1]} \,+\, \mathcal{V}_N^{[2]}$ with

$$\mathcal{V}_N^{[1]}[H](\xi) \;=\; \int_{a_N}^{b_N} \frac{[H(\xi) - H(s)]\,\mathrm{d}s}{(\xi - s)\sqrt{(s - a_N)(b_N - s)}}$$

$$\textit{and} \quad \mathcal{V}_N^{[2]}[H](\xi) \;=\; \int_{a_N}^{b_N} V_N^{[2]}(\xi,\eta) \cdot \mathcal{W}_N[H](\eta)\,\mathrm{d}\eta \quad (5.2.5)$$

and the integral kernel of the operator $\mathcal{V}_N^{[2]}$ *reads:*

$$V_N^{[2]}(\xi,\eta) = \int_{a_N}^{b_N} \frac{S_{\mathrm{reg}}\big[N^{\alpha}(s-\eta)\big] - S_{\mathrm{reg}}\big[N^{\alpha}(\xi-\eta)\big]}{(\xi-s)\sqrt{(s-a_N)(b_N-s)}}\,\mathrm{d}s \quad \textit{with} \quad S_{\mathrm{reg}}(\xi) = S(\xi) - \frac{2\beta}{\xi}\,. \tag{5.2.6}$$

Finally, we have that, for any $\xi \in [a_N\,;b_N]$, $\mathcal{V}_N[V'](\xi) \neq 0$.

Note that the above representation is not completely fit for obtaining a fine bound of the $W_{\ell}^{\infty}(\mathbb{R})$ norm of $\mathcal{U}_N^{-1}[H]$ in the large-N limit. Indeed, we will show in Appendix C that $\mathcal{V}_N[V'](\xi) > c_N > 0$ for N large enough. Unfortunately, the constant $c_N \to 0$ and thus does not provide an optimal bound for the $W_{\ell}^{\infty}(\mathbb{R})$ norm. Gaining a more precise control on c_N (*eg*. its dependence on N) is much harder, but a more precise control is one of the ingredients that are necessary for obtaining sharp N-dependent bounds for the $W_{\ell}^{\infty}(\mathbb{R})$ norm of $\mathcal{U}_N^{-1}[H]$. We shall obtain such a more explicit control on c_N in the course of the proof of Theorem 5.2.2.

Proof Given $H \in (\mathfrak{X}_s \cap \mathcal{C}_{\mathrm{c}}^1)(\mathbb{R})$, let ϕ be the unique solution to the equation $\mathcal{S}_N[\phi](\xi) = H(\xi)$ on $[a_N\,;b_N]$. Reminding the definition of $\mathcal{S}_N$ in (2.4.15), it means that, for $\xi\, \in]a_N\,;b_N[$:

$$\fint_{a_N}^{b_N} \frac{\phi(\eta)\,\mathrm{d}\eta}{(\xi-\eta)\mathrm{i}\pi} = U(\xi)$$

$$\text{where} \qquad U(\xi) = \frac{N^{\alpha}}{2\mathrm{i}\pi\beta}\bigg\{ H(\xi) - \fint_{a_N}^{b_N} S_{\mathrm{reg}}\big[N^{\alpha}(\xi-\eta)\big]\phi(\eta)\,\mathrm{d}\eta \bigg\}\,. \tag{5.2.7}$$

As a consequence, the function

$$F(z) = \frac{1}{q(z)}\int_{a_N}^{b_N} \frac{\phi(\eta)}{z-\eta}\cdot\frac{\mathrm{d}\eta}{2\mathrm{i}\pi} \qquad \text{with} \qquad q(z) = \sqrt{(z-a_N)(z-b_N)} \tag{5.2.8}$$

solves the scalar Riemann–Hilbert problem

- $F \in \mathcal{O}(\mathbb{C}\setminus[a_N\,;b_N])$ and admits $\pm\, L^p\big([a_N\,;b_N]\big)$ boundary values for $p \in]1\,;2[$;
- $F(z) = \mathrm{O}\big(z^{-1}\big)$ when $z \to \infty$;
- $F_+(x) - F_-(x) = U(x)/q_+(x)$ for any $x \in]a_N\,;b_N[$.

Note that the L^p character of the boundary values follows from the fact that both ϕ and the principal value integral are continuous on $[a_N\,;b_N]$. The former follows from Propositions 5.1.4–5.1.6 whereas the latter is a consequence of (5.2.7). By uniqueness of the solution to such a Riemann–Hilbert problem, it follows that

$$F(z) = \int_{a_N}^{b_N} \frac{U(s)}{q_+(s)(s-z)}\,\frac{\mathrm{d}s}{2\mathrm{i}\pi} \qquad \text{for} \qquad z \in \mathbb{C}\setminus[a_N\,;b_N]. \tag{5.2.9}$$

By using that, for $\xi \in]a_N ; b_N[$,

$$-\phi(\xi) = q_+(\xi) \cdot \Big(F_+(\xi) + F_-(\xi) \Big) \qquad \text{and} \qquad \fint_{a_N}^{b_N} \frac{1}{q_+(s) \cdot (s - \xi)} \cdot \frac{\mathrm{d}s}{\mathrm{i}\pi} = 0 , \tag{5.2.10}$$

we obtain that:

$$\phi(\xi) = \frac{\sqrt{N^{2\alpha}(\xi - a_N)(b_N - \xi)}}{2\pi^2 \beta} \mathcal{V}_N[H](\xi) \tag{5.2.11}$$

with the expression of $\mathcal{V}_N$ given by (5.2.5). Further, given any $\xi \in \mathbb{R} \setminus [a_N ; b_N]$, we have:

$$\mathcal{S}_N[\phi](\xi) = \int_{a_N}^{b_N} S_{\mathrm{reg}}\big[N^{\alpha}(\xi - \eta)\big]\phi(\eta)\,\mathrm{d}\eta + \frac{4\mathrm{i}\pi\beta}{N^{\alpha}} q(\xi) F(\xi) . \tag{5.2.12}$$

It then remains to use that, for such ξ's

$$\int_{a_N}^{b_N} \frac{1}{q_+(s)(s - \xi)} \cdot \frac{\mathrm{d}s}{\mathrm{i}\pi} = \frac{1}{q(\xi)} \tag{5.2.13}$$

so as to get the representation

$$\mathcal{S}_N[\phi](\xi) = H(\xi) - \frac{q(\xi)}{\pi} \cdot \mathcal{V}_N[H](\xi) . \tag{5.2.14}$$

We can now go back to the original problem. Let ψ be any solution to $\mathcal{U}_N[\psi] = H$. Due to the integral equation satisfied by the density of equilibrium measure on $[a_N ; b_N]$, it follows that, for any $\xi \in [a_N ; b_N]$ such that $\mathcal{W}_N[V'](\xi) \neq 0$,

$$\psi(\xi) = \frac{\mathcal{W}_N[H](\xi)}{\mathcal{W}_N[V'](\xi)} \tag{5.2.15}$$

and we can conclude thanks to the relation (5.2.11). For $\xi \in \mathbb{R} \setminus [a_N ; b_N]$, we rather have:

$$\psi(\xi) = \frac{\mathcal{S}_N\big[\mathcal{W}_N[H]\big](\xi) - H(\xi)}{\mathcal{S}_N\big[\mathcal{W}_N[V']\big](\xi) - V'(\xi)} \tag{5.2.16}$$

at any point where the denominator does not vanish. It then solely remains to invoke the relation (5.2.14). Note that

$$\mathcal{V}_N[V'](\xi) = \frac{2\pi^2 \beta\, \rho_{\mathrm{eq}}^{(N)}(\xi)}{\sqrt{N^{2\alpha}(\xi - a_N)(b_N - \xi)}} . \tag{5.2.17}$$

It is shown in the proof of Theorem 2.4.2 given in Appendix C, point (ii), that $\rho_{\mathrm{eq}}^{(N)}(\xi) > 0$ for $\xi \in]a_N ; b_N[$ for N large enough and that it vanishes as a square root at the edges. Furthermore, it is also shown in that appendix, Equation (C.0.8), that $V'(\xi) - \mathcal{S}_N\big[\mathcal{W}_N[V']\big](\xi) \neq 0$ on $\mathbb{R} \setminus [a_N ; b_N]$. Thus the denominator in (5.2.4) never vanishes, and thus the equation holds for any $\xi \in \mathbb{R}$ and any $H \in \mathfrak{X}_s \cap \mathcal{C}^1_{\mathrm{c}}(\mathbb{R})$. The result then follows by density of $\mathfrak{X}_s \cap \mathcal{C}^1_{\mathrm{c}}(\mathbb{R})$ in $\mathfrak{X}_s \cap \mathcal{C}^1(\mathbb{R})$. ■

5.2.2 Sharp Weighted Bounds for $\mathcal{U}_N^{-1}$

The aim of the present subsection is to prove one of the most important technical propositions of the paper, namely sharp N-dependent bounds on the $W^{\infty}_{\ell}(\mathbb{R})$ norm of $\mathcal{U}_N^{-1}[H]$. Part of the difficulties of the proof consists in obtaining lower bounds for $\mathcal{W}_N[V']$ in the vicinity of a_N and b_N as well as in gaining a sufficiently precise control on the square root behaviour of $\mathcal{W}_N[H]$ at the edges.

Proposition 5.2.2 below is the key tool for the large-N analysis of the Schwinger–Dyson equations. We insist that although our result is effective in what concerns our purposes, it is *not* optimal. More optimal results can be obtained with respect to local W^{∞}_{ℓ} norms, *viz.* $W^{\infty}_{\ell}(J)$ with J being specific subintervals of $\mathbb{R}$, or with respect to milder ones such as the $W^{p}_{\ell}(\mathbb{R})$ ones. However, obtaining these results demands more efforts on the one hand and, on the other hand, requires much more technical handlings so as to make the best of them when dealing with the Schwinger–Dyson equations. We therefore chose not to venture further in these technicalities.

Before stating the theorem, we remind the expression for the weighted norm (Definition 3.3.1):

$$\mathcal{N}_N^{(\ell)}[\phi] \;=\; \sum_{k=0}^{\ell} \frac{||\phi||_{W^{\infty}_{\ell}(\mathbb{R})}}{N^{\ell\alpha}} \,. \tag{5.2.18}$$

and the *ad hoc* norms on the potential (Definition 3.3.2):

$$\mathfrak{n}_{\ell}[V] \;=\; \frac{\max\Big\{ \prod\limits_{a=1}^{\ell} ||\mathcal{K}_{\kappa}[V']||_{W^{\infty}_{k_a}(\mathbb{R}^n)} \;:\; \sum\limits_{a=1}^{\ell} k_a = 2\ell+1 \Big\}}{\Big\{ \min\Big(1\,,\ \inf_{[a\,;b]}|V''(\xi)|\,,\ |V'(b+\epsilon)-V'(b)|\,,\ |V'(a-\epsilon)-V'(a)|\Big)\Big\}^{\ell+1}} \tag{5.2.19}$$

for some $\epsilon > 0$ small enough but independent of N. We also remind that $\mathcal{K}_{\kappa}[H]$ is an exponential regularisation of H, see Definition 3.1.7.

Proposition 5.2.2 *Let $\ell \geq 0$ be an integer, and C_V, κ be positive constants. There exists a constant $C_{\ell} > 0$ such that for any H and V satisfying*

- $\mathcal{K}_{\kappa/\ell}[H] \in W^{\infty}_{2\ell+1}(\mathbb{R})$ *and* $\mathcal{K}_{\kappa/\ell}[V] \in W^{\infty}_{2\ell+2}(\mathbb{R})$*;*
- $||V||_{W_3^{\infty}([a-\delta;b+\delta])} < C_V$ *for some* $\delta > 0$ *where* (a,b) *are such that* $(a_N, b_N) \underset{N\to+\infty}{\to} (a,b)$*;*
- $H \in \mathfrak{X}_s([a_N\,;b_N])$*;*

we have the following bound:

$$||\mathcal{K}_{\kappa}\big[\mathcal{U}_N^{-1}[H]\big]||_{W_\ell^{\infty}(\mathbb{R})} \le C_\ell \cdot \mathfrak{n}_\ell[V] \cdot N^{(\ell+1)\alpha} \cdot (\ln N)^{2\ell+1} \cdot \mathcal{N}_N^{(2\ell+1)}\big[\mathcal{K}_\kappa[H]\big]. \tag{5.2.20}$$

Proof As discussed in the proof of Proposition 5.2.1, the operator $\mathcal{U}_N^{-1}$ can be recast as

$$\mathcal{U}_N^{-1}[H](\xi) = \frac{\mathcal{W}_N[H](\xi)}{\mathcal{W}_N[V'](\xi)} \cdot \mathbf{1}_{[a_N\,;b_N]}(\xi) + \frac{\mathcal{S}_N\big[\mathcal{W}_N[H]\big](\xi) - H(\xi)}{\mathcal{S}_N\big[\mathcal{W}_N[V']\big](\xi) - V'(\xi)} \cdot \mathbf{1}_{[a_N\,;b_N]^c}(\xi)\,. \tag{5.2.21}$$

Therefore, obtaining sharp bounds on $\mathcal{U}_N^{-1}[H]$ demands to control, with sufficient accuracy, both ratios appearing in the formula above. Observe that the same Proposition 5.2.1 and, in particular, Eqs. (5.2.11)–(5.2.14) ensure that, given $\epsilon > 0$ small enough and H of class $\mathcal{C}^{k+1}$, the functions

$$\xi \mapsto \frac{\mathcal{W}_N[H](\xi)}{q_R(\xi)} \quad \text{and} \quad \xi \mapsto \frac{\mathcal{S}_N\big[\mathcal{W}_N[H]\big](\xi) - H(\xi)}{q_R(\xi)} \tag{5.2.22}$$

with:

$$q_R(\xi) = \sqrt{N^\alpha(b_N - \xi)} = x_R^{1/2} \tag{5.2.23}$$

are respectively $\mathcal{C}^k([b_N - \epsilon\,;b_N])$ and $\mathcal{C}^k([b_N\,;b_N+\epsilon])$. A similar statement holds at the left boundary. Furthermore, the same proposition readily ensures that both functions are $\mathcal{C}^{k+1}$ uniformly away from the boundaries.

The large-N behaviour of both functions in (5.2.22) is not uniform on $\mathbb{R}$ and depends on whether one is in a vicinity of the endpoints a_N, b_N or not. Therefore, we will split the analysis for ξ in one of the four regions, from right to left on the real axis:

$$\mathbb{J}_N^{(R;\mathrm{out})} = [b_N + \epsilon(\ln N)^2 \cdot N^{-\alpha}\,;+\infty[\tag{5.2.24}$$
$$\mathbb{J}_N^{(R;\mathrm{ext})} = [b_N\,;b_N + \epsilon(\ln N)^2 \cdot N^{-\alpha}] \tag{5.2.25}$$
$$\mathbb{J}_N^{(R;\mathrm{in})} = [b_N - \epsilon(\ln N)^2 \cdot N^{-\alpha}\,;b_N] \tag{5.2.26}$$
$$\mathbb{J}_N^{(\mathrm{bk})} = [a_N + \epsilon(\ln N)^2 \cdot N^{-\alpha}\,;b_N - \epsilon(\ln N)^2 \cdot N^{-\alpha}]\,. \tag{5.2.27}$$

Indeed, the behaviour on the three other regions:

$$\mathbb{J}_N^{(L;\mathrm{in})} = [a_N\,;a_N + \epsilon(\ln N)^2 \cdot N^{-\alpha}] \tag{5.2.28}$$
$$\mathbb{J}_N^{(L;\mathrm{ext})} = [a_N - \epsilon(\ln N)^2 \cdot N^{-\alpha}\,;a_N] \tag{5.2.29}$$

$$\mathbb{J}_N^{(L;\mathrm{out})} =]-\infty\,;\, a_N - \epsilon(\ln N)^2 \cdot N^{-\alpha}] \tag{5.2.30}$$

can be deduced by the reflection symmetry

$$\mathcal{W}_N[H](\xi) = -\mathcal{W}_N\big[H^{\wedge}\big](a_N + b_N - \xi) \tag{5.2.31}$$

from the analysis on the local intervals (5.2.24)–(5.2.26).

The proof consists in several steps. First of all, we bound the $W_\ell^\infty(\mathbb{J}_N^{(*)})$ norm of the functions in (5.2.22), this depending on the interval of interest. Also, we obtain *lower* bounds for the same functions with $H \leftrightarrow V'$. Finally, we use the partitioning of $\mathbb{R}$ into the local intervals (5.2.24)–(5.2.26) so as to raise the local bounds into global bounds on $\mathcal{U}_N^{-1}[H]$ issuing from those on $\mathcal{W}_N[H]\cdot q_R^{-1}$ and $\big\{\mathcal{S}_N\big[\mathcal{W}_N[H]\big]-H\big\}\cdot q_R^{-1}$.

Lower and upper bounds on $\mathbb{J}_N^{(R;\mathrm{out})}$

Let us decompose S given in (5.2.2) into:

$$S(x) = S_\infty(x) + (\Delta S)(x), \qquad \text{with} \quad S_\infty(x) = \beta\pi(\omega_1+\omega_2)\mathrm{sgn}(x) \tag{5.2.32}$$

We observe that when $\xi \in \mathbb{J}_N^{(R;\mathrm{out})}$ and $\eta \in [a_N\,;\,b_N]$ one avoids the simple pole in the kernel functions $S[N^\alpha(\xi-\eta)]$ of the integral operator $\mathcal{S}_N$. Besides, the decomposition (5.2.32) has the property that, for any integer $\ell \geq 0$, there exists constants c, $C_\ell > 0$ independent of N such that:

$$\forall \xi \in \mathbb{J}_N^{(R;\mathrm{out})},\ \forall \eta \in [a_N\,;\,b_N], \qquad \big|\partial_\xi^\ell(\Delta S)[N^\alpha(\xi-\eta)]\big| \;\leq\; C_\ell\, N^{\ell\alpha}\, \mathrm{e}^{-c(\ln N)^2}\,. \tag{5.2.33}$$

We have proved in Lemmas 6.1.7 and 6.1.8 that

$$\Big|\int_{a_N}^{b_N} \mathcal{W}_N[H](\xi)\,\mathrm{d}\xi\Big| \;\leq\; C\,||H_{\mathfrak{e}}||_{W_0^\infty(\mathbb{R})}\,, \qquad ||\mathcal{W}_N[H]||_{L^1([a_N\,;b_N])} \;\leq\; C\,||H_{\mathfrak{e}}||_{W_1^\infty(\mathbb{R})} \tag{5.2.34}$$

for some $C > 0$ independent of N. Subsequently:

$$\begin{aligned} ||\mathcal{S}_N\big[\mathcal{W}_N[H]\big]||_{W_\ell^\infty(\mathbb{J}_N^{(R;\mathrm{out})})} &\leq \delta_{\ell 0}\, C\,||H_{\mathfrak{e}}||_{W_0^\infty(\mathbb{R})} \;+\; C_\ell\, N^{\ell\alpha}\,\mathrm{e}^{-c(\ln N)^2}\,||\mathcal{W}_N[H]||_{L^1([a_N\,;b_N])} \\ &\leq \delta_{\ell 0}\, C'\,\mathcal{N}_N^{(0)}\big[\mathcal{K}_\kappa[H]\big] \;+\; C'_\ell\, N^{(\ell+1)\alpha}\,\mathrm{e}^{-c(\ln N)^2}\,(b_N - a_N)\,\mathcal{N}_N^{(1)}\big[\mathcal{K}_\kappa[H]\big]\,. \end{aligned} \tag{5.2.35}$$

We have used: in the first line, the estimates (5.2.34); in the second line, the definition (5.2.18) of the weighted norm, and we have included exponential regularisations via $\mathcal{K}_\kappa$, whose only effect is to change the value of the constant prefactors. Since $(a_N, b_N) \to (a, b)$ in virtue of Corollary 6.2.2, we can write for N large enough:

$$||\mathcal{K}_\kappa\big[\mathcal{S}_N\big[\mathcal{W}_N[H]\big] - H\big]||_{W_\ell^\infty(\mathbb{J}_N^{(R;\mathrm{out})})} \;\leq\; \widetilde{C}_\ell \cdot N^{\ell\alpha}\cdot \mathcal{N}_N^{(\ell)}\big[\mathcal{K}_\kappa[H]\big]\,. \tag{5.2.36}$$

Indeed, a bound from the left-hand side is obtained by adding the W_ℓ^∞ norm of H to (5.2.35), which is itself bounded by a multiple of $N^{\ell\alpha}\mathcal{N}_N^{(\ell)}\big[\mathcal{K}_\kappa[H]\big]$.

Thanks to the decomposition (5.2.32) using that $\mathrm{sgn}(\xi-\eta)=1$ for $\xi\in\mathbb{J}_N^{R;(\mathrm{out})}$ and $\eta\in[a_N\,;b_N]$, as well as the exponential estimate (5.2.33) and the L^1 bound of $\mathcal{W}_N$ from Lemma 6.1.8, we can also write:

$$\mathcal{S}_N\big[\mathcal{W}_N[V']\big](\xi)-V'(\xi)=\underbrace{\pi\beta(\omega_1+\omega_2)}_{=V'(b)}\underbrace{\int_{a_N}^{b_N}\mathcal{W}_N[V'](\xi)\,\mathrm{d}\xi}_{=1}\;-V'(\xi)$$
$$+\,\mathrm{O}\Big(e^{-c(\ln N)^2}\,||V'||_{W_1^\infty([a_N\,;b_N])}\Big)\,. \qquad (5.2.37)$$

The identification of the first term comes from (3.1.16). Further, we have for $|\xi-b|\le\epsilon$ and $\xi\in\mathbb{J}_N^{(R;\mathrm{out})}$:

$$\big|V'(b)-V'(\xi)|\;\ge\;|\xi-b|\cdot\inf_{\xi\in[b\,;b+\epsilon]}|V''(\xi)|\;\ge\;\frac{\epsilon}{2}\frac{(\ln N)^2}{N^\alpha}V''(b)\ge\frac{\epsilon}{2}\frac{V''(b)}{N^\alpha} \qquad (5.2.38)$$

To obtain the last bound we have assumed that ϵ was small enough—but still independent of N—and made use of $|b-b_N|=\mathrm{O}(N^{-\alpha})$ as well as of $||V||_{W_3^\infty([a-\delta\,;b+\delta])}<+\infty$ and N large enough. Finally, it is clear from the strict convexity of V that in the case $|b-\xi|>\epsilon$:

$$|V'(b)-V'(\xi)|\;\ge\;|V'(b+\epsilon)-V'(b)|\ge\frac{\epsilon}{2}\,\frac{V'(b+\epsilon)-V'(b)}{N^\alpha}\,, \qquad (5.2.39)$$

where the last inequality is a trivial one. Therefore, in any case, for N large enough:

$$\big|\mathcal{S}_N\big[\mathcal{W}_N[V']\big]-V'(\xi)\big|\ge\frac{\epsilon}{4N^\alpha}\min\Big\{\inf_{\xi\in[a\,;b]}V''(\xi)\,,\;|V'(b+\epsilon)-V'(b)|\Big\}\,. \qquad (5.2.40)$$

The combination of the numerator upper bound (5.2.36) applied to $H=V'$ (using that the weighted norm is dominated by the W^∞ norm) and the denominator lower bound (5.2.40) implies that, for any $\kappa>0$ such that both sides below are well-defined:

$$\frac{||\mathcal{K}_\kappa\big[\mathcal{S}_N\big[\mathcal{W}_N[V']\big]-V'\big]||_{W_\ell^\infty(\mathbb{J}_N^{(R;\mathrm{out})})}}{\big|\mathcal{S}_N\big[\mathcal{W}_N[V']\big](\xi)\;-\;V'(\xi)\big|}\;\le\;\frac{N^{(\ell+1)\alpha}\cdot C_\ell\cdot||V'||_{W_\ell^\infty(\mathbb{R})}}{\min\Big\{\inf_{\xi\in[a\,;b]}|V''(\xi)|\,,\;|V'(b+\epsilon)-V'(b)|\Big\}}\,. \qquad (5.2.41)$$

Implicitly, we have treated ϵ from (5.2.40) like a constant.

Lower and upper bounds on $\mathbb{J}_N^{(\mathrm{bk})}$

Consider the decomposition of $\mathcal{W}_N$ from (5.1.69):

$$\mathcal{W}_N[H](\xi) \;=\; \mathcal{W}_{\mathrm{bk};k}[H](\xi) \;+\; \Delta_{[k]}\mathcal{W}_{\mathrm{bk};k}[H_{\mathfrak{e}}](\xi) \;+\; \mathcal{W}_R[H_{\mathfrak{e}}](x_R) \\ -\, \mathcal{W}_R\big[H^{\wedge}\big](x_L, b_N + a_N - \xi) \;+\; \mathcal{W}_{\exp}[H_{\mathfrak{e}}](\xi)\;. \qquad (5.2.42)$$

From the expression of $\mathcal{W}_{\mathrm{bk};k}$ in (5.1.40), we have the bound:

$$||\mathcal{W}_{\mathrm{bk};k}[H]||_{W_{\ell}^{\infty}(\mathbb{J}_N^{(\mathrm{bk})})} \;\leq\; c_{k;\ell} \cdot \max_{s \in [\![0\,;\,\ell]\!]} \mathcal{N}_N^{(k-1)}[H^{(s+1)}] \qquad (5.2.43)$$

and recollecting the estimates of the other terms from Propositions 5.1.4 and 5.1.6, we also find:

$$\Big|\Big|\Delta_{[k]}\mathcal{W}_{\mathrm{bk};k}[H_{\mathfrak{e}}] + \mathcal{W}_R[H_{\mathfrak{e}}] - (\mathcal{W}_R)^{\wedge}[H_{\mathfrak{e}}] + \mathcal{W}_{\exp}[H_{\mathfrak{e}}]\Big|\Big|_{W_{\ell}^{\infty}(\mathbb{J}_N^{(\mathrm{bk})})} \;\leq\; c_{\ell}\, N^{-k\alpha}\, ||H_{\mathfrak{e}}^{(k+1)}||_{W_{\ell}^{\infty}(\mathbb{R})}\,, \qquad (5.2.44)$$

with the reflected operator $\mathcal{W}_R^{\wedge}$ as introduced in Definition 5.1.2. We do stress that, in the present context, $H_{\mathfrak{e}}$ denotes a compactly supported extension of H from $[a_N\,;\,b_N]$ to $\mathbb{R}$ that, furthermore, satisfies the same regularity properties as H. All in all, the bounds (5.2.43) and (5.2.44) yield

$$||\mathcal{W}_N[H](\xi)||_{W_{\ell}^{\infty}(\mathbb{J}_N^{(\mathrm{bk})})} \;\leq\; c'_{k;\ell} \cdot \max_{s \in [\![0\,;\,\ell]\!]} \Big\{\mathcal{N}_N^{(k)}[H^{(s+1)}]\Big\}\;. \qquad (5.2.45)$$

Besides, for $k = 1$ we have from (5.1.40):

$$\mathcal{W}_{\mathrm{bk};k}[V'](\xi) \;=\; u_1\, V''(\xi)\;. \qquad (5.2.46)$$

The constant u_1 was introduced in Definition 5.1.5, and according to the expression of $R(\lambda)$ in (4.1.18), it takes the value:

$$u_1 = \frac{1}{2\pi\beta(\omega_1 + \omega_2)} > 0\;. \qquad (5.2.47)$$

So, using the bound (5.2.44) for $k = 1$ and $\ell = 0$ to control the extra terms in $\mathcal{W}_N$ in sup norm, we get

$$\big|\mathcal{W}_N[V'](\xi)\big| \;\geq\; u_1 \inf_{\xi \in [a\,;b]} V''(\xi) \;-\; \frac{C}{N^{\alpha}}||V_{\mathfrak{e}}||_{W_3^{\infty}(\mathbb{R})} \;\geq\; \frac{u_1}{2} \cdot \inf_{\xi \in [a\,;b]} \big\{V''(\xi)\big\} \qquad (5.2.48)$$

where the last lower bound holds for N large enough. The above lower bound leads to

$$\frac{||\mathcal{W}_{\mathrm{bk};k}[V']||_{W_{\ell}^{\infty}(\mathbb{J}_N^{(\mathrm{bk})})}}{\big|\mathcal{W}_N[V'](\xi)\big|} \;\leq\; \frac{C_{\ell} \cdot ||V'||_{W_{k+\ell+1}^{\infty}(\mathbb{J}_N^{(\mathrm{bk})})}}{\inf_{\xi \in [a\,;b]} V''(\xi)}. \qquad (5.2.49)$$

Lower and upper bounds on $\mathbb{J}_N^{(R;\mathrm{in})}$

In virtue of Lemma 5.1.11 and Proposition 5.1.6, given $k \in \mathbb{N}^*$, we have the decomposition

$$\mathcal{W}_N[H](\xi) = \big(\mathcal{W}_{R;k}^{(\mathrm{as})} + \mathcal{W}_{\mathrm{bk};k}^{(\mathrm{as})}\big)[H](x_R) + \Omega_{R;k}[H_{\mathfrak{e}}](x_R,\xi) \tag{5.2.50}$$

$$\begin{aligned}\Omega_{R;k}[H_{\mathfrak{e}}](x_R,\xi) &= \Delta_{[k]}\mathcal{W}_R^{(\mathrm{as})}[H](x_R) + \Delta_{[k]}\mathcal{W}_{\mathrm{bk}}^{(\mathrm{as})}[H](x_R) \\ &\quad - \mathcal{W}_{R;k}[H^{\wedge}](x_L, b_N + a_N - \xi) + \Delta_{[k]}\mathcal{W}_N[H_{\mathfrak{e}}](\xi)\end{aligned} \tag{5.2.51}$$

where $\Delta_{[k]}\mathcal{W}_N[H_{\mathfrak{e}}]$ has been introduced in (5.1.69). We remind from (5.1.106) that:

$$\big(\mathcal{W}_{R;k}^{(\mathrm{as})} + \mathcal{W}_{\mathrm{bk};k}^{(\mathrm{as})}\big)[H](x_R) = H'(b_N)\mathfrak{a}_0(x_R) + \sum_{r=1}^{k} \frac{H^{(r+1)}(b_N)}{N^{r\alpha}} (\mathfrak{a}_0 \cdot \mathfrak{a}_r)(x_R) \tag{5.2.52}$$

For any integers n, ℓ such that $n \geq \ell + 2$, Lemma 5.1.10 applied to $(\ell, m, n) \hookrightarrow (r, \ell+1, n)$ tells us:

$$\begin{aligned}\frac{\mathfrak{a}_0(x)}{\sqrt{x}} &= p_{0;\ell+1,n}(x)\mathrm{e}^{-\varsigma x} + x^{\ell+1/2} f_{0;\ell+1,n}(x), \\ \frac{(\mathfrak{a}_0 \cdot \mathfrak{a}_r)(x)}{\sqrt{x}} &= p_{r;\ell+1,n}(x)\mathrm{e}^{-\varsigma x} + x^{\ell+1/2} f_{r;\ell+1,n}(x)\end{aligned} \tag{5.2.53}$$

for some polynomials $p_{k;\ell+1,n}(x)$ of degree at most $n+k$ and functions $f_{k;\ell+1,n} \in W^{\infty}_{n-(\ell+1)}(\mathbb{R}_+)$. We therefore get:

$$\big|\big|q_R^{-1}\big(\mathcal{W}_{R;k}^{(\mathrm{as})} + \mathcal{W}_{\mathrm{bk};k}^{(\mathrm{as})}\big)[H]\big|\big|_{W_\ell^\infty(\mathbb{J}_N^{(R;\mathrm{in})})} \leq c_{k;\ell} \cdot N^{\ell\alpha} \cdot (\ln N)^{2\ell+1} \cdot \mathcal{N}_N^{(k-1)}[H_{\mathfrak{e}}'] \,. \tag{5.2.54}$$

In this inequality, one power of N^α pops up at each action of the derivative of $x_R = N^\alpha(b_N - \xi)$. Furthermore, by putting together the control of the remainders in Proposition 5.1.6 and Lemma 5.1.11, we get that:

$$\Omega_{R;k}[H_{\mathfrak{e}}](x_R,\xi) = \sum_{m=0}^{k} \Big\{ c_{k;m}^{(0)} x_R^m + c_{k;m}^{(1/2)} x_R^{m+\frac{1}{2}} \Big\} + f_k(x_R) \tag{5.2.55}$$

where, for any $0 \leq \ell \leq k$, the function f_k satisfies:

$$\big|\partial_\xi^\ell\big(x_R^{-1/2} f_k(x_R)\big)\big| \leq C_{k;\ell} \cdot x_R^{k+\frac{1}{2}-\ell} \cdot N^{(\ell-k)\alpha} \cdot \mathcal{N}_N^{(\ell)}[H_{\mathfrak{e}}^{(k+1)}] \cdot \big(\ln(x_R)\,\mathrm{e}^{-Cx_R} + 1\big). \tag{5.2.56}$$

Since the functions $\big(\mathcal{W}^{(\mathrm{as})}_{R;k}+\mathcal{W}^{(\mathrm{as})}_{\mathrm{bk};k}\big)[H]\cdot q_R^{-1}$ and $\mathcal{W}_N[H]\cdot q_R^{-1}$ are smooth on $\mathbb{J}_N^{(R;\mathrm{in})}$, so must be $\Omega_{R;k}[H_{\mathfrak{e}}]\cdot q_R^{-1}$. As a consequence, we necessarily have $c^{(0)}_{k;m}=0$. The properties of the remainders then ensure that, for any $0\le \ell \le k$,

$$\big|c^{(1/2)}_{k;m}\big| \;\le\; C_{k;m}\cdot N^{-k\alpha}\cdot \big|\big|H_{\mathfrak{e}}^{(k+1)}\big|\big|_{W_m^{\infty}(\mathbb{R})}\,. \tag{5.2.57}$$

Thus, all-in-all, by choosing properly the compactly supported regular extension $H_{\mathfrak{e}}$ of H from $[a_N\,;\,b_N]$ to $\mathbb{R}$ we get

$$||q_R^{-1}\cdot \mathcal{W}_N[H]||_{W_\ell^{\infty}(\mathbb{J}_N^{(R;\mathrm{in})})} \;\le\; C_\ell\cdot(\ln N)^{2\ell+1}\cdot N^{(\ell+1)\alpha}\cdot \mathcal{N}_N^{(2\ell+1)}[\mathcal{K}_\kappa[H_{\mathfrak{e}}]] \tag{5.2.58}$$

upon choosing $k=\ell$. This holds for any $\kappa>0$, the right-hand side being possibly $+\infty$.

In what concerns the lower bounds, observe that

$$x_R^{-1/2}\cdot\big(\mathcal{W}^{(\mathrm{as})}_{R;1}+\mathcal{W}^{(\mathrm{as})}_{\mathrm{bk};1}\big)[H](x_R) \;=\; \frac{\mathfrak{a}_0(x_R)}{\sqrt{x_R}}V''(b_N)\bigg(1\,+\,\frac{V^{(3)}(b_N)}{V''(b_N)}\cdot\frac{\mathfrak{a}_1(x_R)}{N^{\alpha}}\bigg) \tag{5.2.59}$$

as well as

$$\begin{aligned}\big|c^{(1/2)}_{1;0}+c^{(1/2)}_{1;1}x_R\,+\,x_R^{-1/2}f_1(x_R)\big| \;\le\; C\cdot\Big\{N^{-\alpha}\cdot(x_R+1)||V_{\mathfrak{e}}''||_{W_1^{\infty}(\mathbb{R})}\\ +N^{-\alpha}\,x_R^{\frac{3}{2}}\big(\ln x_R\,\mathrm{e}^{-Cx_R}+1\big)||V_{\mathfrak{e}}''||_{W_0^{\infty}(\mathbb{R})}\Big\}\,.\end{aligned} \tag{5.2.60}$$

These estimates imply, for N large enough:

$$\bigg|\frac{\mathcal{W}_N[V'](\xi)}{q_R(\xi)}\bigg| \;>\; \frac{\mathfrak{a}_0(x_R)}{\sqrt{x_R}}V''(b_N)\;-\;\frac{(\ln N)^3}{N^{\alpha}}||V_{\mathfrak{e}}||_{W_3^{\infty}(\mathbb{R})}\,. \tag{5.2.61}$$

The function $x\to\mathfrak{a}_0(x)/\sqrt{x}$ is bounded from below on $\mathbb{R}^+$, *cf.* Lemma 5.1.10 and $(a_N,b_N)\to(a,b)$ in virtue of Corollary 6.2.2. As a consequence, for any potential V such that $||V_{\mathfrak{e}}||_{W_3^{\infty}([a\,;b])}<C$, there exists N_0 large enough and $c>0$ such that

$$\bigg|\frac{\mathcal{W}_N[V'](\xi)}{q_R(\xi)}\bigg| \;>\; c\,\inf_{[a\,;b]}\big\{V''(\xi)\big\}\,. \tag{5.2.62}$$

We can deduce from the above bounds that, for any $\xi\in\mathbb{J}_N^{(R;\mathrm{in})}$,

$$\frac{||q_R^{-1}\cdot\mathcal{W}_N[V']||_{W_\ell^{\infty}(\mathbb{J}_N^{(R;\mathrm{in})})}}{q_R^{-1}(\xi)\cdot\mathcal{W}_N[V'](\xi)} \;\le\; C_\ell\cdot(\ln N)^{2\ell+1}\cdot N^{\ell\alpha}\cdot\frac{||V'||_{W^{\infty}_{2\ell+1}(\mathbb{J}_N^{(R;\mathrm{in})})}}{\inf_{[a\,;b]}\big\{V''(\xi)\big\}}. \tag{5.2.63}$$

Lower and upper bounds on $\mathbb{J}_N^{(R;\mathrm{ext})}$

Let us go back to the vector Riemann–Hilbert problem discussed in Lemma 4.1.1. The representation (4.1.10) and the fact that the solution Φ to this vector Riemann–Hilbert problem allows one the reconstruction of the functions ψ_1 and ψ_2 arising in (4.1.10) through (4.1.15). Using the reconstruction formula (4.3.14) with $P_1 = P_2 = 0$ and $z_0 = \infty$ and applying the regularisation trick exactly as in (4.3.67), we get $\xi \in [b_N ; +\infty[$:

$$\mathcal{S}_N\big[\mathcal{W}_N[H]\big](\xi) = N^\alpha \int\limits_{\mathbb{R}+2\mathrm{i}\epsilon} \frac{\mathrm{d}\lambda}{2\pi} \int\limits_{\mathbb{R}+\mathrm{i}\epsilon} \frac{\mathrm{d}\mu}{2\mathrm{i}\pi} \int\limits_{a_N}^{b_N} \mathrm{d}\eta H(\eta) \frac{\mathrm{e}^{\mathrm{i}\lambda N^\alpha(b_N-\xi)-\mathrm{i}\mu N^\alpha(b_N-\eta)}}{\mu-\lambda} \times \Big\{\chi_{21}(\lambda)\chi_{12}(\mu) - \frac{\mu}{\lambda}\cdot\chi_{11}(\mu)\chi_{22}(\lambda)\Big\}. \tag{5.2.64}$$

The local behaviour of the above integral representation can be studied with the set of tools developed throughout this Chapter 5. We do not reproduce this reasoning again. All-in-all, we obtain:

$$||q_R^{-1}\cdot\mathcal{K}_\kappa\big[\mathcal{S}_N\big[\mathcal{W}_N[H]\big]-H\big]||_{W_\ell^\infty(\mathbb{J}_N^{(R;\mathrm{ext})})} \le C_\ell(\ln N)^{2\ell+1}\cdot N^{(\ell+1)\alpha}\cdot\mathcal{N}_N^{(2\ell+1)}\big[\mathcal{K}_\kappa[H]\big] \tag{5.2.65}$$

and, for any $\xi \in \mathbb{J}_N^{(R;\mathrm{ext})}$,

$$\big|q_R^{-1}(\xi)\cdot\big\{\mathcal{S}_N\big[\mathcal{W}_N[V'](\xi)\big]-V'(\xi)\big\}\big| > c\inf_{[a;b]} V''(b_N) \tag{5.2.66}$$

provided that N is large enough. Likewise, we have the bounds:

$$\frac{||q_R^{-1}\mathcal{K}_\kappa\big[\mathcal{S}_N\big[\mathcal{W}_N[V']\big]-V'\big]||_{W_\ell^\infty(\mathbb{J}_N^{(R;\mathrm{ext})})}}{q_R^{-1}(\xi)\cdot\big\{\mathcal{S}_N\big[\mathcal{W}_N[V']\big]-V'\big\}} \le C_\ell\cdot(\ln N)^{2\ell+1}\cdot N^{\ell\alpha}\cdot\frac{||\mathcal{K}_\kappa[V']||_{W_{2\ell+1}^\infty(\mathbb{J}_N^{(R;\mathrm{ext})})}}{\inf_{[a;b]}\big\{V''(\xi)\big\}}. \tag{5.2.67}$$

Synthesis

Let us now write:

$$\mathcal{U}_N^{-1}[H](\xi) = \sum_{A=L,R}\Bigg\{\frac{\mathcal{S}_N\big[\mathcal{W}_N[H]\big](\xi)-H(\xi)}{\mathcal{S}_N\big[\mathcal{W}_N[V']\big](\xi)-V'(\xi)}\cdot\mathbf{1}_{\mathbb{J}_N^{(A;\mathrm{out})}}(\xi)+\frac{\mathcal{W}_N[H](\xi)}{\mathcal{W}_N[V'](\xi)}\cdot\mathbf{1}_{\mathbb{J}_N^{(A;\mathrm{in})}}(\xi) + \frac{q_R^{-1}(\xi)\cdot\big\{\mathcal{S}_N\big[\mathcal{W}_N[H]\big](\xi)-H(\xi)\big\}}{q_R^{-1}(\xi)\cdot\big\{\mathcal{S}_N\big[\mathcal{W}_N[V']\big](\xi)-V'(\xi)\big\}}\cdot\mathbf{1}_{\mathbb{J}_N^{(A;\mathrm{ext})}}(\xi)\Bigg\}+\frac{\mathcal{W}_N[H](\xi)}{\mathcal{W}_N[V'](\xi)}\cdot\mathbf{1}_{\mathbb{J}_N^{(\mathrm{bk})}}(\xi). \tag{5.2.68}$$

The piecewise bounds (5.2.36)–(5.2.41) on $\mathbb{J}_N^{(R;\mathrm{out})}$, (5.2.45)–(5.2.49) on $\mathbb{J}_N^{(\mathrm{bk})}$, (5.2.58)–(5.2.63) on $\mathbb{J}_N^{(\mathrm{bk})}$, (5.2.65)–(5.2.67) on $\mathbb{J}_N^{(R;\mathrm{in})}$, and those which can be deduced by reflection symmetry on the three other segments defined in (5.2.28)–(5.2.30), can now be used together with the Faá di Bruno formula

$$\frac{\mathrm{d}^{\ell}}{\mathrm{d}\xi^{\ell}}\Big(\frac{f}{g}\Big)(\xi) = \sum_{n+m=\ell}\ \sum_{\sum k n_k = n} \frac{\ell!\big(\sum_{k=1}^{n} n_k\big)!}{m!} \cdot \frac{f^{(m)}(\xi)}{g(\xi)} \cdot \prod_{j=1}^{n}\bigg\{\frac{1}{n_j!}\Big(\frac{-g^{(j)}(\xi)}{j!g(\xi)}\Big)^{n_j}\bigg\} \tag{5.2.69}$$

to establish the global bound. Note that, in the intermediate bounds, one should use the obvious property of the exponential regularisation:

$$K_{\kappa}[f_1\cdots f_p] = \prod_{a=1}^{p}\mathcal{K}_{\kappa/p}[f_a]. \tag{5.2.70}$$

The details are left to the reader. ■

Chapter 6
Asymptotic Analysis of Integrals

Abstract In this chapter we carry out the large-N asymptotic analysis of the single and double integrals that arise in the problem. First, in Section 6.1.1, we deal with the one-fold integrals that arise in the characterisation of the image space $\mathfrak{X}_s(\mathbb{R})$ of $H_s\big([a_N\,;\,b_N]\big)$ under the operator $\mathcal{S}_N$. Then, in Section 6.1.2, we evaluate asymptotically in N one-dimensional integrals of $\mathcal{W}_N[H]$ versus test functions G. This provides the first set of results that were necessary in Section 3.4 for a thorough calculation of the large-N expansion of the partition function. Then, in Section 6.2, we build on the obtained large-N expansion of the two types of single integrals so as to characterise the support of the equilibrium measure. Finally, in Section 6.3, we obtain the large-N expansion, up to a vanishing with N remainder, of the double integral (3.4.3) arising in the large-N expansion of the partition function at $\beta = 1$.

6.1 Asymptotic Analysis of Single Integrals

6.1.1 Asymptotic Analysis of the Constraint Functionals $\mathcal{X}_N[H]$

Recall that for any $H \in \mathcal{C}^1([a_N\,;\,b_N])$ the linear form $\mathcal{X}_N[H]$ defined in (3.3.31):

$$\mathcal{X}_N[H] \;=\; \frac{\mathrm{i}N^{\alpha}}{\chi_{11;+}(0)} \int\limits_{\mathbb{R}+\mathrm{i}\epsilon'} \frac{\mathrm{d}\mu}{2\mathrm{i}\pi}\, \chi_{11}(\mu) \int\limits_{a_N}^{b_N} H(\eta)\mathrm{e}^{\mathrm{i}N^{\alpha}\mu(\eta-b_N)}\,\mathrm{d}\eta \qquad (6.1.1)$$

is related to the constraint $\mathscr{I}_{11}[h]$ defined in (4.3.24) where H and h are related by the rescaling (4.1.1):

$$\mathscr{I}_{11}[h] \;=\; -\frac{N^{\alpha}\,\chi_{11;+}(0)}{2\pi\beta}\,\mathcal{X}_N[H]\,, \qquad h(x) = \frac{N^{\alpha}}{2\mathrm{i}\pi\beta}\,H(a_N + N^{-\alpha}x)\,. \quad (6.1.2)$$

G. Borot et al., *Asymptotic Expansion of a Partition Function Related to the Sinh-model*, Mathematical Physics Studies,
DOI 10.1007/978-3-319-33379-3_6

In the following, we shall obtain the large-N expansion of the linear form $\mathcal{X}_N[h]$ introduced in (3.3.31) and defining the hyperplane $\mathfrak{X}_s$ where we inverse operators. We first need to define new constants:

Definition 6.1.1 If $p \geq 0$ is an integer, we define:

$$\urcorner_p = -\frac{R_\downarrow(0)}{2} \int\limits_{\mathbb{R}+\mathrm{i}\epsilon'} \frac{1}{\mu^{p+1} R_\downarrow(\mu)} \cdot \frac{\mathrm{d}\mu}{2\mathrm{i}\pi} = (-1)^{p+1} \frac{R_\downarrow(0)}{2} \int\limits_{\mathbb{R}-\mathrm{i}\epsilon'} \frac{1}{\mu^{p+2} R_\uparrow(\mu)} \cdot \frac{\mathrm{d}\mu}{2\mathrm{i}\pi} . \tag{6.1.3}$$

The equality between the two expressions of $\urcorner_p$ follows from the symmetry (4.1.27).

Lemma 6.1.2 *Let $k \geq 1$ be an integer, and $H \in \mathcal{C}^k\big([a_N\,;\,b_N]\big)$. We have an asymptotic expansion:*

$$\mathcal{X}_N[H] = \sum_{p=0}^{k-1} \frac{\mathrm{i}^p \urcorner_p}{N^{\alpha p}} \Big\{ H^{(p)}(a_N) + (-1)^p H^{(p)}(b_N) \Big\} + \Delta_{[k]}\mathcal{X}_N[H] , \tag{6.1.4}$$

where:

$$\big|\Delta_{[k]}\mathcal{X}_N[H]\big| \leq C\,N^{-k\alpha}\,||H||_{W_k^\infty([a_N\,;b_N])} . \tag{6.1.5}$$

Proof For λ between $\Gamma_\uparrow$ and $\mathbb{R}$, we decompose χ into:

$$\chi(\lambda) = \chi_\uparrow^{(\mathrm{as})}(\lambda) + \chi_\uparrow^{(\mathrm{exp})}(\lambda) . \tag{6.1.6}$$

In terms of the various matrices used Section 4.2.2, the main part is:

$$\begin{aligned} \chi_\uparrow^{(\mathrm{as})}(\lambda) &= \mathcal{R}_\uparrow^{-1}(\lambda) \cdot \big[\upsilon(\lambda)\big]^{-\sigma_3} \cdot M_\uparrow(\lambda) \cdot \left(I_2 + \frac{\sigma^-}{\lambda}\right) \\ &= \begin{pmatrix} -\dfrac{\mathrm{e}^{\mathrm{i}\lambda\overline{x}_N}}{R_\downarrow(\lambda)} + \dfrac{1}{\lambda R_\uparrow(\lambda)} & \dfrac{1}{R_\uparrow(\lambda)} \\ -R_\uparrow(\lambda) & 0 \end{pmatrix} \end{aligned} \tag{6.1.7}$$

and is such that the remainder is exponentially small in N:

$$\chi_\uparrow^{(\mathrm{exp})}(\lambda) = \chi_\uparrow^{(\mathrm{as})}(\lambda)\cdot[\delta\Pi](\lambda) \quad \text{with} \quad [\delta\Pi](\lambda) = \left(I_2 + \frac{\sigma^-}{\lambda}\right)^{-1} \cdot \Pi(\lambda)\cdot P_R(\lambda) - I_2 . \tag{6.1.8}$$

Indeed, the large-N behaviour of θ_R inferred from (4.1.17) and (4.1.34) as well as the estimate (4.2.15) on the matrix $\Pi - I_2$ imply that, for ϵ' fixed but small enough, and uniformly in $\lambda \in \mathbb{R} + \mathrm{i}\tau$, $0 < \tau < \epsilon'$:

$$\big|[\delta\Pi]_{ab}(\lambda)\big| \leq \frac{C\,\mathrm{e}^{-\varkappa_{\epsilon'} N^\alpha}}{1 + |\lambda|} . \tag{6.1.9}$$

Furthermore, a direct calculation shows that

$$[\chi_{\uparrow}^{(\exp)}]_{11}(\lambda) = \left(\frac{1}{\lambda R_{\uparrow}(\lambda)} - \frac{e^{i\lambda \overline{x}_N}}{R_{\downarrow}(\lambda)}\right)[\delta\Pi]_{11}(\lambda) + \frac{[\delta\Pi]_{21}(\lambda)}{R_{\uparrow}(\lambda)}, \tag{6.1.10}$$

and taking into account the large-λ behaviour of $R_{\uparrow/\downarrow}$ given in (4.1.24) and (4.1.25), we also get a uniform bound for $\lambda \in \mathbb{R} + i\tau$, $0 < \tau < \epsilon'$:

$$\left|[\chi_{\uparrow}^{(\exp)}]_{11}(\lambda)\right| \leq \frac{C' e^{-\varkappa_{\epsilon'} N^{\alpha}}}{\sqrt{1+|\lambda|}}. \tag{6.1.11}$$

In particular, this estimate (6.1.11) implies:

$$\frac{1}{\chi_{11;+}(0)} = -\frac{R_{\downarrow}(0)}{2} + O(e^{-\varkappa_{\epsilon'} N^{\alpha}}). \tag{6.1.12}$$

The decomposition (6.1.6) in formula (6.1.1) induces a decomposition:

$$\mathcal{X}_N[H] = \mathcal{X}_N^{(\mathrm{as})}[H] + \mathcal{X}_N^{(\exp)}[H] \tag{6.1.13}$$

where

$$\mathcal{X}^{(\exp)}[H] = \frac{iN^{\alpha}}{\chi_{11;+}(0)} \int\limits_{\widetilde{\mathscr{C}}^{(-)}} \frac{d\mu}{2i\pi} \left[\chi_{\uparrow}^{(\exp)}(\mu)\right]_{11} \int\limits_{a_N}^{b_N} H(\eta) e^{iN^{\alpha}\mu(\eta - b_N)} \cdot d\eta \tag{6.1.14}$$

and $\widetilde{\mathscr{C}}^{(-)}$ is a contour surrounding 0 from above, going to ∞ in $\mathbb{H}^-$ along the rays $te^{-\frac{3i\pi}{4}}$ and $te^{-\frac{i\pi}{4}}$ and such that $\max\left\{\mathrm{Im}(\lambda) : \lambda \in \widetilde{\mathscr{C}}^{(-)}\right\} = \epsilon'$. Note that we could have carried out this contour deformation since $\Pi(\lambda)$ is holomorphic in the domain delimited by $\mathbb{R} + i\epsilon'$ and $\widetilde{\mathscr{C}}^{(-)}$.
Since for $\lambda \in \widetilde{\mathscr{C}}^{(-)}$, we have:

$$\left|\int\limits_{a_N}^{b_N} H(\eta) e^{iN^{\alpha}\lambda(\eta - b_N)} d\eta\right| \leq \frac{C e^{\overline{x}_N \epsilon'}}{|\lambda|} ||H||_{L^{\infty}([a_N;b_N])}, \tag{6.1.15}$$

it is readily seen that

$$\left|\mathcal{X}_N^{(\exp)}[H]\right| \leq C' \cdot N^{\alpha} e^{-\frac{\varkappa_{\epsilon'}}{2} N^{\alpha}} ||H||_{L^{\infty}([a_N;b_N])}. \tag{6.1.16}$$

It thus remains to estimate

$$\mathcal{X}_N^{(\mathrm{as})}[H] = \mathcal{X}_R^{(\mathrm{as})}[H] + \mathcal{X}_R^{(\mathrm{as})}[H^{\wedge}] \tag{6.1.17}$$

where

$$\mathcal{X}_R^{(\mathrm{as})}[H] \;=\; \frac{\mathrm{i}N^{\alpha}}{\chi_{11;+}(0)} \int_{\mathscr{C}_{\mathrm{reg}}^{(-)}} \frac{\mathrm{d}\mu}{2\mathrm{i}\pi} \, \frac{1}{\mu R_{\uparrow}(\mu)} \int_{a_N}^{b_N} H(\eta) \mathrm{e}^{\mathrm{i}N^{\alpha}\mu(\eta-b_N)} \, \mathrm{d}\eta \,, \tag{6.1.18}$$

and the second term arises upon the change of variables $(\mu,\eta)\mapsto(-\mu,a_N+b_N-\eta)$ in the initial expression. The dependence in N is implicit in these new notations. Note that we could deform the contour from $\mathbb{R}+\mathrm{i}\epsilon'$ up to $\mathbb{R}-\mathrm{i}\epsilon'$ or $\mathscr{C}_{\mathrm{reg}}^{(-)}$ since the integrand is holomorphic in the domain swapped in between. Replacing H by its Taylor series with integral remainder at order k, we get:

$$\mathcal{X}_R^{(\mathrm{as})}[H] \;=\; \mathcal{X}_{R;k}^{(\mathrm{as})}[H] \;+\; \Delta_{[k]}\mathcal{X}_R^{(\mathrm{as})}[H] \,. \tag{6.1.19}$$

The first term is:

$$\begin{aligned} \mathcal{X}_{R;k}^{(\mathrm{as})}[H] &= \mathrm{i}N^{\alpha} \sum_{p=0}^{k-1} \frac{H^{(p)}(b_N)}{p!\,\chi_{11;+}(0)} \int_{\mathbb{R}-\mathrm{i}\epsilon'} \frac{\mathrm{d}\mu}{2\mathrm{i}\pi} \, \frac{1}{\mu R_{\uparrow}(\mu)} \int_{-\infty}^{0} \eta^{p} \mathrm{e}^{\mathrm{i}N^{\alpha}\mu\eta} \, \mathrm{d}\eta \\ &= \frac{-2}{R_{\downarrow}(0)\chi_{11;+}(0)} \sum_{p=0}^{k-1} \frac{(-\mathrm{i})^{p}\, \daleth_p \, H^{(p)}(b_N)}{N^{p\alpha}} \end{aligned} \tag{6.1.20}$$

where we have recognised the constants $\daleth_p$ of Definition 6.1.1. The remainder is:

$$\begin{aligned} \Delta_{[k]}\mathcal{X}_R^{(\mathrm{as})}[H] = & \frac{1}{\mathrm{i}\chi_{11;+}(0)} \Bigg\{ \sum_{p=0}^{k-1} \frac{H^{(p)}(b_N)}{p!\,N^{p\alpha}} \int_{\mathscr{C}_{\mathrm{reg}}^{(-)}} \frac{\mathrm{d}\mu}{2\mathrm{i}\pi} \, \frac{1}{\mu R_{\uparrow}(\mu)} \int_{-\infty}^{-\overline{x}_N} \eta^{p} \mathrm{e}^{\mathrm{i}\mu\eta} \, \mathrm{d}\eta \\ & + \int_{\mathscr{C}_{\mathrm{reg}}^{(-)}} \frac{\mathrm{d}\mu}{2\mathrm{i}\pi} \, \frac{N^{\alpha}}{\mu R_{\uparrow}(\mu)} \int_{a_N}^{b_N} \mathrm{d}\eta \, (\eta-b_N)^{k} \int_{0}^{1} \mathrm{d}t \, \frac{(1-t)^{k-1}}{(k-1)!} \mathrm{e}^{\mathrm{i}N^{\alpha}\mu(\eta-b_N)} \, H^{(k)}\big(b_N+t(\eta-b_N)\big) \Bigg\} . \end{aligned} \tag{6.1.21}$$

$\mathcal{X}_{R;k}^{(\mathrm{as})}[H]$ yields the leading terms of the asymptotic expansion announced in (6.1.4). Hence, it remains to bound $\Delta_{[k]}\mathcal{X}_R^{(\mathrm{as})}[H]$. The first line in (6.1.21) is exponentially small and bounded by a term proportional to $||H||_{W_{k-1}^{\infty}([a_N\,;b_N])}$. The second line is bounded by

$$N^{\alpha} \cdot |R_{\downarrow}(0)| \cdot ||H||_{W_k^{\infty}([a_N\,;b_N])} \int_{\mathscr{C}_{\mathrm{reg}}^{(-)}} \frac{|\mathrm{d}\mu|}{2\pi\, k!} \frac{1}{|\mu R_{\uparrow}(\mu)|} \int_{-\infty}^{b_N} \mathrm{d}\eta\, (b_N - \eta)^k \, \mathrm{e}^{-N^{\alpha}\left[\mathrm{Im}\,\mu(\eta - b_N)\right]}$$

$$\leq C\, N^{-k\alpha}\, ||H||_{W_k^{\infty}([a_N\,;b_N])} \,. \tag{6.1.22}$$

It thus solely remains to put all the pieces together. ■

Using these estimates, we obtain the continuity of the linear form $\mathcal{X}_N$ in sup norms:

Corollary 6.1.3 *There exists $C > 0$ independent of N, such that:*

$$\left|\widetilde{\mathcal{X}}_N[H]\right| \leq C\, ||H||_{W_0^{\infty}([a_N\,;b_N])} \,. \tag{6.1.23}$$

Proof We have shown in the proof of Lemma 6.1.2 a decomposition:

$$\mathcal{X}_N^{(\mathrm{as})}[H] = \mathcal{X}_R^{(\mathrm{as})}[H] + \mathcal{X}_R^{(\mathrm{as})}[H^{\wedge}] + \mathcal{X}_N^{(\mathrm{exp})}[H] \,. \tag{6.1.24}$$

$\mathcal{X}_R^{(\mathrm{as})}[H]$ is given in (6.1.18). It has $\chi_{11;+}(0)$ as prefactor, and we have seen in (6.1.12) that this quantity takes the non-zero value $-2/R_{\downarrow}(0)$ up to exponential small (in N) corrections. So, we have the bound:

$$\left|\mathcal{X}_R^{(\mathrm{as})}[H]\right| \leq \frac{|R_{\downarrow}(0)|}{2} \cdot ||H||_{W_0^{\infty}([a_N\,;b_N])} \cdot \int_{\mathscr{C}_{\mathrm{reg}}^{(-)}} \frac{1}{|\mu||\mathrm{Im}\,\mu|\, R_{\uparrow}(\mu)|} \frac{|\mathrm{d}\mu|}{2\pi} \tag{6.1.25}$$

where the inverse power of $|\mathrm{Im}\,\mu|$ and the loss of the prefactor N^{α} resulted from integrating the decaying exponential $|e^{\mathrm{i}N^{\alpha}\mu(\eta - b_N)}|$ over $[a_N\,;b_N]$, given that $\mathrm{Im}\,\mu < 0$ for $\mu \in \mathscr{C}_{\mathrm{reg}}^{(-)}$. We conclude by combining this estimate with (6.1.16) which shows that the remainder is exponentially small. ■

6.1.2 Asymptotic Analysis of Simple Integrals

In the present subsection, we obtain the large-N asymptotic expansion of one-dimensional integrals involving $\mathcal{W}_N[H]$. This provides the first set of results that were necessary in Section 3.4 for a thorough calculation of the large-N expansion of the partition function.

Definition 6.1.4 If G and H are two functions on $[a_N\,;b_N]$, we define:

$$\mathfrak{I}_{\mathrm{s}}\big[G, H\big] = \int_{a_N}^{b_N} G(\xi) \cdot \mathcal{W}_N[H](\xi)\, \mathrm{d}\xi \tag{6.1.26}$$

where the $\mathcal{W}_N$ is the operator defined in (2.4.17).

To write the large N-expansion of $\mathfrak{I}_{\mathrm{s}}$, we need to introduce some more constants:

Definition 6.1.5 If $s, \ell \geq 0$ are integers, we set:

$$\daleth_{s,\ell} = \int_0^{+\infty} x^s\, \mathfrak{b}_\ell(x)\, \mathrm{d}x \tag{6.1.27}$$

where the function $\mathfrak{b}_\ell$ has been introduced in Definition 5.1.9.

Proposition 6.1.6 *Let $k \geq 1$ be an integer, $G \in \mathcal{C}^{k-1}([a_N\,;b_N])$ and $H \in \mathcal{C}^{k+1}([a_N\,;b_N])$. We have the asymptotic expansion:*

$$\begin{aligned}\mathfrak{I}_{\mathrm{s}}\big[G,H\big] = u_1 \int_{a_N}^{b_N} G(\xi)\cdot H'(\xi)\,\mathrm{d}\xi + \sum_{p=1}^{k-1} \frac{1}{N^{\alpha p}} \Bigg\{ u_{p+1} \int_{a_N}^{b_N} G(\xi) H^{(p+1)}(\xi)\,\mathrm{d}\xi \\ + \sum_{\substack{s+\ell=p-1 \\ s,\ell\geq 0}} \frac{\daleth_{s,\ell}}{s!} \Big[(-1)^s H^{(\ell+1)}(b_N)\cdot G^{(s)}(b_N) + (-1)^\ell H^{(\ell+1)}(a_N) G^{(s)}(a_N) \Big] \Bigg\} \\ + \Delta_{[k]} \mathfrak{I}_{\mathrm{s}}\big[G,H\big]\,.\end{aligned} \tag{6.1.28}$$

where we remind that u's are the constants appearing in Definition 5.1.5. The remainder is bounded as

$$\Big|\Delta_{[k]}\mathfrak{I}_{\mathrm{s}}\big[G,H\big]\Big| \leq C\,N^{-k\alpha}\,||G||_{W^{\infty}_{k-1}([a_N\,;b_N])}\,||H_{\mathfrak{e}}||_{W^{\infty}_{k+1}([a_N\,;b_N])} \tag{6.1.29}$$

for some constant $C > 0$ independent of N, G and H.

Note that the leading asymptotics of $\mathfrak{I}_{\mathrm{s}}\big[G,H\big]$, *i.e.* up to the $\mathrm{o}(1)$ remainder, correspond precisely to the contribution obtained by replacing the integral kernel $S\big(N^\alpha(\xi-\eta)\big)$ of $\mathcal{S}_N$ by the sign function—which corresponds to the almost sure pointwise limit of $S\big(N^\alpha(\xi-\eta)\big)$, see (2.4.15)—and then inverting the formal limiting operator. The corrections, however, are already more complicated as they stem from the fine behaviour at the boundaries.

Proof Recall from Propositions 5.1.4 and 5.1.6 that $\mathcal{W}_N[H]$ decomposes as

$$\begin{aligned}\mathcal{W}_N[H](\xi) = \mathcal{W}_{R;k}[H](x_R,\xi) + \mathcal{W}_{\mathrm{bk};k}[H](\xi) - \mathcal{W}_{R;k}[H^\wedge](x_L, a_N+b_N-\xi) \\ + \Delta_{[k]}\mathcal{W}_N[H_{\mathfrak{e}}](\xi)\end{aligned} \tag{6.1.30}$$

where

$$\Big|\Big|\Delta_{[k]}\mathcal{W}_N[H_{\mathfrak{e}}]\Big|\Big|_{L^\infty([a_N\,;b_N])} \leq C\,N^{-k\alpha}\,||H_{\mathfrak{e}}^{(k+1)}||_{L^\infty(\mathbb{R})}\,. \tag{6.1.31}$$

This leads to the decomposition

$$\mathfrak{I}_{\mathrm{s}}[G,H] = \mathfrak{I}_{\mathrm{s};k}^{(\mathrm{bk})}[G,H] + \mathfrak{I}_{\mathrm{s};k}^{(\partial)}[G,H] - \mathfrak{I}_{\mathrm{s};k}^{(\partial)}[G^{\wedge},H^{\wedge}] + \Delta_{[k]}\mathfrak{I}_{\mathrm{s}}[G,H_{\mathfrak{e}}] \tag{6.1.32}$$

where:

$$\mathfrak{I}_{\mathrm{s};k}^{(\mathrm{bk})}[G,H] = \int_{a_N}^{b_N} G(\xi)\cdot \mathcal{W}_{\mathrm{bk};k}[H](\xi)\,\mathrm{d}\xi\ ,$$

$$\mathfrak{I}_{\mathrm{s};k}^{(\partial)}[G,H] = \frac{1}{N^{\alpha}}\int_{0}^{\overline{x}_N} G\big(b_N - N^{-\alpha}x\big)\cdot \mathcal{W}_{R;k}[H]\big(x, b_N - N^{-\alpha}x\big)\,\mathrm{d}x\ ,$$

$$\Delta_{[k]}\mathfrak{I}_{\mathrm{s}}[G,H_{\mathfrak{e}}] = \int_{a_N}^{b_N} G(\xi)\cdot \Delta_{[k]}\mathcal{W}_N[H_{\mathfrak{e}}](\xi)\,\mathrm{d}\xi\ . \tag{6.1.33}$$

Clearly from the estimate (6.1.31), there exist a constant $C' > 0$ such that:

$$\big|\Delta_{[k]}\mathfrak{I}_{\mathrm{s}}[G,H_{\mathfrak{e}}]\big| \leq C'\,N^{-k\alpha}\cdot ||G||_{L^{\infty}([a_N;b_N])}\cdot ||H_{\mathfrak{e}}^{(k+1)}||_{L^{\infty}(\mathbb{R})}\ . \tag{6.1.34}$$

The asymptotic expansion of $\mathfrak{I}_{\mathrm{s};k}^{(\mathrm{bk})}$ follows readily from the expression (5.1.40) for $\mathcal{W}_{\mathrm{bk};k}[H]$. It produces the first line of (6.1.28). As a consequence, it remains to focus on $\mathfrak{I}_{\mathrm{s};k}^{(\partial)}$. Recall from Proposition 5.1.11 the decomposition

$$\mathcal{W}_{R;k}[H]\big(x, b_N - N^{-\alpha}x\big) = \mathcal{W}_{R;k}^{(\mathrm{as})}[H](x) + \Delta_{[k]}\mathcal{W}_{R}^{(\mathrm{as})}[H](x) \tag{6.1.35}$$

and especially the bounds (5.1.103)–(5.1.105) on the remainder, which imply:

$$\big|\Delta_{[k]}\mathcal{W}_{R}^{(\mathrm{as})}[H](x)\big| \leq C\,\mathrm{e}^{-\varsigma x}\,x^{k+1}\,\ln x\cdot N^{-k\alpha}\cdot ||H_{\mathfrak{e}}||_{W_{k+1}^{\infty}(\mathbb{R})}\ . \tag{6.1.36}$$

The contribution of the first term of (6.1.35) involves the functions $\mathfrak{b}_{\ell}$. It remains to replace G by its Taylor series with integral remainder of appropriate order so as to get

$$\mathfrak{I}_{\mathrm{s};k}^{(\partial)}[G,H] = \sum_{p=0}^{k-1}\frac{1}{N^{(p+1)\alpha}}\sum_{\substack{s+\ell=p\\ s,\ell\geq 0}}\frac{(-1)^s}{s!}H^{(\ell+1)}(b_N)\cdot G^{(s)}(b_N)\int_{0}^{\overline{x}_N} x^s\,\mathfrak{b}_{\ell}(x)\,\mathrm{d}x$$
$$+\ \Delta_{[k]}\mathfrak{I}_{\mathrm{s}}^{(\partial)}[G,H] \tag{6.1.37}$$

where

$$\Delta_{[k]}\mathfrak{I}_{\mathrm{s}}^{(\partial)}[G,H] = \frac{1}{N^{k\alpha}}\sum_{\ell=0}^{k-1}\frac{H^{(\ell+1)}(b_N)}{(k-\ell-2)!}\int_0^{\overline{x}_N}\mathrm{d}x\,\mathfrak{b}_\ell(x)\,(-x)^{k-\ell-1}$$
$$\times\int_0^1 \mathrm{d}t\,(1-t)^{k-2-\ell}G^{(k-\ell-1)}(b_N-N^{-\alpha}tx) \qquad (6.1.38)$$
$$+\frac{1}{N^\alpha}\int_0^{\overline{x}_N} G(b_N-N^{-\alpha}x)\cdot\Delta_{[k]}\mathcal{W}_R^{(\mathrm{as})}[H](x)\,\mathrm{d}x\,. \qquad (6.1.39)$$

Clearly from (6.1.36), there exists $C''>0$ such that:

$$\big|\Delta_k\mathfrak{I}_{\mathrm{s};k}^{(\partial)}[G,H]\big| \le C''\,N^{-k\alpha}\,||H_{\mathfrak{e}}||_{W^{\infty}_{k+1}(\mathbb{R})}\cdot||G||_{W^{\infty}_{k-1}([a_N\,;b_N])}\,. \qquad (6.1.40)$$

Moreover, one can extend the integration in (6.1.37) from $[0\,;\overline{x}_N]$ up to $\mathbb{R}^+$, this for the price of exponentially small corrections in N. Adding up all the pieces leads to (6.1.28). ■

In the case when $G=1$, *i.e.* to estimate the magnitude of the total integral of $\mathcal{W}_N[H]$, we can obtain slightly better bounds, solely involving the sup norm.

Lemma 6.1.7 *There exists* $C>0$ *independent of* N *such that, for any* $H\in\mathcal{C}^1([a_N\,;b_N])$,

$$\Big|\int_{a_N}^{b_N}\mathcal{W}_N[H](\xi)\,\mathrm{d}\xi\Big| \le C\,||H_{\mathfrak{e}}||_{W_0^{\infty}(\mathbb{R})}\,. \qquad (6.1.41)$$

Proof Recall from Propositions 5.1.4 the decomposition:

$$\mathcal{W}_N[H](\xi) = \mathcal{W}_R[H_{\mathfrak{e}}](x_R,\xi) + \mathcal{W}_{\mathrm{bk}}[H_{\mathfrak{e}}](\xi)$$
$$-\mathcal{W}_R[H_{\mathfrak{e}}^{\wedge}](x_L,a_N+b_N-\xi) + \mathcal{W}_{\exp}[H](\xi)\,. \qquad (6.1.42)$$

We focus on the integral of each of the terms taken individually. We have:

$$\int_{a_N}^{b_N}\mathcal{W}_{\mathrm{bk}}[H_{\mathfrak{e}}](\xi)\,\mathrm{d}\xi = \frac{N^\alpha}{2\pi\beta}\int_{\mathbb{R}}\mathrm{d}y\,J(y)\int_0^{N^{-\alpha}y}\big[H_{\mathfrak{e}}(b_N+t) - H_{\mathfrak{e}}(a_N+t)\big]\mathrm{d}t\,, \qquad (6.1.43)$$

thus leading to

$$\Big|\int_{a_N}^{b_N}\mathcal{W}_{\mathrm{bk}}[H_{\mathfrak{e}}](\xi)\,\mathrm{d}\xi\Big| \le C\,||H_{\mathfrak{e}}||_{W_0^{\infty}(\mathbb{R})}\,. \qquad (6.1.44)$$

Next, we have:

$$\int_{a_N}^{b_N} \mathcal{W}_R[H_\mathfrak{e}](x_R,\xi)\,\mathrm{d}\xi = -\frac{N^\alpha}{2\pi\beta}\int_{\overline{x}_N}^{+\infty}\mathrm{d}y\,J(y)\int_{a_N}^{b_N}\mathrm{d}\xi\,\big[H_\mathfrak{e}(\xi+N^{-\alpha}y)-H_\mathfrak{e}(\xi)\big]$$
$$-\frac{N^\alpha}{2\pi\beta}\int_0^{\overline{x}_N}\mathrm{d}y\,J(y)\int_0^{N^{-\alpha}y}\big[H_\mathfrak{e}(b_N+t)-H_\mathfrak{e}(b_N-N^{-\alpha}y+t)\big]\mathrm{d}t$$
$$+\frac{N^\alpha}{2\mathrm{i}\pi\beta}\int_{\mathscr{C}_{\mathrm{reg}}^{(+)}}\frac{\mathrm{d}\lambda}{2\mathrm{i}\pi}\int_{\mathscr{C}_{\mathrm{reg}}^{(-)}}\frac{\mathrm{d}\mu}{2\mathrm{i}\pi}\,\frac{1}{(\mu-\lambda)R_\downarrow(\mu)R_\uparrow(\lambda)}$$
$$\times\left\{\frac{\mathrm{e}^{\mathrm{i}\lambda\overline{x}_N}-1}{\lambda}\int_{a_N}^{b_N}H_\mathfrak{e}(\eta)\mathrm{e}^{-\mathrm{i}\mu y_R}\mathrm{d}\eta+\frac{1}{\mu}\int_{a_N}^{b_N}H_\mathfrak{e}(\xi)\mathrm{e}^{\mathrm{i}\lambda x_R}\mathrm{d}\xi\right\}. \tag{6.1.45}$$

The exponential decay of J at $+\infty$ ensures that the first two lines of (6.1.45) are indeed bounded by $C\,||H_\mathfrak{e}||_{W_0^\infty(\mathbb{R})}$ for some N-independent $C>0$. The last line is bounded similarly by using

$$\forall\lambda\in\mathscr{C}_{\mathrm{reg}}^{(\pm)},\qquad \left|\int_{a_N}^{b_N}H_\mathfrak{e}(\xi)\mathrm{e}^{\pm\mathrm{i}\lambda N^\alpha(b_N-\xi)}\,\mathrm{d}\xi\right| \le \frac{C'\,||H_\mathfrak{e}||_{W_0^\infty(\mathbb{R})}}{|\lambda|N^\alpha}. \tag{6.1.46}$$

It thus solely remains to focus on the exponentially small term $\mathcal{W}_{\exp}[H]$. In fact, we only discuss the operator $\mathcal{W}_N^{(++)}$ as all others can be treated in a similar fashion. Thanks to the bound (4.2.15) for $\Pi(\lambda)-I_2$ and the expression (4.2.19) of the matrix Ψ in terms of Π, we have:

$$\Psi(\lambda) = \begin{pmatrix}1 & 0\\ 1/\lambda & 1\end{pmatrix} + \mathrm{O}\Big(\frac{\mathrm{e}^{-\varkappa_\epsilon N^\alpha}}{1+|\lambda|}\Big) \tag{6.1.47}$$

which is valid for λ uniformly away from the jump contour Σ_Ψ (see Figure 4.1). Therefore, using the definition (5.1.8) of $\mathcal{W}_N^{(++)}$:

$$\left|\int_{a_N}^{b_N}\mathcal{W}_N^{(++)}[H_\mathfrak{e}](\xi)\,\mathrm{d}\xi\right| \le C''\,||H_\mathfrak{e}||_{W_0^\infty(\mathbb{R})}\,\mathrm{e}^{-\varkappa_\epsilon N^\alpha}\int_{\mathscr{C}_{\mathrm{reg}}^{(+)}}\frac{|\mathrm{d}\lambda\mathrm{d}\mu|}{(2\pi)^2}\,\frac{1}{|\lambda-\mu|\,|R_\downarrow(\lambda)R_\downarrow(\mu)\,\lambda|}. \tag{6.1.48}$$

Adding up all the intermediate bounds readily leads to the claim. ■

By a slight modification of the method leading to Lemma 6.1.7, we can likewise control the $L^1([a_N\,;b_N])$ norm of $\mathcal{W}_N$ in terms of the W_1^{∞} norm of (an extension of) H.

Lemma 6.1.8 *For any $H \in \mathcal{C}^1([a_N\,;b_N])$ it holds*

$$||\mathcal{W}_N[H]||_{L^1([a_N\,;b_N])} \;\le\; C\,||H_{\mathfrak{e}}||_{W_1^{\infty}(\mathbb{R})}$$

and

$$||\mathcal{W}_{\exp}[H]||_{L^1([a_N\,;b_N])} \;\le\; C\,\mathrm{e}^{-C'N^{\alpha}}\,||H_{\mathfrak{e}}||_{W_1^{\infty}(\mathbb{R})}\,. \tag{6.1.49}$$

6.2 The Support of the Equilibrium Measure

In the present subsection we build on the previous analysis so as to prove the existence of the endpoints (a_N, b_N) of the support of the equilibrium measure and thus the fact that

$$\rho_{\mathrm{eq}}^{(N)}(\xi) \;=\; \mathbf{1}_{[a_N\,;b_N]}(\xi)\cdot\mathcal{W}_N[V'](\xi)\,\mathrm{d}\xi\;, \tag{6.2.1}$$

where $\mathcal{W}_N$ is as defined in (2.4.17).

Lemma 6.2.1 *There exists a unique sequence (a_N, b_N)—defining the support of the Lebesgue-continuous equilibrium measure which corresponds to the unique solution to the minimisation problem (2.4.8)–(2.4.9). The sequences a_N and b_N are bounded in N.*

Proof The existence and uniqueness of the solution to the minimisation problem (2.4.8)–(2.4.9) is obtained through a straightforward generalisation of the proof arising in the random matrix case, see *e.g.* [1].

The endpoint of the support of the equilibrium measure should be chosen in such a way that, on the one hand, the density of equilibrium measure admits at most a square root behaviour at the endpoints and, on the other hand, that it indeed defines a probability measure. In other words, the endpoints are to be chosen so that the two constraints are satisfied

$$\mathcal{X}_N[V'] \;=\; 0 \qquad\text{and}\qquad \mathfrak{I}_s[1, V'] = \int_{a_N}^{b_N} \mathcal{W}_N[V'](\xi)\,\mathrm{d}\xi \;=\; 1\,. \tag{6.2.2}$$

The asymptotic expansion of $\mathcal{X}_N[V']$ and $\mathfrak{I}_s[1, V']$ is given, respectively, in Lemma 6.1.2 and Proposition 6.1.6. However, the control on the remainder obtained there does depend on a_N and b_N. Should a_N or b_N be unbounded in N this could brake the *a priori* control on the remainder. Still, observe that if (a_N, b_N) solve the system of equations (6.2.2) then $\xi \mapsto \mathcal{W}_N[V'](\xi)$ with $\mathcal{W}_N$ associated with the support $[a_N\,;b_N]$ provides one with a solution to the minimisation problem of $\mathcal{E}_N$

defined in (2.4.7). By uniqueness of solutions to this minimisation problem, it thus corresponds to the density of equilibrium measure. As a consequence, there exists at most one solution (a_N, b_N) to the system of equations (6.2.2).

Assume that the sequence a_N and b_N are bounded in N. Then, the leading asymptotic expansion of the two functionals in (6.2.2) yields

$$\begin{cases} V'(b_N) + V'(a_N) = \quad \mathrm{O}(N^{-\alpha}) \\ V'(b_N) - V'(a_N) = u_1^{-1} + \mathrm{O}(N^{-\alpha}) \end{cases} \quad viz. \quad \begin{pmatrix} 1 & 1 \\ 1 & -1 \end{pmatrix} \cdot \begin{pmatrix} V'(b_N) - V'(b) \\ V'(a_N) - V'(a) \end{pmatrix} = \mathrm{O}(N^{-\alpha}) \,. \tag{6.2.3}$$

Note that the control on the remainder follows from the fact that $|a_N|$ and $|b_N|$ are bounded by an N-independent constant. Also, (a, b) appearing above corresponds to the unique solution to the system

$$V'(b) + V'(a) = 0 \quad \text{and} \quad V'(b) - V'(a) = u_1^{-1} \,. \tag{6.2.4}$$

We do stress that the existence and uniqueness of this solution is ensured by the strict convexity of V.

The smoothness of the remainder in (a_N, b_N) away from 0, the control on its magnitude (guaranteed by the boundedness of a_N and b_N) as well as the strict convexity of V and the invertibility of the matrix occurring in (6.2.3) ensure the existence of solutions (a_N, b_N) by the implicit function theorem, this provided that N is large enough. Hence, since a solution to (6.2.2) with a_N and b_N bounded in N does exists, by uniqueness of the solutions to (6.2.2), it is the one that defines the endpoints of the support of the equilibrium measure. ■

Corollary 6.2.2 *Let the pair (a, b) correspond to the unique solution to the system*

$$V'(b) + V'(a) = 0 \quad and \quad V'(b) - V'(a) = u_1^{-1} \,. \tag{6.2.5}$$

Then the endpoints (a_N, b_N) of the support of the equilibrium measure admit the asymptotic expansion

$$a_N = \sum_{\ell=0}^{k-1} \frac{a_{N:\ell}}{N^{\ell\alpha}} + \mathrm{O}\big(N^{-k\alpha}\big) \quad and \quad b_N = \sum_{\ell=0}^{k-1} \frac{b_{N:\ell}}{N^{\ell\alpha}} + \mathrm{O}\big(N^{-k\alpha}\big) \,, \tag{6.2.6}$$

where $a_{N:0} = a$ and $b_{N:0} = b$.

Note that the existence and uniqueness of solutions to the system (6.2.5) follows from the strict convexity of the potential V.

Proof The invertibility of the matrix occurring in (6.2.3) as well as the strict convexity of the potential V ensure that a_N and b_N admit the expansion (6.2.6) for $k = 1$, *viz.* up to $\mathrm{O}\big(N^{-\alpha}\big)$ corrections. Now suppose that this expansion holds up to $\mathrm{O}\big(N^{-(k-1)\alpha}\big)$. It follows from Lemma 6.1.2 and Proposition 6.1.6 that the asymptotic expansion of $\mathcal{X}_N[V']$ and $\mathfrak{I}_s[1, V']$ up to $\mathrm{O}\big(N^{-k\alpha}\big)$ can be recast as

$$\begin{pmatrix} \mathcal{X}_N[V'] \cdot \daleth_0^{-1} \\ \mathfrak{J}_s[1, V'] \cdot u_1^{-1} \end{pmatrix} = \begin{pmatrix} V'(b_N) + V'(a_N) \; + \; \mathcal{B}_{1;k-1}[V'] \; + \; \daleth_0^{-1} \cdot \Delta_{[k]}\mathcal{X}_N[V'] \\ V'(b_N) - V'(a_N) \; + \; \mathcal{B}_{2;k-1}[V'] + u_1^{-1} \cdot \Delta_{[k]}\mathfrak{J}_s[1, V'] \end{pmatrix}. \tag{6.2.7}$$

In this expression, we have $\big|\daleth_0^{-1} \cdot \Delta_{[k]}\mathcal{X}_N[V']\big| + \big|u_1^{-1} \cdot \Delta_{[k]}\mathfrak{J}_s[1, V']\big| \leq CN^{-k\alpha}$ since a_N and b_N are bounded uniformly in N, while

$$\begin{pmatrix} \mathcal{B}_{1;k-1}[V'] \\ \mathcal{B}_{2;k-1}[V'] \end{pmatrix} = \sum_{p=1}^{k-1} \frac{1}{N^{p\alpha}} \begin{pmatrix} \mathrm{i}^p \cdot \daleth_p \daleth_0^{-1} \cdot \big(V^{(p+1)}(a_N) + (-1)^p V^{(p+1)}(b_N)\big) \\ \big(u_{p+1} + \daleth_{0,p-1}\big)u_1^{-1} \cdot V^{(p+1)}(b_N) - \big(u_{p+1} + (-1)^p \daleth_{0,p-1}\big)u_1^{-1} \cdot V^{(p+1)}(a_N) \end{pmatrix}. \tag{6.2.8}$$

We remind that $\daleth_p$ was introduced in Definition 6.1.1, u_p in Definition 5.1.5, and $\daleth_{0,p}$ in Definition 6.1.5.

Since both $\mathcal{B}_{1;k-1}[V']$ and $\mathcal{B}_{2;k-1}[V']$ have $N^{-\alpha}$ as a prefactor, by composition of asymptotic expansions, there exist functions

$$B_{p;\ell}\big(b_{N;1}, \ldots, b_{N;\ell-1} \mid a_{N;1}, \ldots, a_{N;\ell-1}\big),$$

indexed by $p \in \{1, 2\}$ and $\ell \in [\![1 \,;\, k-1]\!]$, independent of k, such that

$$\begin{pmatrix} \mathcal{B}_{1;k-1}[V'] \\ \mathcal{B}_{2;k-1}[V'] \end{pmatrix} = \sum_{\ell=1}^{k-1} \frac{1}{N^{\ell\alpha}} \begin{pmatrix} B_{1;\ell}\big(b_{N;1}, \ldots, b_{N;\ell-1} \mid a_{N;1}, \ldots, a_{N;\ell-1}\big) \\ B_{2;\ell}\big(b_{N;1}, \ldots, b_{N;\ell-1} \mid a_{N;1}, \ldots, a_{N;\ell-1}\big) \end{pmatrix} + \mathrm{O}\big(N^{-\alpha k}\big). \tag{6.2.9}$$

As a consequence, we have the relation:

$$\begin{pmatrix} 1 & 1 \\ 1 & -1 \end{pmatrix} \begin{pmatrix} V'(b_N) - V'(b) \\ V'(a_N) - V'(a) \end{pmatrix} = \sum_{\ell=1}^{k-1} \frac{-1}{N^{\ell\alpha}} \begin{pmatrix} B_{1;\ell}\big(b_{N;1}, \ldots, b_{N;\ell-1} \mid a_{N;1}, \ldots, a_{N;\ell-1}\big) \\ B_{2;\ell}\big(b_{N;1}, \ldots, b_{N;\ell-1} \mid a_{N;1}, \ldots, a_{N;\ell-1}\big) \end{pmatrix} + \mathrm{O}\big(N^{-k\alpha}\big). \tag{6.2.10}$$

This implies the existence of an asymptotic expansion of a_N and b_N up to a remainder of the order $\mathrm{O}\big(N^{-k\alpha}\big)$. ■

6.3 Asymptotic Evaluation of the Double Integral

In this section we study the large-N asymptotic expansion for the double integral in:

Definition 6.3.1

$$\mathfrak{I}_{\mathrm{d}}[H,V] \;=\; \int_{a_N}^{b_N} \mathcal{W}_N \circ \widetilde{\mathcal{X}}_N\Big[\partial_{\xi}\big\{S\big(N^{\alpha}(\xi-*)\big)\cdot \mathcal{G}_N\big[\widetilde{\mathcal{X}}_N[H],V\big](\xi,*)\big\}\Big](\xi)\,\mathrm{d}\xi \;, \tag{6.3.1}$$

with

$$\mathcal{G}_N\big[H,V\big](\xi,\eta) \;=\; \frac{\mathcal{W}_N[H](\xi)}{\mathcal{W}_N[V'](\xi)} \;-\; \frac{\mathcal{W}_N[H](\eta)}{\mathcal{W}_N[V'](\eta)}\;. \tag{6.3.2}$$

We remind that $*$ indicates the variable on which the operator $\mathcal{W}_N$ acts. The asymptotic analysis of the double integral $\mathfrak{I}_{\mathrm{d}:\beta}$ arising in the $\beta \neq 1$ large-N asymptotics (*cf.* (3.4.4)) can be carried out within the setting of the method developed in this section. However, in order to keep the discussion minimal, we shall not present this calculation here.

In order to carry out the large-N asymptotic analysis of $\mathfrak{I}_{\mathrm{d}}[H,V]$, it is convenient to write down a decomposition for $\mathcal{G}_N\big[H,V\big]$ ensuing from the decomposition of $\mathcal{W}_N$ that has been described in Propositions 5.1.4 and 5.1.6. We omit the proof since it consists of straightforward algebraic manipulations.

Lemma 6.3.2 *The function $\mathcal{G}_N\big[H,V\big](\xi,\eta)$ can be recast as*

$$\begin{aligned}\mathcal{G}_N[H,V](\xi,\eta) \;&=\; \mathcal{G}_{\mathrm{bk};k}[H,V](\xi,\eta) \;+\; \mathcal{G}^{(\mathrm{as})}_{R;k}[H,V](x_R,y_R;\xi,\eta)\\ &-\; \mathcal{G}^{(\mathrm{as})}_{R;k}\big[H^{\wedge},V^{\wedge}\big](x_L,y_L;a_N+b_N-\xi,a_N+b_N-\eta) + \;\Delta_{[k]}\mathcal{G}_N\big[H,V\big](\xi,\eta)\;.\end{aligned} \tag{6.3.3}$$

The functions arising in this decomposition read

$$\mathcal{G}_{\mathrm{bk};k}[H,V](\xi,\eta) \;=\; \frac{\mathcal{W}_{\mathrm{bk};k}[H](\xi)}{\mathcal{W}_{\mathrm{bk};k}[V'](\xi)} \;-\; (\xi\leftrightarrow\eta)\;, \tag{6.3.4}$$

and

$$\begin{aligned}\mathcal{G}^{(\mathrm{as})}_{R;k}[H,V](x,y;\xi,\eta) \;=\; &\left\{\frac{\mathcal{W}^{(\mathrm{as})}_{R;k}[H](x)}{\mathcal{W}_{\mathrm{bk};k}[V'](\xi)} \;-\; \frac{\mathcal{W}^{(\mathrm{as})}_{R;k}[V'](x)}{\mathcal{W}_{\mathrm{bk};k}[V'](\xi)}\cdot\frac{(\mathcal{W}^{(\mathrm{as})}_{\mathrm{bk};k}+\mathcal{W}^{(\mathrm{as})}_{R;k})[H](x)}{(\mathcal{W}^{(\mathrm{as})}_{\mathrm{bk};k}+\mathcal{W}^{(\mathrm{as})}_{R;k})[V'](x)}\right\}\\ &-\;\begin{pmatrix}\xi\leftrightarrow\eta\\ x\leftrightarrow y\end{pmatrix}.\end{aligned} \tag{6.3.5}$$

Finally, the remainder $\Delta_{[k]}\mathcal{G}_N$ takes the form

$$\begin{aligned}&\Delta_{[k]}\mathcal{G}_N[H,V](\xi,\eta)\\ &=\; \frac{1}{\mathcal{W}_{\mathrm{bk};k}[V'](\xi)}\left\{\Delta_{[k]}\mathcal{W}_N[H](\xi) \;-\; \Delta_{[k]}\mathcal{W}_N[V'](\xi)\cdot\frac{\mathcal{W}_N[H](\xi)}{\mathcal{W}_N[V'](\xi)}\right\} \;-\; \big(\xi\leftrightarrow\eta\big)\\ &+\;\Delta_{[k]}\mathcal{G}^{(\mathrm{as})}_N[H,V](x_R,y_R;\xi,\eta)\\ &-\;\Delta_{[k]}\mathcal{G}^{(\mathrm{as})}_N\big[H^{\wedge},V^{\wedge}\big](x_L,y_L;a_N+b_N-\xi,a_N+b_N-\eta)\;.\end{aligned} \tag{6.3.6}$$

The reminder $\Delta_{[k]}\mathcal{W}_N$ of order k has been introduced in (5.1.69), while

$$\Delta_{[k]}\mathcal{G}_N^{(\mathrm{as})}[H,V](x,y;\xi,\eta)=\frac{1}{\mathcal{W}_{\mathrm{bk};k}[V'](\xi)}\Big\{\Delta_{[k]}\mathcal{W}_R^{(\mathrm{as})}[H](x)-\Delta_{[k]}\mathcal{W}_R^{(\mathrm{as})}[V'](x)\cdot\frac{\mathcal{W}_N[H](\xi)}{\mathcal{W}_N[V'](\xi)}$$
$$-\Big[\big(\Delta_{[k]}\mathcal{W}_N\big)_R[H](\xi)-\big(\Delta_{[k]}\mathcal{W}_N\big)_R[V'](\xi)\cdot\frac{\mathcal{W}_N[H](\xi)}{\mathcal{W}_N[V'](\xi)}\Big]\frac{\mathcal{W}_{R;k}^{(\mathrm{as})}[V'](x)}{(\mathcal{W}_{\mathrm{bk};k}^{(\mathrm{as})}+\mathcal{W}_{R;k}^{(\mathrm{as})})[V'](x)}\Big\}-\begin{pmatrix}\xi\leftrightarrow\eta\\ x\leftrightarrow y\end{pmatrix}. \tag{6.3.7}$$

The local right boundary remainder arising above is defined as

$$\big(\Delta_{[k]}\mathcal{W}_N\big)_R \;=\; \mathcal{W}_N \;-\; \mathcal{W}_{R;k}^{(\mathrm{as})} \;-\; \mathcal{W}_{\mathrm{bk};k}^{(\mathrm{as})}\,. \tag{6.3.8}$$

Note that the two terms $\mathcal{G}_{R;k}^{(\mathrm{as})}$ present in (6.3.3) correspond to the parts of $\mathcal{G}_N$ that localise at the right and left boundary. The way in which they appear is reminiscent of the symmetry satisfied by $\mathcal{W}_N$:

$$\mathcal{W}_N[H](a_N+b_N-\xi) \;=\; -\mathcal{W}_N[H^{\wedge}](\xi)\,. \tag{6.3.9}$$

Lemma 6.3.3 *The double integral $\mathfrak{J}_{\mathrm{d}}[H,V]$ can be recast as*

$$\begin{aligned}\mathfrak{J}_{\mathrm{d}}[H,V] \;=\;& \mathfrak{J}_{\mathrm{d;bk}}\Big[\mathcal{G}_{\mathrm{bk};k}[H,V]\Big] \;+\; \mathfrak{J}_{\mathrm{d;bk}}\Big[\mathcal{G}_{R;k}^{(\mathrm{as})}[H,V] \;+\; \mathcal{G}_{R;k}^{(\mathrm{as})}[H^{\wedge},V^{\wedge}]\Big]\\ &+\; \mathfrak{J}_{\mathrm{d};R}\Big[\big(\mathcal{G}_{\mathrm{bk};k}+\mathcal{G}_{R;k}^{(\mathrm{as})}\big)[H,V] \;+\; \big(\mathcal{G}_{\mathrm{bk};k}+\mathcal{G}_{R;k}^{(\mathrm{as})}\big)[H^{\wedge},V^{\wedge}]\Big]\\ &+\; \Delta_{[k]}\mathfrak{J}_{\mathrm{d}}\Big[\widetilde{\mathcal{X}}_N[H],V\Big]\,.\end{aligned} \tag{6.3.10}$$

The bulk part of the double integral is described by the functional

$$\mathfrak{J}_{\mathrm{d;bk}}[F] \;=\; \frac{-N^{2\alpha}}{4\pi\beta}\int\limits_{[a_N\,;b_N]^2} J\big(N^{\alpha}(\xi-\eta)\big)\cdot\big(\partial_{\xi}\,-\,\partial_{\eta}\big)\big\{S\big(N^{\alpha}(\xi-\eta)\big)F(\xi,\eta)\big\}\,\mathrm{d}\xi\mathrm{d}\eta\,. \tag{6.3.11}$$

The local (right) part of the double integral is represented as

$$\mathfrak{J}_{\mathrm{d};R}[F] \;=\; -\frac{N^{2\alpha}}{2\pi\beta}\int\limits_{\mathscr{C}_{\mathrm{reg}}^{(+)}}\frac{\mathrm{d}\lambda}{2\mathrm{i}\pi}\int\limits_{\mathscr{C}_{\mathrm{reg}}^{(-)}}\frac{\mathrm{d}\mu}{2\mathrm{i}\pi}\int\limits_{a_N}^{b_N}\frac{\mathrm{d}\xi\;\mathrm{e}^{\mathrm{i}\lambda N^{a}(b_N-\xi)}}{(\mu-\lambda)R_{\downarrow}(\lambda)R_{\uparrow}(\mu)}$$
$$\times\int\limits_{a_N}^{b_N}\mathrm{d}\eta\,\mathrm{e}^{-\mathrm{i}\mu N^{\alpha}(b_N-\eta)}\,\partial_{\xi}\big\{S\big(N^{\alpha}(\xi-\eta)\big)F(\xi,\eta)\big\}\,. \tag{6.3.12}$$

Finally, $\Delta_{[k]}\mathfrak{I}_{\mathrm{d}}$ *represents the remainder which decomposes as*

$$\Delta_{[k]}\mathfrak{I}_{\mathrm{d}}[H,V] = \sum_{p=1}^{4} \Delta_{[k]}\mathfrak{I}_{\mathrm{d};p}[H,V] \tag{6.3.13}$$

$$\Delta_{[k]}\mathfrak{I}_{\mathrm{d};1}[H,V] = \int_{a_N}^{b_N} \mathcal{W}_{\mathrm{exp}}\Big[\partial_\xi\big\{S\big(N^\alpha(\xi-*)\big)\cdot\big(\mathcal{G}_N-\Delta_{[k]}\mathcal{G}_N\big)\big[H,V\big](\xi,*)\big\}\Big](\xi)\,\mathrm{d}\xi \tag{6.3.14}$$

$$\Delta_{[k]}\mathfrak{I}_{\mathrm{d};2}[H,V] = \int_{a_N}^{b_N} \mathcal{W}_N\Big[\partial_\xi\big\{S\big(N^\alpha(\xi-*)\big)\cdot\Delta_{[k]}\mathcal{G}_N\big[H,V\big](\xi,*)\big\}\Big](\xi)\,\mathrm{d}\xi \tag{6.3.15}$$

$$\Delta_{[k]}\mathfrak{I}_{\mathrm{d};3}[H,V] = -\int_{a_N}^{b_N} \mathcal{W}_N[1](\xi)\cdot\mathcal{X}_N\Big[\partial_\xi\big\{S\big(N^\alpha(\xi-*)\big)\cdot\mathcal{G}_N\big[H,V\big](\xi,*)\big\}\Big](\xi)\,\mathrm{d}\xi\,. \tag{6.3.16}$$

$$\Delta_{[k]}\mathfrak{I}_{\mathrm{d};4}[H,V] = -\mathfrak{I}_{\mathrm{d};R}\Big[\big(\mathcal{G}^{(\mathrm{as})}_{R;k}[H,V]\big)^{\wedge} + \big(\mathcal{G}^{(\mathrm{as})}_{R;k}[H^{\wedge},V^{\wedge}]\big)^{\wedge}\Big] \tag{6.3.17}$$

where $\mathcal{W}_{\mathrm{exp}}$ *is as defined in (5.1.32).*

Proof We first invoke the Definition 3.3.5 of the operator $\widetilde{\mathcal{X}}_N$ so as to recast $\mathfrak{I}_{\mathrm{d}}[H,V]$ as an integral involving solely $\mathcal{W}_N$, and another one containing the action of $\mathcal{X}_N$. Then, in the first integral, we decompose the operator $\mathcal{W}_N$ arising in the "exterior" part of the double integral $\mathfrak{I}_{\mathrm{d}}[H,V]$ as $\mathcal{W}_N = (\mathcal{W}_R^{(0)} + \mathcal{W}_{\mathrm{bk}}^{(0)} + \mathcal{W}_L^{(0)} + \mathcal{W}_{\mathrm{exp}})$, *cf.* (5.1.32). Then, it remains to observe that

$$\int_{a_N}^{b_N} \mathcal{W}_L^{(0)}\Big[\partial_\xi\big\{S\big(N^\alpha(\xi-*)\big)\cdot\mathcal{G}_N\big[H,V\big](\xi,*)\big\}\Big](x_L)\,\mathrm{d}\xi$$

$$= \int_{a_N}^{b_N} \mathcal{W}_R^{(0)}\Big[\partial_\xi\big\{S\big(N^\alpha(\xi-*)\big)\cdot\mathcal{G}_N\big[H^{\wedge},V^{\wedge}\big](\xi,*)\big\}\Big](x_R)\,\mathrm{d}\xi \tag{6.3.18}$$

and that

$$-\,\mathfrak{I}_{\mathrm{d;bk}}\Big[\big(\mathcal{G}^{(\mathrm{as})}_{R;k}[H^{\wedge},V^{\wedge}]\big)^{\wedge}\Big] = \mathfrak{I}_{\mathrm{d;bk}}\Big[\mathcal{G}^{(\mathrm{as})}_{R;k}[H^{\wedge},V^{\wedge}]\Big]\,. \tag{6.3.19}$$

Putting all these results together, and using that the functions $\mathcal{G}_{\mathrm{bk};k}[H,V]$ and $\mathcal{G}^{(\mathrm{as})}_{R;k}[H,V]$ solely involve derivatives of H, implies:

$$\mathcal{G}_{\mathrm{bk};k}\big[\widetilde{\mathcal{X}}_N[H], V\big] = \mathcal{G}_{\mathrm{bk};k}[H, V] \quad \text{and} \quad \mathcal{G}^{(\mathrm{as})}_{R;k}\big[\widetilde{\mathcal{X}}_N[H], V\big] = \mathcal{G}^{(\mathrm{as})}_{R;k}[H, V] \,, \tag{6.3.20}$$

we obtain the desired decomposition of the double integral. ■

6.3.1 The Asymptotic Expansion Related to $\mathfrak{J}_{\mathrm{d;bk}}$ and $\mathfrak{J}_{\mathrm{d};R}$

Once again, we need to introduce new constants:

Definition 6.3.4 If $\ell \geq 0$ is an integer, we set:

$$\beth_{2\ell} = \int_{\mathbb{R}} \frac{J(u)\, u^{2\ell}}{4\pi\beta\,(2\ell)!}\big[uS'(u) + S(u)\big]\mathrm{d}u \quad \textit{and} \quad \beth_{2\ell+1} = \int_{\mathbb{R}} \frac{J(u)\, S(u)\, u^{2(\ell+1)}}{4\pi\beta\,(2\ell+1)!}\mathrm{d}u \tag{6.3.21}$$

where the function J comes from Definition 5.1.3 and S is the kernel of $\mathcal{S}_N$ and appears lately in (5.2.2).

They are useful in the following:

Lemma 6.3.5 *Assume* $F \in \mathcal{C}^{2k+2}([a_N\,;b_N]^2)$ *and antisymmetric viz.* $F(\xi,\eta) = -F(\eta,\xi)$. *We have the asymptotic expansion:*

$$\mathfrak{J}_{\mathrm{d;bk}}[F] = -N^{\alpha}\cdot\beth_0\cdot\mathcal{T}_{\mathrm{even}}[F](0) - \sum_{\ell=1}^{k}\frac{1}{N^{(2\ell-1)\alpha}}\Big\{\beth_{2\ell}\cdot\mathcal{T}^{(2\ell)}_{\mathrm{even}}[F](0) + \beth_{2\ell-1}\cdot\mathcal{T}^{(2\ell-1)}_{\mathrm{odd}}[F](0)\Big\} + \mathrm{O}\big(N^{-2k\alpha}\big) \tag{6.3.22}$$

in terms of the integral transforms:

$$\mathcal{T}_{\mathrm{even}}[F](s) = \frac{1}{s}\int_{2a_N-|s|}^{2b_N-|s|} F\big[(v+s)/2, (v-s)/2\big]\,\mathrm{d}v$$

$$\text{and} \quad \mathcal{T}_{\mathrm{odd}}[F](s) = \int_{2a_N-|s|}^{2b_N-|s|} \partial_s\big\{s^{-1}\,F\big[(v+s)/2, (v-s)/2\big]\big\}\,\mathrm{d}v\,. \tag{6.3.23}$$

The integral transforms $\mathcal{T}_{\mathrm{even}}$, $\mathcal{T}_{\mathrm{odd}}$ can be slightly simplified in the case of specific examples of the function F. In particular, if F takes the form $F(\xi,\eta) = g(\xi) - g(\eta)$ for some sufficiently regular function g, then we have:

$$\mathcal{T}_{\text{even}}[F](0) = \int_{2a_N}^{2b_N} g'(v/2)\,\mathrm{d}v = 2\big[g(b_N) - g(a_N)\big]\,. \qquad (6.3.24)$$

Proof We first implement the change of variables

$$\begin{cases} u = N^{\alpha}(\xi - \eta) \\ v = \xi + \eta \end{cases} \quad i.e. \quad \begin{cases} \xi = (v + N^{-\alpha}u)/2 \\ \eta = (v - N^{-\alpha}u)/2 \end{cases} \qquad (6.3.25)$$

in the integral representation for $\mathfrak{I}_{\text{d;bk}}[F]$. This recasts the integral as

$$\begin{aligned} \mathfrak{I}_{\text{d;bk}}\big[F\big] &= -\frac{N^{2\alpha}}{4\pi\beta} \int_{-\overline{x}_N}^{\overline{x}_N} du\, J(u) \int_{2a_N+|u|N^{-\alpha}}^{2b_N-|u|N^{-\alpha}} \partial_u\Big\{S(u)\cdot F\Big[\frac{v+uN^{-\alpha}}{2}, \frac{v-uN^{-\alpha}}{2}\Big]\Big\}\,\mathrm{d}v \\ &= -\frac{N^{\alpha}}{4\pi\beta}\int_{-\overline{x}_N}^{\overline{x}_N} \bigg(J(u)\big[uS'(u) + S(u)\big]\mathcal{T}_{\text{even}}[F](uN^{-\alpha}) \\ &\qquad + N^{-\alpha} J(u)\cdot uS(u)\cdot \mathcal{T}_{\text{odd}}[F](uN^{-\alpha})\bigg)\mathrm{d}u \end{aligned} \qquad (6.3.26)$$

Both $J(u)\cdot uS(u)$ and $J(u)\big[uS'(u) + S(u)\big]$ decay exponentially fast at infinity. Hence, the expansion (6.3.22) readily follows by using the Taylor expansion with integral remainder for the functions $\mathcal{T}_{\text{even/odd}}[F](uN^{-\alpha})$ around $u = 0$, and the parity properties of $\mathcal{T}_{\text{even/odd}}[F]$. ∎

Lemma 6.3.6 *Let $F(x, y; \xi, \eta)$ be such that*

- $F(x, y; \xi, \eta) = -F(y, x; \eta, \xi)$;
- *the map* $(x, y; \xi, \eta) \mapsto F(x, y; \xi, \eta)$ *is* $C^3\big(\mathbb{R}^+ \times \mathbb{R}^+ \times [a_N ; b_N]^2\big)$;
- *F—and any combination of partial derivatives of total order at most 3—decays exponentially fast in x, y uniformly in* $(\xi, \eta) \in [a_N ; b_N]$, *viz.*

$$\max\Big\{\big|\partial_x^{p_1}\partial_y^{p_2}\partial_\xi^{p_3}\partial_\eta^{p_4} F(x, y; \xi, \eta)\big| \; : \; \sum_{a=1}^{4} p_a \le 3\Big\} \le C\,\mathrm{e}^{-c\min(x,y)}\,. \qquad (6.3.27)$$

- *the following asymptotic expansion holds uniformly in* $(x, y) \in [0 ; \epsilon N^{\alpha}]$, *for some* $\epsilon > 0$ *and with a differentiable remainder in the sense of* (6.3.27).

$$F\big(x, y; b_N - N^{-\alpha}x, b_N - N^{-a}\,y\big) = \sum_{\ell=1}^{k} \frac{f_\ell(x, y)}{N^{\ell\alpha}} + \mathrm{O}\Big(\frac{C_k\,\mathrm{e}^{-c\min(x,y)}}{N^{(k+1)\alpha}}\Big), \qquad (6.3.28)$$

where $f_\ell \in \mathcal{C}^3(\mathbb{R}^+ \times \mathbb{R}^+)$ *for* $\ell \in [\![1 ; k]\!]$ *while*

$$\max\Big\{ \big|\partial_x^p \partial_y^q f_\ell(x,y)\big| \;:\; p+q \le 3 \quad \text{and} \quad \ell \in [\![1 ; k]\!] \Big\} \;\le\; C_k \, \mathrm{e}^{-c \min(x,y)} \,. \tag{6.3.29}$$

Then, denoting $F_N(\xi,\eta) = F\big(N^\alpha(b_N-\xi), N^\alpha(b_N-\eta); \xi,\eta\big)$, *we have an asymptotic expansion:*

$$\mathfrak{I}_{\mathrm{d;bk}}[F_N] = -\sum_{\ell=1}^{k} \frac{1}{N^{(\ell-1)\alpha}} \int_{\mathbb{R}} \frac{\mathrm{d}u \, J(u)}{4\pi\beta} \int_{|u|}^{+\infty} \mathrm{d}v \, \partial_u \Big\{ S(u) \cdot f_\ell\big[(v-u)/2, (v+u)/2\big] \Big\} + \mathrm{O}\Big(\frac{1}{N^{k\alpha}}\Big) \,. \tag{6.3.30}$$

Note that, necessarily, f_ℓ are antisymmetric functions of (x,y).

Proof The change of variables

$$\begin{cases} u = N^\alpha(\xi-\eta) \\ v = N^\alpha\big(2b_N - \xi - \eta\big) \end{cases} \qquad i.e. \qquad \begin{cases} \xi = b_N - N^{-\alpha}(v-u)/2 \\ \eta = b_N - N^{-\alpha}(v+u)/2 \end{cases} \tag{6.3.31}$$

recasts the integral as

$$\mathfrak{I}_{\mathrm{d;bk}}[F] = -N^\alpha \int_{-\overline{x}_N}^{\overline{x}_N} \frac{\mathrm{d}u \, J(u)}{4\pi\beta} \int_{|u|}^{2\overline{x}_N-|u|} \mathrm{d}v \, \partial_u \Big\{ S(u) \cdot F\Big[\frac{v-u}{2}, \frac{v+u}{2}; b_N - \frac{v-u}{2N^\alpha}, b_N - \frac{v+u}{2N^\alpha}\Big]\Big\} \,. \tag{6.3.32}$$

At this stage, we can limit all the domains of integration to $|u|, |v| \le \epsilon N^\alpha$, this for the price of exponentially small corrections. Then, we insert the asymptotic expansion (6.3.28) and extend the domains of integration up to $+\infty$ this, again, for the price of exponentially small corrections, and we get the claim. ■

Very similarly, but under slightly different assumptions on the function F, we have the large-N asymptotic expansion of the right edge double integral.

Lemma 6.3.7 *Let* $F(x,y;\xi,\eta)$ *be such that*

- $F(x,y;\xi,\eta) = -F(y,x;\eta,\xi)$;
- *the map* $(x,y;\xi,\eta) \mapsto F(x,y;\xi,\eta)$ *is* $\mathcal{C}^3\big(\mathbb{R}^+ \times \mathbb{R}^+ \times [a_N ; b_N]^2\big)$;
- F *decays exponentially fast in* x, y *this uniformly in* $(\xi,\eta) \in [a_N ; b_N]$ *and for any combination of partial derivatives of total order at most 3, viz.:*

$$\max\Big\{ \big|\partial_x^{p_1}\partial_y^{p_2}\partial_\xi^{p_3}\partial_\eta^{p_4} F(x,y;\xi,\eta)\big| \;:\; \sum_{a=1}^{4} p_a \le 3 \Big\} \;\le\; C\, \mathrm{e}^{-c\min(x,y)} \,. \tag{6.3.33}$$

- *the following asymptotic expansion holds uniformly in* $(x, y) \in [0 ; \epsilon N^{\alpha}]$, *for some* $\epsilon > 0$ *and with a differentiable remainder in the sense of* (6.3.33).

$$F\big(x, y; b_N - N^{-\alpha}x, b_N - N^{-\alpha}y\big) = \sum_{\ell=1}^{k} \frac{f_\ell(x, y)}{N^{\ell\alpha}} + \mathrm{O}\bigg(\frac{C_k\,(x^k + y^k + 1)}{N^{(k+1)\alpha}}\bigg) , \tag{6.3.34}$$

where $f_\ell \in \mathcal{C}^3(\mathbb{R}^+ \times \mathbb{R}^+)$ *for* $\ell \in [\![1 ; k]\!]$ *while*

$$\max\Big\{ \big|\partial_x^p \partial_y^q f_\ell(x, y)\big| \ : \ p+q \le 3 \ \ and \ \ \ell \in [\![1 ; k]\!] \Big\} \ \le \ C_k\,(x^k + y^k + 1) \, . \tag{6.3.35}$$

Then, we have the following asymptotic expansions

$$\mathfrak{I}_{\mathrm{d};R}[F_N] = \sum_{\ell=1}^{k} \frac{1}{N^{(\ell-1)\alpha}} \int\limits_{\mathscr{C}^{(+)}_{\mathrm{reg}}} \frac{\mathrm{d}\lambda}{2\mathrm{i}\pi} \int\limits_{\mathscr{C}^{(-)}_{\mathrm{reg}}} \frac{\mathrm{d}\mu}{2\mathrm{i}\pi} \frac{(2\pi\beta)^{-1}}{(\lambda-\mu)R_{\downarrow}(\lambda)R_{\uparrow}(\mu)}$$
$$\times \int\limits_{0}^{+\infty} \mathrm{e}^{\mathrm{i}\lambda x - \mathrm{i}\mu y}\, \partial_x\big\{S(x-y)\cdot f_\ell(x, y)\big\} \mathrm{d}x\,\mathrm{d}y \ + \ \mathrm{O}\bigg(\frac{1}{N^{\alpha k}}\bigg) . \tag{6.3.36}$$

The function F_N *occurring above is as defined in the previous Lemma.*

Proof The change of variables $x = N^{\alpha}(b_N - \xi)$ and $y = N^{\alpha}(b_N - \eta)$ recasts the integral in the form

$$\mathfrak{I}_{\mathrm{d};R}\big[F\big] = N^{\alpha} \int\limits_{\mathscr{C}^{(+)}_{\mathrm{reg}}} \frac{\mathrm{d}\lambda}{2\mathrm{i}\pi} \int\limits_{\mathscr{C}^{(-)}_{\mathrm{reg}}} \frac{\mathrm{d}\mu}{2\mathrm{i}\pi} \frac{(2\pi\beta)^{-1}}{(\lambda-\mu)R_{\downarrow}(\lambda)R_{\uparrow}(\mu)}$$
$$\times \int\limits_{0}^{\overline{x}_N} \mathrm{e}^{\mathrm{i}\lambda x - \mathrm{i}\mu y}\, \partial_x\Big\{S(x-y)\cdot F\big(x, y; b_N - N^{-\alpha}x, b_N - N^{-\alpha}y\big)\Big\}\mathrm{d}x\mathrm{d}y \, .$$

We can then conclude exactly as in the proof of Lemma 6.3.6. ■

6.3.2 Estimation of the Remainder $\Delta_{[k]}\mathfrak{I}_{\mathrm{d}}[H, V]$

Lemma 6.3.8 *Let* $k \ge 1$ *be an integer. Given* $C_V > 0$, *assume* V *strictly convex, smooth enough and* $||V_{\mathfrak{e}}||_{W_3^{\infty}(\mathbb{R})} < C_V$. *There exists* $C > 0$ *such that, for any* $H \in \mathfrak{X}_s(\mathbb{R})$ *smooth enough, the remainder integral* $\Delta_{[k]}\mathfrak{I}_{\mathrm{d}}[H, V]$ *satisfies:*

$$\big|\Delta_{[k]}\mathfrak{I}_{\mathrm{d}}[H, V]\big| \ \le \ C\,N^{-(k-5)\alpha} \cdot \mathfrak{n}_{k+4}[V_{\mathfrak{e}}] \cdot ||H_{\mathfrak{e}}||_{W^{\infty}_{\max\{k,5\}+4}(\mathbb{R})} \, . \tag{6.3.37}$$

Proof It follows from Lemma 5.1.10 and 5.1.11, as well as $V''(b_N) \neq 0$ by strict convexity, that:

$$x \mapsto \frac{\left(\mathcal{W}_{\mathrm{bk};k}^{(\mathrm{as})} + \mathcal{W}_{R;k}^{(\mathrm{as})}\right)[H](x)}{\left(\mathcal{W}_{\mathrm{bk};k}^{(\mathrm{as})} + \mathcal{W}_{R;k}^{(\mathrm{as})}\right)[V'](x)} \tag{6.3.38}$$

is smooth at $x = 0$. As a consequence, the function

$$\begin{aligned}(\xi, \eta) &\mapsto \mathcal{G}_{\mathrm{bk};k}\big[H, V\big](\xi, \eta) + \mathcal{G}_{R;k}^{(\mathrm{as})}\big[H, V\big](x_R, y_R; \xi, \eta) \\ &= \frac{\Delta_{[k]}\mathcal{W}_{\mathrm{bk}}^{(\mathrm{as})}\big[H\big](\xi)}{\mathcal{W}_{\mathrm{bk};k}\big[V'\big](\xi)} + \frac{\mathcal{W}_{\mathrm{bk};k}^{(\mathrm{as})}\big[V'\big](x_R)}{\mathcal{W}_{\mathrm{bk};k}\big[V'\big](\xi)} \cdot \frac{\left(\mathcal{W}_{\mathrm{bk};k}^{(\mathrm{as})} + \mathcal{W}_{R;k}^{(\mathrm{as})}\right)[H](x_R)}{\left(\mathcal{W}_{\mathrm{bk};k}^{(\mathrm{as})} + \mathcal{W}_{R;k}^{(\mathrm{as})}\right)[V'](x_R)} - \big(\xi \leftrightarrow \eta\big)\end{aligned} \tag{6.3.39}$$

is smooth in (ξ, η). Furthermore, it follows from Theorem 5.2.1 that $(\xi, \eta) \mapsto \mathcal{G}_N[H, V](\xi, \eta)$ is smooth on $[a_N\,; b_N]$ as well. Since $\mathcal{G}_{R;k}^{(\mathrm{as})}\big[H, V\big](x_R, y_R; \xi, \eta)$ is smooth in ξ—respectively η—as soon as the latter variable is away from b_N or a_N, it follows that $(\xi, \eta) \mapsto \Delta_{[k]}\mathcal{G}_N[H, V](\xi, \eta)$ is smooth as well.

The remainder $\Delta_{[k]}\mathcal{G}_N$ described in (5.2.36) involves the remainders $\Delta_{[k]}\mathcal{W}_{R/\mathrm{bk}}^{(\mathrm{as})}$ studied in Lemma 5.1.11, and $(\Delta_{[k]}\mathcal{W}_N)_R$ defined in (6.3.8) and for which Proposition 5.1.6 and Lemma 5.1.11 also provide estimates. Using the properties of the $\mathfrak{a}$s obtained in Lemma 5.1.10 and involved in the asymptotic expansion of the quantities of interest, it shows the existence of constants $c_{\ell;k}^{(0)}, c_{\ell;k}^{(1/2)}$ and of functions $f_{m;k} \in W_m^{\infty}\big(\mathbb{R}^+\big)$ bounded uniformly in N and satisfying $f_{m;k}(x) = \mathrm{O}\big(x^{m+1/2}\big)$ such that

$$\Delta_{[k]}\mathcal{G}_N[H, V](\xi, \eta) = \frac{1}{N^{k\alpha}} \sum_{\ell=0}^{m} \left(c_{\ell;k}^{(0)} x_R^{\ell} + c_{\ell;k}^{(1/2)} x_R^{\ell-1/2}\right) + \frac{f_{m;k}(x_R)}{N^{k\alpha}} - \big(x_R \leftrightarrow y_R\big)\,, \tag{6.3.40}$$

for $(x_R, y_R) \in [0\,; \epsilon]^2$. Since $\Delta_{[k]}\mathcal{G}_N[H, V](\xi, \eta)$ is smooth, we necessarily have that $c_{\ell;k}^{(1/2)} = 0$ for $\ell \in [\![\, 0\,;\, m\,]\!]$. The representation (6.3.40) thus ensures that

$$\max_{0 \leq \ell + p \leq n} \max_{\substack{(x_R, y_R) \\ \in [0\,;\epsilon]^2}} \left|\partial_\xi^\ell \partial_\eta^p \Delta_{[k]}\mathcal{G}_N[H, V](\xi, \eta)\right| \leq \frac{C_n}{N^{(k-n)\alpha}} \cdot \mathfrak{n}_{n+k}[V] \cdot ||H_{\mathfrak{e}}||_{W_{n+1+k}^{\infty}(\mathbb{R})}\,. \tag{6.3.41}$$

Here, the explicit control on the dependence of the bound on V and H issues from the control on the remainders entering in the expression for $\Delta_{[k]}\mathcal{G}_N[H, V]$.

Similar types of bounds can, of course, be obtained for $(x_L, y_L) \in [0\,; \epsilon]^2$. Finally, as soon as a variable, be it ξ or η, is uniformly (in N) away from an immediate neighbourhood of the endpoints a_N and b_N, we can use more crude expressions for the remainders so as to bound derivatives of the remainder $\Delta_{[k]}\mathcal{G}_N[H, V]$. This does not spoil (6.3.41) and we conclude:

$$\max_{0\leq \ell+p\leq n} \max_{\substack{(\xi,\eta) \\ \in [a_N\,;b_N]^2}} \big|\partial_\xi^\ell \partial_\eta^p \Delta_{[k]}\mathcal{G}_N[H, V](\xi,\eta)\big| \leq \frac{C_n}{N^{(k-n)\alpha}} \cdot \mathfrak{n}_{n+k}[V] \cdot ||H_{\mathfrak{e}}||_{W^\infty_{n+1+k}(\mathbb{R})} . \tag{6.3.42}$$

Having at disposal such a control on the remainder $\Delta_{[k]}\mathcal{G}_N[H, V]$, we are in position to bound the double integral of interest. The latter decomposes into a sum of four terms

$$\Delta_{[k]}\mathfrak{I}_{\mathrm{d}}[H, V] = \sum_{p=1}^{4} \Delta_{[k]}\mathfrak{I}_{\mathrm{d};p}[H, V] \tag{6.3.43}$$

that have been defined in (6.3.14)–(6.3.17).

6.3.2.1 Bounding $\Delta_{[k]}\mathfrak{I}_{\mathrm{d};1}[H, V]$

Let

$$\begin{aligned} \tau(\xi,\eta) &= \partial_\xi\Big\{S\big(N^\alpha(\xi-\eta)\big)\cdot \mathcal{G}_N\big[H, V\big](\xi,\eta)\Big\} \\ \text{and}\quad \Delta_{[k]}\tau(\xi,\eta) &= \partial_\xi\Big\{S\big(N^\alpha(\xi-\eta)\big)\cdot \Delta_{[k]}\mathcal{G}_N\big[H, V\big](\xi,\eta)\Big\} . \end{aligned} \tag{6.3.44}$$

Observe that given $(\xi,\eta)\mapsto f(\xi,\eta)$ sufficiently regular, we have the decomposition:

$$\int_{a_N}^{b_N} \mathcal{W}_{\mathrm{exp}}\big[f(\xi,*)\big](\xi)\,\mathrm{d}\xi = \int_{a_N}^{b_N} \mathcal{W}_{\mathrm{exp}}\big[f(a_N,*)\big](\xi)\,\mathrm{d}\xi + \int_{a_N}^{b_N}\mathrm{d}\xi \int_{a_N}^{\xi}\mathrm{d}\eta\, \mathcal{W}_{\mathrm{exp}}\big[\partial_\eta f(\eta,*)\big](\xi) . \tag{6.3.45}$$

The latter ensures that

$$\begin{aligned} \Big|\int_{a_N}^{b_N} \mathcal{W}_{\mathrm{exp}}\big[f(\xi,*)\big](\xi)\,\mathrm{d}\xi\Big| &\leq \big|\big|\mathcal{W}_{\mathrm{exp}}\big[f(a_N,*)\big]\big|\big|_{L^1([a_N\,;b_N])} \\ &\quad + (b_N-a_N) \sup_{\eta\in[a_N\,;b_N]} \big|\big|\mathcal{W}_{\mathrm{exp}}\big[\partial_\eta f(\eta,*)\big]\big|\big|_{L^1([a_N\,;b_N])} . \end{aligned} \tag{6.3.46}$$

The two terms can be estimated directly using the L^1 bound (6.1.49) obtained in Lemma 6.1.8. For the first one:

$$\big|\big|\mathcal{W}_{\text{exp}}\big[f(a_N,*)\big]\big|\big|_{L^1([a_N;b_N])} \leq C_1 e^{-C_2 N^{\alpha}} ||f_{\mathfrak{e}}(a_N,*)||_{W_1^{\infty}(\mathbb{R})} \leq C_1 e^{-C_2 N^{\alpha}} ||f||_{W_1^{\infty}(\mathbb{R}^2)} \tag{6.3.47}$$

for some $C_1, C_2 > 0$ independent of N and f, and likewise for the second term. But the $W_p^{\infty}(\mathbb{R}^2)$ norm of $f_{\mathfrak{e}}$ is also bounded by a constant times the $W_p^{\infty}([a_N;b_N]^2)$ norm of f, and we can make the constant depends only on the compact support of the extension $f_{\mathfrak{e}}$. Therefore:

$$\big|\big|\mathcal{W}_{\text{exp}}\big[f(a_N,*)\big]\big|\big|_{L^1([a_N;b_N])} \leq C_1' e^{-C_2' N^{\alpha}} ||f||_{W_1^{\infty}([a_N;b_N]^2)} \tag{6.3.48}$$

for some $C_1', C_2' > 0$. Taking $f = \tau - \Delta_{[k]}\tau$ to match the definition (6.3.14) of $\Delta_{[k]}\mathfrak{I}_{\mathrm{d};1}$, this implies:

$$\big|\Delta_{[k]}\mathfrak{I}_{\mathrm{d};1}[H,V]\big| \leq C_1' e^{-C_2 N^{\alpha}} \Big\{||\tau||_{W_2^{\infty}([a_N;b_N]^2)} + ||\Delta_{[k]}\tau||_{W_2^{\infty}([a_N;b_N]^2)}\Big\}\,. \tag{6.3.49}$$

It solely remains to bound the $W_2^{\infty}([a_N;b_N]^2)$ norm of τ and $\Delta_{[k]}\tau$. We remind that, for $\xi \in [a_N;b_N]$, we have from the definition (6.3.2) and the expression of $\mathcal{U}_N^{-1}$ given in (5.2.21):

$$\mathcal{G}_N[H,V](\xi) = \mathcal{U}_N^{-1}[H](\xi) - \mathcal{U}_N^{-1}[H](\eta)\,. \tag{6.3.50}$$

By invoking the mean value theorem and the estimate of Proposition 5.2.2 for W_ℓ^{∞} norm of $\mathcal{U}_N^{-1}[H]$, we obtain:

$$\begin{aligned} ||\tau||_{W_\ell^{\infty}([a_N;b_N]^2)} &\leq C N^{\alpha} \left|\left|(\xi,\eta) \mapsto \frac{\mathcal{G}_N\big[H,V\big](\xi,\eta)}{\xi-\eta}\right|\right|_{W_{\ell+1}^{\infty}([a_N;b_N]^2)} \\ &\leq C' N^{\alpha} \,\big|\big|\mathcal{U}_N^{-1}[H]\big|\big|_{W_{\ell+2}^{\infty}([a_N;b_N])} \qquad (6.3.51) \\ &\leq C_\ell' \cdot (\ln N)^{2\ell+5} \cdot N^{(\ell+4)\alpha} \cdot \mathfrak{n}_{\ell+2}[V] \cdot \mathcal{N}_N^{(2\ell+5)}\big[\mathcal{K}_\kappa[H]\big] \qquad (6.3.52) \\ &\leq C_\ell'' \cdot (\ln N)^{2\ell+5} \cdot N^{(\ell+4)\alpha} \cdot \mathfrak{n}_{\ell+2}[V] \cdot ||H_{\mathfrak{e}}||_{W_{2\ell+5}^{\infty}(\mathbb{R})} \qquad (6.3.53) \end{aligned}$$

where the last step comes from domination of the weighted norm by the W^{∞} norm of the same order—and the exponential regularisation can easily be traded for a compactly supported extension up to increasing the constant prefactor. Similarly, in virtue of the bounds (6.3.42), we get:

$$||\Delta_{[k]}\tau||_{W_\ell^{\infty}([a_N;b_N]^2)} \leq C' \cdot N^{(\ell+3-k)\alpha} \cdot \mathfrak{n}_{k+\ell+2}[V] \cdot ||H_{\mathfrak{e}}||_{W_{k+\ell+3}^{\infty}(\mathbb{R})}\,. \tag{6.3.54}$$

Putting these two estimates back in (6.3.49) with $\ell = 2$, we see that:

$$\big|\Delta_{[k]}\mathfrak{I}_{\mathrm{d};1}[H,V]\big| \leq C_1' \cdot N^{6\alpha} \cdot e^{-C_2 N^{\alpha}} \mathfrak{n}_{k+4}[V] \cdot ||H_{\mathfrak{e}}||_{W_{\max\{k,5\}+4}^{\infty}(\mathbb{R})}\,, \tag{6.3.55}$$

which is exponentially small when $N \to \infty$.

6.3.2.2 Bounding $\Delta_{[k]}\mathfrak{I}_{d;2}[H, V]$

$\Delta_{[k]}\mathfrak{I}_{d;2}[H, V]$ has been defined in (6.3.15) and can be bounded by repeating the previous handlings. Indeed, using (6.1.49) on the L^1 norm of $\mathcal{W}_N$ and then following the previous steps, one finds:

$$\left|\Delta_{[k]}\mathfrak{I}_{d;2}[H, V]\right| \leq ||\Delta_{[k]}\tau||_{W_2^\infty([a_N ; b_N]^2)} \tag{6.3.56}$$

with $\Delta_{[k]}\tau$ defined in (6.3.44) and bounded in W_ℓ^∞ norms in (6.3.54). Hence, we find:

$$\left|\Delta_{[k]}\mathfrak{I}_{d;2}[H, V]\right| \leq C_1' \cdot N^{(5-k)\alpha} \cdot \mathfrak{n}_{k+4}[V] \cdot ||H_\mathfrak{e}||_{W_{k+5}^\infty(\mathbb{R})} \,. \tag{6.3.57}$$

6.3.2.3 Bounding $\Delta_{[k]}\mathfrak{I}_{d;3}[H, V]$

This quantity is defined in (6.3.16), and it follows from the explicit expression for $\mathcal{W}_N[1](\xi)$ given in (4.3.72) that

$$\begin{aligned}\left|\Delta_{[k]}\mathfrak{I}_{d;3}[H, V]\right| &\leq C\,N^\alpha ||\tau||_{W_0^\infty([a_N ; b_N]^2)} \cdot \left|\chi_{12;+}(0)\right| \cdot |b_N - a_N| \\ &\times \sup_{\xi \in [a_N ; b_N]} \left| \int\limits_{\mathbb{R}+\mathrm{i}\epsilon'} \frac{\chi_{11}(\lambda)}{\lambda} \mathrm{e}^{-\mathrm{i}N^\alpha\lambda(\xi - a_N)} \cdot \frac{\mathrm{d}\lambda}{2\mathrm{i}\pi} \right|\end{aligned} \tag{6.3.58}$$

where τ is as defined in (6.3.44). The decomposition (6.1.6) for χ and its properties show the existence of $C, C' > 0$ such that:

$$\forall \lambda \in \mathbb{R} + \mathrm{i}\epsilon', \qquad |\chi_{11}(\lambda)| \leq C\,|\lambda|^{-1/2}, \qquad \text{and} \qquad |\chi_{12}(\lambda)| \leq C'\,\mathrm{e}^{-N^\alpha \varkappa_{\epsilon'}} \,. \tag{6.3.59}$$

Hence, by invoking the bounds (6.3.53) satisfied by τ, we get:

$$\left|\Delta_{[k]}\mathfrak{I}_{d;3}[H, V]\right| \leq C'' \cdot N^{5\alpha} \cdot (\ln N)^5 \cdot \mathrm{e}^{-N^\alpha[\varkappa_{\epsilon'} - \epsilon'(b_N - a_N)]} \cdot \mathfrak{n}_2[V] \cdot ||H_\mathfrak{e}||_{W_5^\infty(\mathbb{R})} \,. \tag{6.3.60}$$

Since $\varkappa_{\epsilon'} > 0$ is bounded away from 0 when $\epsilon' \to 0$ according to its definition (4.2.7), we also have $\varkappa_\epsilon - \epsilon' x_N > 0$ uniformly in N for some choice of ϵ' small enough but independent of N.

6.3.2.4 Bounding $\Delta_{[k]}\mathfrak{I}_{d;4}[H, V]$

This quantity is defined in (6.3.17), and it involves integration of:

$$\tau_L(\xi, \eta) = \partial_\xi \Big\{ S\big(N^\alpha(\xi - \eta)\big) \cdot \mathcal{G}_{R;k}^{(\mathrm{as})}\big[H, V\big](x_L, y_L; b_N + a_N - \xi, b_N + a_N - \eta) \Big\} \tag{6.3.61}$$

where $\mathcal{G}^{(\mathrm{as})}_{R;k}$ was defined in (6.3.5). It only involves the operators $\mathcal{W}^{(\mathrm{as})}_{bk;k}$ and $\mathcal{W}^{(\mathrm{as})}_{R;k}$, whose expression is given in Lemma 5.1.11. Let us fix $\epsilon > 0$. Straightforward manipulations show that, for $(\xi,\eta) \in [a_N+\epsilon\,;b_N]^2$, we have:

$$\begin{aligned} \big|\tau_L(\xi,\eta)\big| &\le CN^{3\alpha}\mathrm{e}^{-C'\min(x_L,y_L)}\cdot \mathfrak{n}_{k+1}[V]\cdot ||H_{\mathrm{e}}||_{W_k^{\infty}(\mathbb{R})} \\ &\le \widetilde{C}\cdot N^{3\alpha}\cdot \mathrm{e}^{-\epsilon C' N^{\alpha}}\cdot \mathfrak{n}_{k+1}[V]\cdot ||H_{\mathrm{e}}||_{W_k^{\infty}(\mathbb{R})} \end{aligned} \tag{6.3.62}$$

which is thus exponentially small in N. Similar steps show that, for

$$\begin{aligned} (\xi,\eta)\in &\Big\{[a_N+\epsilon\,;b_N]\times[a_N\,;a_N+\epsilon]\Big\}\cup\Big\{[a_N\,;a_N+\epsilon]\times[a_N+\epsilon\,;b_N]\Big\} \\ &\cup\Big\{[a_N\,;a_N+\epsilon]\times[a_N\,;a_N+\epsilon]\Big\}\,, \end{aligned} \tag{6.3.63}$$

we have:

$$\big|\tau_L(\xi,\eta)\big| \le C\,N^{3\alpha}\mathfrak{n}_{k+1}[V]\,||H||_{W_k^{\infty}(\mathbb{R})}\,. \tag{6.3.64}$$

Here, the exponential decay in N will come after integration of τ_L as it appears in (6.3.17). Indeed, given $\operatorname{Im}\lambda > 0$ and $\operatorname{Im}\mu < 0$ we have:

$$\begin{aligned} &\Bigg|\int_{a_N}^{b_N}\mathrm{e}^{\mathrm{i}\lambda x_R}\mathrm{e}^{-\mathrm{i}\mu y_R}\tau_L(\xi,\eta)\,\mathrm{d}\xi\,\mathrm{d}\eta\Bigg| \\ &\le C\,N^{3\alpha}\mathrm{e}^{-\epsilon C'N^{\alpha}}\mathfrak{n}_{k+1}[V]\,||H||_{W_k^{\infty}(\mathbb{R})}\int_{a_N+\epsilon}^{b_N}\mathrm{e}^{-|\operatorname{Im}\lambda|N^{\alpha}(b_N-\xi)-|\operatorname{Im}\mu|N^{\alpha}(b_N-\eta)}\,\mathrm{d}\eta\,\mathrm{d}\xi \\ &+ CN^{3\alpha}\mathfrak{n}_{k+1}[V]\,||H||_{W_k^{\infty}(\mathbb{R})}\Bigg\{\int_{a_N+\epsilon}^{b_N}\mathrm{d}\xi\int_{a_N}^{a_N+\epsilon}\mathrm{d}\eta + \int_{a_N}^{a_N+\epsilon}\mathrm{d}\xi\int_{a_N+\epsilon}^{b_N}\mathrm{d}\eta \\ &\qquad + \int_{a_N}^{a_N+\epsilon}\int_{a_N}^{a_N+\epsilon}\mathrm{d}\xi\,\mathrm{d}\eta\Bigg\}\,\mathrm{e}^{-|\operatorname{Im}\lambda|N^{\alpha}(b_N-\xi)-|\operatorname{Im}\mu|N^{\alpha}(b_N-\eta)} \\ &\le \mathfrak{n}_{k+1}[V]\,||H||_{W_k^{\infty}(\mathbb{R})}\cdot\frac{\widetilde{C}\,N^{3\alpha}\mathrm{e}^{-\widetilde{C}'N^{\alpha}}}{|\lambda\cdot\mu|}\,. \end{aligned} \tag{6.3.65}$$

Note that, above, we have used that for $\lambda \in \mathscr{C}^{(+)}_{\mathrm{reg}}$ and $\mu \in \mathscr{C}^{(-)}_{\mathrm{reg}}$, we can bound:

$$|\operatorname{Im}\lambda|^{-1} \le c_1|\lambda|^{-1}\,, \qquad |\operatorname{Im}\mu|^{-1} \le c_1|\mu|^{-1} \tag{6.3.66}$$

for some constant $c_1 > 1$. Hence, all in all, we have:

$$\Big|\mathfrak{I}_{\mathrm{d};R}\Big[\big(\mathcal{G}_{R;k}^{(\mathrm{as})}[H,V]\big)^{\wedge}\Big]\Big| \leq C'' N^{3\alpha} \mathrm{e}^{-\widetilde{C}' N^{\alpha}} \cdot \int\limits_{\mathscr{C}_{\mathrm{reg}}^{(+)}} |\mathrm{d}\lambda| \int\limits_{\mathscr{C}_{\mathrm{reg}}^{(-)}} |\mathrm{d}\mu| \cdot \frac{\mathfrak{n}_{k+1}[V]\, ||H||_{W_k^{\infty}(\mathbb{R})}}{|\mu-\lambda| \cdot |\lambda R_{\downarrow}(\lambda) R_{\uparrow}(\mu)\mu|}$$
$$\leq C''' N^{-3\alpha}\, \mathrm{e}^{-\widetilde{C}' N^{\alpha}} \cdot \mathfrak{n}_{k+1}[V]\, ||H||_{W_k^{\infty}(\mathbb{R})} \,. \qquad (6.3.67)$$

Then, putting together all of the results for each $\Delta_{[k]}\mathfrak{I}_{\mathrm{d};p}$ for $p \in [\![\, 1\,;\, 4\,]\!]$ entails the global bound (6.3.37). ■

6.3.3 Leading Asymptotics of the Double Integral

We need to introduce two new quantities before writing down the asymptotic expansion of the double integral $\mathfrak{I}_{\mathrm{d}}$.

Definition 6.3.9 We define the function:

$$\mathfrak{c}(x) = \frac{\mathfrak{b}_1(x) - \mathfrak{b}_0(x)\mathfrak{a}_1(x)}{u_1} \qquad (6.3.68)$$

and the constant:

$$\aleph_0 = -\int\limits_{\mathbb{R}} \frac{\mathrm{d}u\, J(u)}{4\pi\beta} \int\limits_{|u|}^{+\infty} \mathrm{d}v\; \partial_u\Big\{ S(u)\cdot\Big(\mathfrak{c}\big[\tfrac{v-u}{2}\big] - \mathfrak{c}\big[\tfrac{v+u}{2}\big]\Big)\Big\}$$
$$+ \frac{1}{2\pi\beta} \int\limits_{\mathscr{C}_{\mathrm{reg}}^{(+)}} \frac{\mathrm{d}\lambda}{2\mathrm{i}\pi} \int\limits_{\mathscr{C}_{\mathrm{reg}}^{(-)}} \frac{\mathrm{d}\mu}{2\mathrm{i}\pi}\, \frac{1}{(\lambda-\mu) R_{\downarrow}(\lambda) R_{\uparrow}(\mu)}$$
$$\times \int\limits_{0}^{+\infty} \mathrm{e}^{\mathrm{i}\lambda x - \mathrm{i}\mu y}\, \partial_x\Big\{ S(x-y)\big[\mathfrak{c}(x) - \mathfrak{c}(y)\big] - x + y \Big\} \mathrm{d}x\, \mathrm{d}y \qquad (6.3.69)$$

Proposition 6.3.10 *We have the large-N behaviour:*

$$\mathfrak{I}_{\mathrm{d}}[H,V] = -2\mathfrak{I}_0 \cdot N^{\alpha} \cdot \Big\{ \frac{H'(b_N)}{V''(b_N)} - \frac{H'(a_N)}{V''(a_N)} \Big\} + \aleph_0 \cdot \Big\{ \Big(\frac{H'}{V''}\Big)'(b_N) + \Big(\frac{H'}{V''}\Big)'(a_N) \Big\}$$
$$+ \Delta\mathfrak{I}_{\mathrm{d}}[H,V] \qquad (6.3.70)$$

and the remainder is bounded as:

$$\Delta\mathfrak{I}_{\mathrm{d}}[H,V] \leq \frac{C}{N^{\alpha}} \cdot \mathfrak{n}_{10}[V_{\mathfrak{e}}] \cdot ||H_{\mathfrak{e}}||_{W_{11}^{\infty}(\mathbb{R})} \,. \qquad (6.3.71)$$

Proof We first need to introduce two universal sequences of polynomials $\mathcal{P}_\ell\big(\{x_p\}_1^\ell\big)$ and $\mathcal{Q}_\ell\big(\{y_p\}_1^\ell;\{a_p\}_1^\ell\big)$. Given formal power series

$$f(z) = 1 + \sum_{\ell\geq 1} f_\ell\, z^\ell \qquad \text{and} \qquad g(z) = 1 + \sum_{\ell\geq 1} g_\ell\, z^\ell \tag{6.3.72}$$

they are defined to be the coefficients arising in the formal power series

$$\frac{1}{f(z)} = 1 + \sum_{\ell\geq 1} \mathcal{P}_\ell\big(\{f_p\}_1^\ell\big)\, z^\ell \quad \text{and} \quad \frac{g(z)}{f(z)} = 1 + \sum_{\ell\geq 1} \mathcal{Q}_\ell\big(\{g_p\}_1^\ell;\{f_p\}_1^\ell\big)\, z^\ell\,. \tag{6.3.73}$$

Note that

$$\mathcal{Q}_\ell\big(\{g_p\}_1^\ell;\{f_p\}_1^\ell\big) = \sum_{\substack{r+s=\ell\\ r,s\geq 0}} g_r\cdot \mathcal{P}_s\big(\{f_p\}_1^s\big)\,, \tag{6.3.74}$$

where we agree upon the convention $\mathcal{P}_0 = 1$ and $g_0 = 1$. This notation is convenient to write down the large-N expansion of $\mathcal{G}_{\mathrm{bk};k}$—defined in (6.3.4)—ensuing from the large N-expansion of $\mathcal{W}_{\mathrm{bk};k}$ provided by Lemma 5.1.11. We find, uniformly in $(\xi,\eta)\in[a_N\,;\,b_N]^2$:

$$\mathcal{G}_{\mathrm{bk};k}[H,V](\xi,\eta) = \sum_{\ell=0}^{k-1} \frac{\mathfrak{G}_{\mathrm{bk};\ell}[H,V](\xi,\eta)}{N^{\alpha\ell}} + \mathrm{O}\big(N^{-k\alpha}\big) \tag{6.3.75}$$

where

$$\mathfrak{G}_{\mathrm{bk};\ell}[H,V](\xi,\eta) = \mathfrak{g}_{\mathrm{bk};\ell}[H,V](\xi) - \mathfrak{g}_{\mathrm{bk};\ell}[H,V](\eta) \tag{6.3.76}$$

with

$$\mathfrak{g}_{\mathrm{bk};\ell}[H,V](\xi) = \frac{H'(\xi)}{V''(\xi)}\cdot \mathcal{Q}_\ell\bigg(\bigg\{\frac{H^{(\ell+1)}(\xi)}{H'(\xi)}\,\frac{u_{\ell+1}}{u_1}\bigg\}_\ell;\bigg\{\frac{V^{(\ell+2)}(\xi)}{V''(\xi)}\,\frac{u_{\ell+1}}{u_1}\bigg\}_\ell\bigg)\,. \tag{6.3.77}$$

Also, in the case of a localisation of the variables around b_N, we have:

$$\begin{aligned}&\mathcal{G}_{\mathrm{bk};k}[H,V](b_N - N^{-\alpha}x, b_N - N^{-a}y)\\ &\quad = \frac{H'(b_N)}{V''(b_N)}\sum_{\ell=1}^{k} N^{-\ell\alpha}\mathcal{Q}_\ell\bigg(\bigg\{\frac{H^{(p+1)}(b_N)}{H'(b_N)}\,\frac{\mathfrak{u}_p(x)}{u_1}\bigg\};\bigg\{\frac{V^{(p+2)}(b_N)}{V''(b_N)}\,\frac{\mathfrak{u}_p(x)}{u_1}\bigg\}\bigg)\\ &\qquad - (x\leftrightarrow y) + \mathrm{O}\bigg(\frac{x^k+y^k+1}{N^{(k+1)\alpha}}\bigg)\,.\end{aligned} \tag{6.3.78}$$

Finally, we also have the expansion,

$$\mathcal{G}^{(\mathrm{as})}_{R;k}[H,V](x,y;b_N-N^{-\alpha}x,b_N-N^{-\alpha}y) = \sum_{\ell=1}^{k}\frac{\mathfrak{G}_{R;\ell}[H,V](x,y)}{N^{\alpha\ell}} + \mathrm{O}\Big(\frac{\mathrm{e}^{-c\min(x,y)}}{N^{(k+1)\alpha}}\Big) \tag{6.3.79}$$

where $\mathfrak{G}_{R;\ell}[H,V](x,y) = \mathfrak{g}_{R;\ell}[H,V](x) - \mathfrak{g}_{R;\ell}[H,V](y)$ and

$$\begin{aligned}\mathfrak{g}_{R;\ell}[H,V](x) &= \frac{1}{u_1V''(b_N)}\sum_{\substack{m+s=\ell\\ m,s\geq 0}}\mathcal{P}_m\bigg(\bigg\{\frac{V^{(q+2)}(b_N)}{V''(b_N)}\,\frac{\mathfrak{u}_q(x)}{u_1}\bigg\}_q\bigg)\cdot\frac{H^{(s+1)}(b_N)}{H'(b_N)}\cdot\mathfrak{b}_s(x)\\ &\quad -\frac{H'(b_N)}{u_1\big[V''(b_N)\big]^2}\sum_{\substack{m+s+p=\ell\\ m,s,p\geq 0}}\mathcal{P}_m\bigg(\bigg\{\frac{V^{(q+2)}(b_N)}{V''(b_N)}\,\frac{\mathfrak{u}_q(x)}{u_1}\bigg\}_q\bigg)\\ &\quad\times\mathcal{Q}_p\bigg(\bigg\{\frac{H^{(q+1)}(b_N)}{H'(b_N)}\,\mathfrak{a}_q(x)\bigg\}_q;\bigg\{\frac{V^{(q+2)}(b_N)}{V''(b_N)}\,\mathfrak{a}_q(x)\bigg\}_q\bigg)\cdot\frac{V^{(s+2)}(b_N)}{V''(b_N)}\cdot\mathfrak{b}_s(x)\,.\end{aligned} \tag{6.3.80}$$

We can now come back to the double integral $\mathfrak{I}_{\mathrm{d}}$. It has been decomposed in Lemma 6.3.3. If we want a remainder $\Delta_{[k]}\mathfrak{I}_{\mathrm{d}}$ decaying with N, we should take $k=6$ in Lemma 6.3.8. Then, up to $\mathrm{O}(N^{-\alpha})$, we are thus left with operators $\mathfrak{I}_{\mathrm{d;bk}}$ and $\mathfrak{I}_{\mathrm{d};R}$, and Lemmas 6.3.5 and 6.3.7 describe for us their asymptotic expansion knowing the asymptotic expansion of the functions to which they are applied. Here, they are applied to the various functions involving $\mathcal{G}_{\mathrm{bk};k}$ and $\mathcal{G}^{(\mathrm{as})}_{\mathrm{R};k}$ whose expansion has been described in (6.3.78) and (6.3.79). As these expression shows, in order to get $\mathfrak{I}_{\mathrm{d}}$ up to $\mathrm{O}(N^{-\alpha})$, one just need the expressions of $\mathfrak{g}_{\mathrm{bk};0}[H,V](\xi)$ from (6.3.77) and $\mathfrak{g}_{R;1}[H,V](x)$ from (6.3.80). These only involve the universal polynomials $\mathcal{P}_1$ and $\mathcal{Q}_1$, whose expression follows from their definitions in (6.3.73):

$$\mathcal{P}_1\big(\{f_1\}\big) = -f_1 \qquad \mathcal{Q}_1\big(\{g_1\};\{f_1\}\big) = g_1 - f_1\,. \tag{6.3.81}$$

Therefore, we get

$$\mathfrak{g}_{\mathrm{bk};1}[H,V](\xi) = \frac{H'(\xi)}{V''(\xi)} \qquad\text{and}\qquad \mathfrak{g}_{R;1}[H,V](x) = \frac{\mathfrak{b}_1(x)-\mathfrak{a}_1(x)\mathfrak{b}_0(x)}{u_1}\cdot\Big(\frac{H'}{V''}\Big)'(b_N) \tag{6.3.82}$$

and we recognize in the prefactor of the second equation the function $\mathfrak{c}(x)$ of Definition 6.3.9. Finally, we remind that we take the remainder at order $k=6$. The claim then follows upon recognising the constant $\aleph_0$ from Definition 6.3.4 in the computation of the leading term by Lemma 6.3.7. ■

Reference

1. Deift, P.A.: Orthogonal Polynomials and Random Matrices. A Riemann-Hilbert Approach. Courant Lecture Notes, vol. 3. New-York University (1999)

Appendix A
Several Theorems and Properties of Use to the Analysis

Theorem A.0.11 (Hunt et al. [1]) *The Hilbert transform, defined as an operator*

$$\mathcal{H} \ : \ L^2(\mathbb{R}, w(x)\mathrm{d}x) \ \to \ L^2(\mathbb{R}, w(x)\mathrm{d}x)$$

is bounded if and only if there exists a constant $C > 0$ *such that, for any interval* $I \subseteq \mathbb{R}$*:*

$$\left\{ \frac{1}{|I|} \int\limits_I w(x)\mathrm{d}x \right\} \cdot \left\{ \frac{1}{|I|} \int\limits_I \frac{\mathrm{d}x}{w(x)} \right\} \ < \ C \, . \tag{A.0.1}$$

In particular, the operators "upper/lower boundary values"

$$\mathcal{C}_{\pm} \ : \ \mathcal{F}\big[H_s(\mathbb{R})\big] \ \to \ \mathcal{F}\big[H_s(\mathbb{R})\big]$$

are bounded if and only if $|s| < 1/2$.

A less refined version of this theorem takes the form:

Proposition A.0.12 *For any* $\gamma > 0$*, the shifted Cauchy operators* $\mathcal{C}_{\gamma} : f \mapsto \mathcal{C}_{\gamma}[f]$ *with* $\mathcal{C}_{\gamma}[f](\lambda) = \mathcal{C}[f](\lambda + \mathrm{i}\gamma)$ *are continuous on* $\mathcal{F}\big[H_s(\mathbb{R})\big]$ *with* $|s| < 1/2$.

Theorem A.0.13 (Calderon [2]) *Let* Σ *be a non-self intersecting Lipschitz curve in* $\mathbb{C}$ *and* C_{Σ} *the Cauchy transform on* $L^2(\Sigma, \mathrm{d}s)$*:*

$$\forall f \in L^2(\Sigma, \mathrm{d}s) \qquad \mathcal{C}_{\Sigma}[f](z) \ = \ \int\limits_{\Sigma} \frac{f(s)}{s - z} \cdot \frac{\mathrm{d}s}{2\mathrm{i}\pi} \in \mathcal{O}\big(\mathbb{C} \setminus \Sigma\big) \, . \tag{A.0.2}$$

For any $f \in L^2(\Sigma, \mathrm{d}s)$*,* $\mathcal{C}_{\Sigma}[f]$ *admits* $L^2(\Sigma, \mathrm{d}s)$ $\pm$ *boundary values* $\mathcal{C}_{\Sigma;\pm}[f]$*. The operators* $\mathcal{C}_{\Sigma;\pm}[f]$ *are continuous operators on* $L^2(\Sigma, \mathrm{d}s)$ *which, furthermore, satisfy* $\mathcal{C}_{\Sigma;+} - \mathcal{C}_{\Sigma;-} = \mathrm{id}$.

G. Borot et al., *Asymptotic Expansion of a Partition Function Related to the Sinh-model*, Mathematical Physics Studies,
DOI 10.1007/978-3-319-33379-3

Theorem A.0.14 (Paley and Wiener [3]) *Let* $u \in L^2(\mathbb{R}^{\pm})$. *Then* $\mathcal{F}[u]$ *is the* $L^2(\mathbb{R})$ *boundary value on* $\mathbb{R}$ *of a function* $\widehat{u}$ *that is holomorphic on* $\mathbb{H}^{\pm}$, *and there exists a constant* $C > 0$ *such that:*

$$\forall \mu > 0, \qquad \int_{\mathbb{R}} \left|\widehat{u}(\lambda \pm \mathrm{i}\mu)\right|^2 \cdot \mathrm{d}\lambda \; < \; C\,. \tag{A.0.3}$$

Reciprocally, every holomorphic function on $\widehat{u}$ *on* $\mathbb{H}^{\pm}$ *that satisfies the bounds* (A.0.3) *and admits* $L^2(\mathbb{R}) \pm$ *boundary values* $\widehat{u}_{\pm}$ *on* $\mathbb{R}$, *is the Fourier transform of a function* $u \in L^2(\mathbb{R}^{\pm})$, *viz.* $\widehat{u}(z) = \mathcal{F}[u](z)$, $z \in \mathbb{H}^{\pm}$.

Appendix B
Proof of Theorem 2.1.1

We denote by $\mathfrak{p}_N$ the rescaled probability density on $\mathbb{R}^N$ associated with $\mathfrak{z}_N$, namely

$$\mathfrak{p}_N(\boldsymbol{\lambda}) = \frac{N^{\alpha_q N}}{\mathfrak{z}_N[W]} \prod_{a<b}^{N} \Big\{ \sinh\big[\pi \omega_1 N^{\alpha_q}(\lambda_a - \lambda_b)\big] \sinh\big[\pi \omega_2 N^{\alpha_q}(\lambda_a - \lambda_b)\big] \Big\}^{\beta} \cdot \prod_{a=1}^{N} \mathrm{e}^{-W(N^{\alpha_q}\lambda_a)}$$

with

$$\alpha_q = \frac{1}{q-1} .$$

To obtain the above probability density, we have rescaled in the variables in (1.5.27) as $y_a = N^{\alpha_q}\lambda_a$ with the value of α_q guided by the heuristic arguments that followed the statement of Theorem 2.1.1. We shall denote by $\mathcal{P}_N$ the probability measure on $\mathcal{M}^1(\mathbb{R})$ induced by $\mathfrak{p}_N$, *viz*. the measurable sets in $\mathcal{M}^1(\mathbb{R})$ are generated by the Borel σ-algebra for the weak topology, and for any open subset in $\mathcal{M}^1(\mathbb{R})$, we have:

$$\mathcal{P}_N\big[O\big] = \int_{\{L_N^{(\boldsymbol{\lambda})} \in O\}} \mathfrak{p}_N(\boldsymbol{\lambda})\, \mathrm{d}^N\boldsymbol{\lambda} . \tag{B.0.1}$$

The strategy of the proof consists in proving that $\mathcal{P}_N$ is exponentially tight and then establishing a weak large deviation principle, namely upper and lower bounding $\mathcal{P}_N$ on balls of shrinking radii this for balls relatively to the bounded Lipschitz topology, see *e.g.* [4].

G. Borot et al., *Asymptotic Expansion of a Partition Function Related to the Sinh-model*, Mathematical Physics Studies,
DOI 10.1007/978-3-319-33379-3

Exponential Tightness

Lemma B.0.15 *The sequence of measures $\mathcal{P}_N$ is exponentially tight, i.e.:*

$$\limsup_{L\to+\infty}\limsup_{N\to\infty} N^{-(2+\alpha_q)} \ln \mathcal{P}_N\big[K_L^c\big] = -\infty \tag{B.0.2}$$

where $K_L = \big\{\mu \in \mathcal{M}^1(\mathbb{R}) \; : \; \int_{\mathbb{R}} |x|^q \,\mathrm{d}\mu(x) \leq L\big\}$.

Proof By the monotone convergence theorem,

$$\int_{\mathbb{R}} |x|^q \,\mathrm{d}\mu(x) = \sup_{M\in\mathbb{N}} \int_{\mathbb{R}} \min(|x|^q\,, M)\,\mathrm{d}\mu(x). \tag{B.0.3}$$

The left-hand side is lower semi-continuous as a supremum of a continuous family of functionals on $\mathcal{M}_1(\mathbb{R})$. Thus, K_L is closed as a level set of a lower semi-continuous function. For any $\mu \in K_L$, we have by Chebyshev inequality:

$$\mu\big[[-M\,;\,M]^c\big] \leq \frac{1}{M^q}\int_{[-M\,;M]^c} |x|^q \,\mathrm{d}\mu(x) \leq \frac{L}{M^q}\,. \tag{B.0.4}$$

As a consequence,

$$K_L \subseteq \bigcap_{M\in\mathbb{N}} \left\{\mu \in \mathcal{M}^1(\mathbb{R}) \; : \; \mu\big[[-M\,;\,M]^c\big] \leq \frac{L}{M^q}\right\}. \tag{B.0.5}$$

The right-hand side is uniformly tight, by construction and is closed as an intersection of level sets of lower semi-continuous functions on $\mathcal{M}^1(\mathbb{R})$. Thence by Prokhorov theorem, it is compact. As K_L is closed, it must be as well compact.

We now estimate $\mathcal{P}_N\big[K_L^c\big]$. We start by a rough estimate for the partition function. It follows by Jensen inequality applied to the probability measure of $\mathbb{R}^N$

$$\prod_{a=1}^{N} \frac{\mathrm{e}^{-W(\lambda_a)}\mathrm{d}\lambda_a}{\int_{\mathbb{R}} \mathrm{e}^{-W(\lambda)}\mathrm{d}\lambda}\,, \tag{B.0.6}$$

that

$$\ln\big[\mathfrak{z}_N[W]\big] \geq N \ln\Big[\int \mathrm{e}^{-W(\lambda)}\mathrm{d}\lambda\Big]$$

$$+ \int_{\mathbb{R}^N} \sum_{a<b} \beta \ln\big\{\sinh\big[\pi\omega_1(\lambda_a-\lambda_b)\big]\sinh\big[\pi\omega_2(\lambda_a-\lambda_b)\big]\big\} \prod_{a=1}^{N} \frac{\mathrm{e}^{-W(\lambda_a)}\mathrm{d}\lambda_a}{\int \mathrm{e}^{-W(\lambda)}\mathrm{d}\lambda}$$

$$\geq N \ln\Big[\int e^{-W(\lambda)} d\lambda\Big]$$

$$+\frac{\beta N(N-1)}{2}\int_{\mathbb{R}^2} \ln\big\{\sinh\big[\pi\omega_1(\lambda_1-\lambda_2)\big]\sinh\big[\pi\omega_2(\lambda_1-\lambda_2)\big]\big\}\frac{e^{-W(\lambda_1)-W(\lambda_2)}d\lambda_1 d\lambda_2}{\Big(\int e^{-W(\lambda)}d\lambda\Big)^2} \tag{B.0.7}$$

As a consequence, $\mathfrak{z}_N[W] \geq e^{-N^2\kappa}$ for some $\kappa \in \mathbb{R}$. It now remains to estimate the integral arising from the integration over K_L^c. Using that $|\sinh(\lambda)| \leq e^{|\lambda|}$ we get:

$$\begin{aligned}\prod_{a<b}^{N}\Big\{\sinh\big[\pi\omega_1 N^{\alpha_q}(\lambda_a-\lambda_b)\big]\sinh\big[\pi\omega_2 N^{\alpha_q}(\lambda_a-\lambda_b)\big]\Big\}^{\beta} &\leq \prod_{a<b}^{N}\exp\Big\{\pi\beta(\omega_1+\omega_2)N^{\alpha_q}|\lambda_a-\lambda_b|\Big\}\\ \leq \prod_{a<b}^{N}\exp\Big\{\pi\beta(\omega_1+\omega_2)N^{\alpha_q}\big(|\lambda_a|+|\lambda_b|\big)\Big\} &\leq \prod_{a=1}^{N}\exp\Big\{\pi\beta(\omega_1+\omega_2)N^{\alpha_q+1}|\lambda_a|\Big\}.\end{aligned} \tag{B.0.8}$$

Hence,

$$\mathcal{P}_N\big[K_L^c\big] \leq e^{\kappa N^2} N^{\alpha_q N}\int_{\{L_N^{(\lambda)}\in K_L^c\}}\prod_{a=1}^{N}\exp\Big\{\pi\beta(\omega_1+\omega_2)N^{\alpha_q+1}|\lambda_a| - W(N^{\alpha_q}\lambda_a)\Big\}\cdot d^N\boldsymbol{\lambda} \tag{B.0.9}$$

Since $|\xi|^{1-q} \underset{|\xi|\to+\infty}{\longrightarrow} 0$ there exists a constant $C \in \mathbb{R}$ such that

$$\forall \xi \in \mathbb{R}, \qquad \pi\beta(\omega_1+\omega_2)|\xi| \leq \frac{c_q\,|\xi|^q}{2} + C\,. \tag{B.0.10}$$

Likewise it follows from (2.1.2) that given any $\epsilon > 0$ there exists $\tau_\epsilon \in \mathbb{R}^+$ such that

$$\forall \xi \in \mathbb{R}, \qquad -c_q(1+\epsilon)\,|\xi|^q - \tau_\epsilon \leq -W(\xi) \leq -c_q(1-\epsilon)\,|\xi|^q + \tau_\epsilon\,. \tag{B.0.11}$$

In the following, ϵ will be taken small. Taking into account that $q\alpha_q = \alpha_q + 1$, (B.0.10) and the upper bound of (B.0.11) lead to:

$$\begin{aligned}\mathcal{P}_N\big[K_L^c\big] \leq e^{\kappa N^2} N^{\alpha_q N}\int_{\{L_N^{(\lambda)}\in K_L^c\}}\prod_{a=1}^{N}\exp\Big\{&N^{\alpha_q+1}C + \frac{c_q}{2}N^{\alpha_q+1}\,|\lambda_a|^q\\ &+\tau_\epsilon - c_q(1-\epsilon)N^{q\alpha_q}\,|\lambda_a|^q\Big\}\cdot d^N\boldsymbol{\lambda}\end{aligned}$$

$$
\begin{aligned}
&\leq N^{\alpha_q N} e^{\kappa N^2 + C N^{2+\alpha_q} + \tau_\epsilon N} \int_{\left\{L_N^{(\lambda)} \in K_L^c\right\}} \left(\prod_{a=1}^{N} e^{-\epsilon c_q N^{\alpha_q+1} |\lambda_a|^q} \right) \\
&\quad \times \exp\left\{ -\frac{c_q(1-4\epsilon)}{2} N^{2+\alpha_q} \int_{\mathbb{R}} |x|^q \, dL_N^{(\lambda)}(x) \right\} \cdot d^N \lambda \\
&\leq N^{\alpha_q N} e^{C' N^{2+\alpha_q} + \tau_\epsilon N - [c_q(1-4\epsilon)/2] L N^{2+\alpha_q}} \left(\int_{\mathbb{R}} e^{-\epsilon c_q |\lambda_a|^q} d\lambda \right)^N
\end{aligned}
\tag{B.0.12}
$$

for some constant $C' > C$ and N large enough. As a consequence,

$$
\limsup_{N \to +\infty} N^{-(2+\alpha_q)} \ln \mathcal{P}_N\left[K_L^c\right] \leq C - L c_q (1-4\epsilon)/2 ,
$$

and this upper bound goes to $-\infty$ when $L \to +\infty$. ■

Lower Bound

In the following we focus on the renormalised measure on $\mathcal{M}_1(\mathbb{R})$ defined as $\overline{\mathcal{P}}_N = \mathfrak{z}_N[W] \cdot \mathcal{P}_N$. We will now derive a lower bound for the $\overline{\mathcal{P}}_N$ volume of small Vasershtein balls, in terms of the energy functional $\mathcal{E}_{(\text{ply})}$ of (2.1.1), namely:

$$
\mathcal{E}_{(\text{ply})}[\mu] = \int E(\xi, \eta) \, d\mu(\xi) d\mu(\eta),
$$

$$
E(\xi, \eta) = \frac{c_q}{2} \left(|\xi|^q + |\eta|^q \right) - \frac{\beta\pi(\omega_1 + \omega_2)}{2} |\xi - \eta| . \tag{B.0.13}
$$

Lemma B.0.16 *Let $B_\delta(\mu)$ be the ball in $\mathcal{M}^1(\mathbb{R})$ centred at μ and of radius δ with respect to D_V (1.6.1). Then, for any $\mu \in \mathcal{M}^1(\mathbb{R})$, it holds*

$$
\liminf_{\delta \to 0} \liminf_{N \to \infty} N^{-(2+\alpha_q)} \ln \overline{\mathcal{P}}_N\left[B_\delta(\mu)\right] \geq -\mathcal{E}_{(\text{ply})}[\mu] . \tag{B.0.14}
$$

Proof Let $\mu \in \mathcal{M}^1(\mathbb{R})$ and $\delta > 0$. If $\int |x|^q \, d\mu(x) = +\infty$, then $\mathcal{E}_{(\text{ply})}[\mu] = +\infty$ and there is nothing to prove. Thus we may assume from the very beginning that $\int |x|^q \, d\mu(x) < +\infty$. If $M > 0$ is large enough, we have $\mu([-M \,;\, M]) \neq 0$, and we can introduce:

$$
\mu_M = \frac{\mathbf{1}_{[-M\,;\,M]} \cdot \mu}{\mu\big([-M \,;\, M]\big)} \tag{B.0.15}
$$

which is now a compactly supported measure. We will obtain the lower bound for $\overline{\mathcal{P}}_N\left[B_\delta(\mu)\right]$ by restricting to configurations close enough to the classical positions of μ_M, and only at the end, see how the estimate behaves when $M \to \infty$. For any given integer N, we define:

$$\forall a \in [\![1, N]\!], \qquad x_a^{N,M} = \inf\left\{x \in \mathbb{R} \;:\; \int_{-\infty}^{x} \mathrm{d}\mu_M \geq \frac{a}{N+1}\right\}. \tag{B.0.16}$$

When $N \to \infty$, $L_N^{(x^{N,M})}$ approximates μ_M for the Vasershtein distance, so there exists N_δ such that, for any $N \geq N_\delta$, we have the inclusion:

$$\begin{aligned} \Omega_\delta &:= \left\{\boldsymbol{\lambda} \in \mathbb{R}^N \;:\; \forall a \in [\![1, N]\!],\ \left|\lambda_a - x_a^{N,M}\right| < \delta/2\right\} \\ &\subseteq \left\{\boldsymbol{\lambda} \in \mathbb{R}^N \;:\; D_V\big(\mu_M, L_N^{(\boldsymbol{\lambda})}\big) < \delta\right\}. \end{aligned} \tag{B.0.17}$$

Subsequently:

$$\begin{aligned} \overline{\mathcal{P}}_N\big[B_\delta(\mu_M)\big] \geq\ & N^{N\alpha_q} \int_{\Omega_\delta} \prod_{a<b}^{N} \Big\{ \sinh\big[\pi\omega_1 N^{\alpha_q}(\lambda_a - \lambda_b)\big] \sinh\big[\pi\omega_2 N^{\alpha_q}|\lambda_a - \lambda_b|\big]\Big\}^{\beta} \\ & \times \prod_{a=1}^{N} \mathrm{e}^{-W(N^{\alpha_q}\lambda_a)} \cdot \mathrm{d}^N\boldsymbol{\lambda}\,. \end{aligned} \tag{B.0.18}$$

It follows from the lower bound

$$|\sinh(x)| \geq \frac{\mathrm{e}^{|x|}}{2}\frac{|x|}{1+|x|} \tag{B.0.19}$$

from the lower bound for W in (B.0.11), and $q\alpha_q = \alpha_q + 1$, that:

$$\begin{aligned} \overline{\mathcal{P}}_N\big[B_\delta(\mu)\big] \geq\ & \frac{e^{N(\alpha_q \ln N + \tau_\epsilon)}}{2^{\beta N(N-1)}} \int_{\Omega_\delta} \exp\left\{\pi\beta(\omega_1 + \omega_2)N^{\alpha_q}\sum_{a<b}^{N}|\lambda_a - \lambda_b| - N^{\alpha_q+1}c_q(1+\epsilon)\sum_{a=1}^{N}|\lambda_a|^q\right\} \\ & \times \prod_{a<b}^{N}\Big\{g_N(\lambda_a - \lambda_b)\Big\}^{\beta} \cdot \mathrm{d}^N\boldsymbol{\lambda}\,. \end{aligned} \tag{B.0.20}$$

where we have set

$$g_N(\lambda) = \frac{\pi\omega_1 N^{\alpha_q}|\lambda|}{1 + \pi\omega_1 N^{\alpha_q}|\lambda|} \cdot \frac{\pi\omega_2 N^{\alpha_q}|\lambda|}{1 + \pi\omega_2 N^{\alpha_q}|\lambda|} \tag{B.0.21}$$

Now, we would like to replace λ_a by $x_a^{N,M}$. Since the configurations $\boldsymbol{\lambda} \in \Omega_\delta$ satisfy $|x_a^{N,M} - \lambda_a| < \delta/2$, we have:

$$\sum_{a<b}|\lambda_a - \lambda_b| \geq -N(N-1)\frac{\delta}{2} + \sum_{a<b}(x_b^{N,M} - x_a^{N,M})\,. \tag{B.0.22}$$

Since $q > 1$, we also deduce from the mean value theorem:

$$|\lambda_a|^q \le |x_a^{N,M}|^q + \frac{q\delta}{2}\big(|x_a^{N,M}| + \delta/2\big)^{q-1} \tag{B.0.23}$$

and thus

$$\begin{aligned} -(1+\epsilon)|\lambda_a|^q &= -(1+\epsilon)\,|x_a^{N,M}|^q + \frac{\delta}{c_q}\,h_{\epsilon,\delta}(x_a^{N,M}) \\ h_{\epsilon,\delta}(x) &= \frac{qc_q}{2}(1+\epsilon)\cdot\big(|x| + \delta/2\big)^{q-1}. \end{aligned} \tag{B.0.24}$$

These inequalities yield the lower bound:

$$\overline{\mathcal{P}}_N\big[B_\delta(\mu)\big] \ge \exp\bigg\{C\,N^2 - N^{2+\alpha_q}\bigg(\delta\Big\{C' + \int h_{\epsilon,\delta}(\xi)\mathrm{d}L_N^{(\boldsymbol{x}^{N,M})}(\xi)\Big\} + \mathcal{E}_{(\mathrm{ply})}[L_N^{(\boldsymbol{x}^{N,M})}] + \epsilon c_q\int |\xi|^q\,\mathrm{d}L_N^{(\boldsymbol{x}_{N,M})}(\xi)\bigg)\bigg\}\cdot G_{N,\delta} \tag{B.0.25}$$

for some irrelevant, N and δ independent, constants $C, C' > 0$. Furthermore, the factor $G_{N,\delta}$ reads

$$G_{N,\delta} = \int_{\Omega_\delta^{\mathrm{ord}}} \prod_{a>b}^{N} \{g_N(\lambda_b - \lambda_a)\}^\beta\cdot\mathrm{d}^N\boldsymbol{\lambda} \tag{B.0.26}$$

in which $\Omega_\delta^{\mathrm{ord}} = \Omega_\delta \cap \{\boldsymbol{\lambda}\in\mathbb{R}^N\ :\ \lambda_1 < \cdots < \lambda_N\}$.

To find a lower bound for $G_{N,\delta}$, we can restrict further to configurations such that $u_a = \lambda_a - x_a^{N,M}$ increases with $a \in [\![1, N]\!]$, and satisfies $|u_1| < \delta/(2N)$ and $|u_{a+1} - u_a| \le \delta/2N$ for any $a \in [\![1, N-1]\!]$. Using that $\xi \mapsto g_N(\xi)$ is increasing on $\mathbb{R}_+$, we have:

$$G_{N,\delta} \ge \int_{[-\delta/2,\delta/2]^N} \prod_{a=1}^{N-1}\{g_N(u_{a+1}-u_a)\}^{\beta(N-a)}\cdot\mathrm{d}^N u \ge \int_{[0,\delta/2N]^N}\prod_{a=2}^{N}\{g_N(v_a)\}^{\beta(N-a+1)}\cdot\mathrm{d}^N\boldsymbol{v}. \tag{B.0.27}$$

Now, using an arithmetic-geometric upper bound for the denominator in $g_N(v)$, we can write:

$$\forall v\in[0,\delta/2N],\qquad g_N(v) \ge \frac{N^{\alpha_q+1}\pi\sqrt{\omega_1\omega_2}}{2\delta}\cdot|v|^2 \ge \widetilde{C}N^{\alpha_q+1}\cdot|v|^2 \tag{B.0.28}$$

for some irrelevant $C' > 0$ independent of δ provided that $\delta < 1$. So, we arrive to:

$$G_{N,\delta} \ge \frac{\delta}{2N}\cdot\big(\widetilde{C}\,N^{\alpha_q-1}\big)^{\beta N(N-1)/2}\cdot\prod_{a=2}^{N-1}\frac{(\delta/2N)^{2\beta(N-a+1)}}{2\beta(N-a+1)+1} \ge e^{\widetilde{C}'\,N^2\ln N} \tag{B.0.29}$$

for some $\widetilde{C}' > 0$ independent of δ. Hence, ultimately

$$\overline{\mathcal{P}}_N\big[B_\delta(\mu)\big] \geq e^{\widetilde{C}'' N^2 \ln N} \exp\bigg\{ - N^{2+\alpha_q}\bigg[\delta\Big(C' + \int h_{\epsilon,\delta}(\xi)\,\mathrm{d}L_N^{(\boldsymbol{x}^{N,M})}(\xi)\Big) + \mathcal{E}_{(\mathrm{ply})}\big[L_N^{(\boldsymbol{x}^{N,M})}\big] + \epsilon c_q \int |\xi|^q\,\mathrm{d}L_N^{(\boldsymbol{x}^{N,M})}(\xi)\bigg]\bigg\} \qquad \text{(B.0.30)}$$

To establish the desired result (B.0.14), we only need to focus on the last exponential. If ϕ is a $\mathcal{C}^1$ function of p real variables, we denote:

$$\phi^{[M]}(\xi) = \min\Big[\phi(\xi)\,;\;||\phi||_{L^\infty([-M\,;\,M]^p)}\Big] \qquad \text{(B.0.31)}$$

which has the advantage of being bounded and Lipschitz. Since μ_M is supported on $[-M\,;\,M]$, so must be the classical positions $x_a^{N,M}$, and we can apply the truncation to all the functions against which $L_N^{(\boldsymbol{x}^{N,M})}$ is integrated. In particular, we make appear the truncated functional:

$$\mathcal{E}_{(\mathrm{ply})}^{[M]}[\mu] \;=\; \int E^{[M]}(\xi,\eta)\,\mathrm{d}\mu(\xi)\mathrm{d}\mu(\eta)\,. \qquad \text{(B.0.32)}$$

The advantage is that now, all functions to be integrated are Lipschitz bounded. Since, $D_V\big(\mu_M, L_N^{(\boldsymbol{x}^{N,M})}\big) \to 0$ when $N \to \infty$, we get:

$$\liminf_{N\to\infty} \frac{\ln \overline{\mathcal{P}}_N\big[B_\delta(\mu)\big]}{N^{2+\alpha_q}} \;\geq\; -\delta\Big(C + \int h_{\epsilon,\delta}(\xi)\,\mathrm{d}\mu_M(\xi)\Big) - \mathcal{E}_{(\mathrm{ply})}^{[M]}[\mu_M] - \epsilon c_q \int \big\{\max(|\xi|, M)\big\}^q\,\mathrm{d}\mu_M(\xi)\,. \qquad \text{(B.0.33)}$$

The right-hand is an affine function of ϵ, and at this stage, we can send ϵ to 0:

$$\liminf_{N\to\infty} N^{-(2+\alpha_q)} \ln \overline{\mathcal{P}}_N\big[B_\delta(\mu_M)\big] \geq -\delta\Big(C + \int h_{0,\delta}(\xi)\,\mathrm{d}\mu_M(\xi)\Big) - \mathcal{E}_{(\mathrm{ply})}[\mu_M]\,. \qquad \text{(B.0.34)}$$

Now, for any fixed δ, there exists M_δ such that, for any $M \geq M_\delta$, $D_V(\mu, \mu_M) \leq \delta$, and consequently:

$$\liminf_{N\to\infty} N^{-(2+\alpha_q)} \ln \overline{\mathcal{P}}_N\big[B_{2\delta}(\mu)\big] \geq -\delta\Big(C + \int h_{0,\delta}(\xi)\,\mathrm{d}\mu_M(\xi)\Big) - \mathcal{E}_{(\mathrm{ply})}[\mu_M]\,. \qquad \text{(B.0.35)}$$

We could replace $\mathcal{E}_{(\text{ply})}^{[M]}$ by $\mathcal{E}_{(\text{ply})}$ here because μ_M is supported on $[-M\,;\,M]$. Now, we can consider sending $M \to \infty$. Since we have the bound:

$$\forall \xi, \eta \in \mathbb{R}, \qquad E(\xi,\eta) \le C'\left(1 + |\xi|^q + |\eta|^q\right), \qquad h_{0,\delta} \le C'\left(1 + |\xi|^q\right) \quad \text{(B.0.36)}$$

and we assumed that $\int |\xi|^q \,\mathrm{d}\mu(\xi) < +\infty$, we get by dominated convergence:

$$\liminf_{N\to\infty} N^{-(2+\alpha_q)} \ln \overline{\mathcal{P}}_N\big[B_{2\delta}(\mu)\big] \ge -\delta\left(C + \int h_{0,\delta}(\xi)\,\mathrm{d}\mu(\xi)\right) - \mathcal{E}_{(\text{ply})}[\mu]\,. \quad \text{(B.0.37)}$$

Last but not least, sending $\delta \to 0$, the first term disappears and we find:

$$\liminf_{\delta\to 0}\liminf_{N\to\infty} N^{-(2+\alpha_q)} \ln \overline{\mathcal{P}}_N\big[B_{2\delta}(\mu)\big] \ge -\mathcal{E}_{(\text{ply})}[\mu]\,. \quad \text{(B.0.38)}$$

∎

Upper Bound

In this paragraph, we complete our estimate by an upper bound on the probability of small Vasershtein balls:

Lemma B.0.17

$$\limsup_{\delta\to 0}\limsup_{N\to\infty} N^{-(2+\alpha_q)} \ln \overline{\mathcal{P}}_N\big[B_{\delta}(\mu)\big] \le -\mathcal{E}_{(\text{ply})}[\mu] \quad \text{(B.0.39)}$$

Proof Let $\mu \in \mathcal{M}^1(\mathbb{R})$. In order to establish an upper bound, we use that $|\sinh(x)| \le \mathrm{e}^{|x|}$ and the upper bound in (B.0.11) for the potential W. This makes appear again the function $\mathcal{E}_{(\text{ply})}$ of (B.0.13):

$$\overline{\mathcal{P}}_N\big[B_\delta(\mu)\big] \le e^{N(\alpha_q \ln N + \tau_\epsilon)}$$
$$\times \int_{L_N^{(\boldsymbol{\lambda})}\in B_\delta(\mu)} \exp\left\{-N^{2+\alpha_q}\left(-2c_q\epsilon\int |\xi|^q\,\mathrm{d}L_N^{(\boldsymbol{\lambda})} + \mathcal{E}_{(\text{ply})}[L_N^{(\boldsymbol{\lambda})}]\right)\right\} \times \prod_{a=1}^{N} e^{-N^{\alpha_q+1} c_q\epsilon|\lambda_a|^q}\cdot \mathrm{d}^N\boldsymbol{\lambda} \quad \text{(B.0.40)}$$

where we have put aside one exponential decaying with rate ϵ to ensure later convergence of the integral. If $M > 0$, let us define the truncated functional:

$$\mathcal{E}_{(\text{ply})}^{\{M,\epsilon\}}[\mu] = \int E^{\{M,\epsilon\}}(\xi,\eta)\,\mathrm{d}\mu(\xi)\mathrm{d}\mu(\eta),$$

$$E^{\{M,\epsilon\}} = \min\big[M\,;\,E(\xi,\eta) - c_q\epsilon\left(|\xi|^q + |\eta|^q\right)\big]\,. \quad \text{(B.0.41)}$$

Since $E^{\{M,\epsilon\}}$ is a Lipschitz function bounded by M, with Lipschitz constant bounded by $O(M^{1-1/q})$, we deduce the following bounds when the event $L_N^{(\lambda)} \in B_\delta(\mu)$ is realised:

$$\left|\mathcal{E}_{(\mathrm{ply})}^{\{M,\epsilon\}}[L_N^{(\lambda)}] - \mathcal{E}_{(\mathrm{ply})}^{\{M,\epsilon\}}[\mu]\right| \leq C\,\delta\, M\,, \tag{B.0.42}$$

for some constant $C > 0$ independent of N, δ and ϵ. Therefore:

$$\overline{\mathcal{P}}_N\big[B_\delta(\mu)\big] \leq \exp\left\{C'N\ln N + N^{\alpha_q+2}\left(CM\cdot\delta - \mathcal{E}_{(\mathrm{ply})}^{\{M,\epsilon\}}[\mu]\right)\right\}\cdot\left(\int_{\mathbb{R}} \mathrm{e}^{-c_q\epsilon\,|\lambda|^q}\,\mathrm{d}\lambda\right)^N \tag{B.0.43}$$

It follows that:

$$\limsup_{N\to\infty} N^{-(2+\alpha_q)}\ln\overline{\mathcal{P}}_N\big[B_\delta(\mu)\big] \leq CM\cdot\delta - \mathcal{E}_{(\mathrm{ply})}^{\{M,\epsilon\}}[\mu] \tag{B.0.44}$$

We observe that $-E^{\{M,\epsilon\}}$ is an increasing function of ϵ. We can now let $\epsilon \to 0$ by applying the monotone convergence theorem:

$$\limsup_{N\to\infty} N^{-(2+\alpha_q)}\ln\overline{\mathcal{P}}_N\big[B_\delta(\mu)\big] \leq C\,M\cdot\delta - \mathcal{E}_{(\mathrm{ply})}^{\{M,0\}}[\mu]\,. \tag{B.0.45}$$

Then, sending $\delta \to 0$ erases the first term, and finally letting $M \to \infty$ using again monotone convergence:

$$\limsup_{\delta\to 0}\limsup_{N\to\infty} N^{-(2+\alpha_q)}\ln\overline{\mathcal{P}}_N\big[B_\delta(\mu)\big] \leq -\mathcal{E}_{(\mathrm{ply})}[\mu]\,, \tag{B.0.46}$$

Notice that monotone convergence proves this last inequality even in the case where $\mathcal{E}_{(\mathrm{ply})}[\mu] = +\infty$. ∎

B.0.4 Partition Function and Equilibrium Measure

By applying the reasoning described in [5], to the lower bounds (Lemma B.0.16) and upper bounds (Lemma B.0.17), along with the property of exponential tightness (Lemma B.0.15), we deduce that $\mathcal{E}_{(\mathrm{ply})}$ is a good rate function for large deviations, *i.e.*

$$\begin{array}{ll} \text{for any open set } \Omega \subseteq \mathcal{M}^1(\mathbb{R}), & \liminf\limits_{N\to+\infty} N^{-(2+\alpha_q)}\ln\overline{\mathcal{P}}_N[\Omega] \geq -\inf\limits_{\mu\in\Omega}\mathcal{E}_{(\mathrm{ply})}[\mu]\,, \\ \text{for any closed set } F \subseteq \mathcal{M}^1(\mathbb{R}) & \limsup\limits_{N\to+\infty} N^{-(2+\alpha_q)}\ln\overline{\mathcal{P}}_N[F] \leq -\inf\limits_{\mu\in F}\mathcal{E}_{(\mathrm{ply})}[\mu]\,. \end{array} \tag{B.0.47}$$

These two estimates, taken for $\Omega = F = \mathcal{M}^1(\mathbb{R})$, lead to

$$\lim_{N\to\infty} N^{-(2+\alpha_q)}\ln\mathfrak{Z}_N[W] = -\inf_{\mu\in\mathcal{M}^1(\mathbb{R})}\mathcal{E}_{(\mathrm{ply})}[\mu]\,. \tag{B.0.48}$$

The proof of the statements relative to the existence, uniqueness and characterisation of the minimiser of $\mathcal{E}_{(\mathrm{ply})}$ is identical to those for the usual logarithmic energy [6]—and even simpler since there is no logarithmic singularity here. The minimiser is denoted $\mu_{\mathrm{eq}}^{(\mathrm{ply})}$ and it is characterised by the existence of a constant $C_{\mathrm{eq}}^{(\mathrm{ply})}$ such that:

$$c_q\,|\xi|^q - \pi\beta(\omega_1+\omega_2)\int |\xi-\eta|\,\mathrm{d}\mu_{\mathrm{eq}}^{(\mathrm{ply})}(\eta) = C_{\mathrm{eq}}^{(\mathrm{ply})} \quad \text{for } \xi\,, \quad \mu_{\mathrm{eq}}^{(\mathrm{ply})} \text{ everywhere} \tag{B.0.49}$$

$$c_q\,|\xi|^q - \pi\beta(\omega_1+\omega_2)\int |\xi-\eta|\,\mathrm{d}\mu_{\mathrm{eq}}^{(\mathrm{ply})}(\eta) \geq C_{\mathrm{eq}}^{(\mathrm{ply})} \quad \text{for any } \xi \in \mathbb{R} \tag{B.0.50}$$

The construction of the solution of this regular integral equation is left as an exercise to the reader. We only give the final result in the announcement of Theorem 2.4.1. Actually, the fact that (2.1.4) is a solution can be checked directly by integration by parts, and we can conclude by uniqueness.

■

Appendix C
Properties of the N-Dependent Equilibrium Measure

We give here elements for the proof of Theorem 2.4.2, which establishes the main properties of the minimiser of:

$$\mathcal{E}_N[\mu] = \frac{1}{2} \int \bigg(V(\xi) + V(\eta) - \frac{\beta}{N^{\alpha}} \ln\bigg\{ \prod_{p=1}^{2} \sinh\big[\pi N^{\alpha} \omega_p(\xi - \eta)\big] \bigg\} \bigg) \mathrm{d}\mu(\xi)\mathrm{d}\mu(\eta) \,. \tag{C.0.1}$$

among probability measure μ on $\mathbb{R}$, with N considered as a fixed parameter. As for any probability measures μ, ν and $\alpha \in [0, 1]$,

$$\mathcal{E}_N[\alpha\mu + (1-\alpha)\nu] - \alpha\mathcal{E}_N[\mu] - (1-\alpha)\mathcal{E}_N[\nu] = -\alpha(1-\alpha)\mathfrak{D}^2\big[\mu - \nu, \mu - \nu\big] \,,$$

with $\mathfrak{D}$ as in 3.1.2, $\mathcal{E}_N$ is strictly convex, and the standard arguments of potential theory [6, 7] show that it admits a unique minimiser, denoted $\mu_{\mathrm{eq}}^{(N)}$. More precisely, one can prove that $\mu_{\mathrm{eq}}^{(N)}$ has a continuous density $\rho_{\mathrm{eq}}^{(N)}$ (as soon as V is $\mathcal{C}^2$) and is supported on a compact of $\mathbb{R}$ (since the potential here is confining for any given value of N) a priori depending on N, see e.g. [8, Lemma 2.4]. What we really need to justify in our case is that:

- (0) the support of $\mu_{\mathrm{eq}}^{(N)}$ is contained in a compact independent of N;
- (i) $\mu_{\mathrm{eq}}^{(N)}$ is supported on a segment;
- (ii) $\rho_{\mathrm{eq}}^{(N)}$ does not vanish on the interior of this segment and vanishes like a square root at the edges.

As a preliminary, we recall that the characterisation of the equilibrium measure is obtained by writing that $\mathcal{E}_N[\mu_{\mathrm{eq}}^{(N)} + \epsilon\nu] \geq \mathcal{E}_N[\mu_{\mathrm{eq}}^{(N)}]$ for all $\epsilon > 0$, all measures ν with zero mass and such that $\mu_{\mathrm{eq}}^{(N)} + \epsilon\nu$ is non-negative. The resulting condition can be formulated in terms of the effective potential introduced in (3.1.16):

G. Borot et al., *Asymptotic Expansion of a Partition Function Related to the Sinh-model*, Mathematical Physics Studies,
DOI 10.1007/978-3-319-33379-3

$$V_{N;\text{eff}}(\xi) \;=\; U_{N;\text{eff}}(\xi) - \inf_{\mathbb{R}} U_{N;\text{eff}}, \qquad U_N(\xi) = V(\xi) \;-\; \int s_N(\xi - \eta)\,\mathrm{d}\mu_{\text{eq}}^{(N)}(\eta) \tag{C.0.2}$$

with the two-point interaction kernel:

$$s_N(\xi) \;=\; \frac{\beta}{2N^{\alpha}} \ln\Big[\sinh\big(\pi\omega_1 N^{\alpha}\xi\big)\,\sinh\big(\pi\omega_2 N^{\alpha}\xi\big)\Big]. \tag{C.0.3}$$

The equilibrium measure is characterised by the condition:

$$V_{N;\text{eff}}(\xi) \geq 0\,, \qquad \text{with equality } \mu_{\text{eq}}^{(N)} \text{ almost everywhere}. \tag{C.0.4}$$

Proof of (0). Let $m_N > 0$ such that the support of $\mu_{\text{eq}}^{(N)}$ is contained in $[-m_N, m_N]$. For $|\xi| > 2m_N$, we have an easy lower bound:

$$\begin{aligned} \Big| \int s_N(\xi - \eta)\,\mathrm{d}\mu_{\text{eq}}^{(N)}(\eta) \Big| &\geq \frac{\beta}{2N^{\alpha}} \ln\Big[\sinh\big(\pi\omega_1 N^{\alpha} m_N\big)\,\sinh\big(\pi\omega_2 N^{\alpha} m_N\big)\Big] \\ &\geq \frac{\beta\pi(\omega_1 + \omega_2)}{2}\, m_N + \mathrm{O}(1) \end{aligned} \tag{C.0.5}$$

where the remainder is bounded uniformly when $N \to \infty$ and $m_N \to \infty$. By the growth assumption on the potential, there exists constant $C, C' > \epsilon > 0$ such that

$$V(\xi) \geq C|\xi|^{1+\epsilon} + C' \tag{C.0.6}$$

Therefore, we can choose $m := 2m_N$ large enough and independent of N such that $V_{N;\text{eff}}(\xi) > 0$ for any $|\xi| > m$. This guarantees that the support of $\mu_{\text{eq}}^{(N)}$ is included in the compact $[-m\,;\,m]$ for any N.

Proof of (i).

We observe that $-s_N$ is strictly convex:

$$s_N''(\xi) \;=\; -\frac{\beta N^{a}}{2} \sum_{p=1}^{2} \frac{(\pi\omega_p)^2}{\big(\sinh \pi\omega_p \xi\big)^2} \;<\; 0\,. \tag{C.0.7}$$

Since V is assumed strictly convex and $\mu_{\text{eq}}^{(N)}$ is a positive measure, it implies that $V_{N;\text{eff}}$ is strictly convex. Therefore, the locus where it reaches its minimum must be a segment. So, there exists $a_N < b_N$ such that $[a_N\,;\,b_N]$ is the support of $\mu_{\text{eq}}^{(N)}$. This strict convexity also ensures that

$$V_{N;\text{eff}}'(\xi) \;>\; 0 \quad \text{for any} \quad \xi > b_N, \quad V_{N;\text{eff}}'(\xi) \;<\; 0 \quad \text{for any} \quad \xi < a_N\,. \tag{C.0.8}$$

Proof of (ii).

This piece of information is enough so as to build the representation:

$$\rho_{\mathrm{eq}}^{(N)}(\xi) = \mathcal{W}_N[V'] \cdot \mathbf{1}_{[a_N\,;b_N]}(\xi) \tag{C.0.9}$$

for the equilibrium measure. Indeed, we constructed $\mathcal{W}_N[H]$ in Section 4.3.4 so that it provides the unique solution to:

$$\forall \xi \in]a_N\,; b_N[, \qquad \fint_{a_N}^{b_N} s_N[N^a(\xi-\eta)]\,\mathrm{d}\mu_{\mathrm{eq}}^{(N)}(\eta) \;=\; V'(\xi) \tag{C.0.10}$$

which extends continuously on $[a_N\,; b_N]$, and this was only possible when $\mathcal{X}_N[V'] = 0$ in terms of the linear form introduced in Definition 3.3.5. Since the equilibrium measure exists, this imposes the constraint:

$$\mathcal{X}_N[V'] = 0\,. \tag{C.0.11}$$

Besides, since the total mass of (C.0.9) must be 1, we must also have:

$$\int_{a_N}^{b_N} \mathcal{W}_N[V'] = 1\,. \tag{C.0.12}$$

At this stage, we can use Corollary 6.2.2, which shows that (C.0.11) and (C.0.12) determine uniquely the large-N asymptotic expansion of a_N and b_N, in particular there exists $a < b$ such that $(a_N, b_N) \to (a, b)$ with rate of convergence $N^{-\alpha}$. Besides, the leading behaviour at $N \to \infty$ of $\mathcal{W}_N$ is described by Propositions 5.1.4 and 5.1.6. It follows from the reasonings outlined in the proof of Proposition 5.2.2 that

$$\rho_{\mathrm{eq}}^{(N)}(\xi) = \mathcal{W}_N[V'](\xi)$$
$$= \begin{cases} \frac{V''(\xi)}{2\pi\beta(\omega_1+\omega_2)} + \mathrm{O}(N^{-\alpha}) & \xi \in \Big[a_N + \frac{(\ln N)^2}{N^\alpha}\,; b_N - \frac{(\ln N)^2}{N^\alpha}\Big] \\ V''(b_N)\,\mathfrak{a}_0\big(N^\alpha(b_N-\xi)\big) + \mathrm{O}\Big(\frac{(\ln N)^3}{N^\alpha}\sqrt{N^\alpha(b_N-\xi)}\Big) & \xi \in [b_N - (\ln N)^2\cdot N^{-\alpha}\,; b_N] \\ V''(a_N)\,\mathfrak{a}_0\big(N^\alpha(\xi-a_N)\big) + \mathrm{O}\Big(\frac{(\ln N)^3}{N^\alpha}\sqrt{N^\alpha(\xi-a_N)}\Big) & \xi \in [a_N\,; a_N + (\ln N)^2\cdot N^{-\alpha}] \end{cases} \tag{C.0.13}$$

Therefore, for N large enough, $\rho_{\mathrm{eq}}^{(N)}(\xi) > 0$ on $[a_N\,; b_N]$. The vanishing like a square root at the edges then follows from the properties of the $\mathfrak{a}$s as established in Lemma 5.1.10. In fact, one even has

$$\lim_{\xi \to b_N^-} \frac{\rho_{\mathrm{eq}}^{(N)}(\xi)}{\sqrt{b_N - \xi}} = N^{\alpha/2} \Big(V''(b_N) \cdot \lim_{x \to 0} x^{-1/2} \mathfrak{a}_0(x) \; + \; \mathrm{O}(N^{-\alpha}) \Big\} = \frac{N^{\alpha/2} V''(b)}{\pi \beta \sqrt{\pi(\omega_1 + \omega_2)}} + \mathrm{O}(N^{-\alpha/2}) \, . \tag{C.0.14}$$

■

Appendix D
The Gaussian Potential

In this appendix we focus on the case of a Gaussian potential and establish two results. On the one hand, we establish in Lemma D.0.18 that, for N large enough, there exists a unique sequence of Gaussian potential $V_{G;N} = g_N \lambda^2 + t_N \lambda$ such that their associated equilibrium measure has support $\sigma_{\mathrm{eq}}^{(N)} = [a_N ; b_N]$. On the other hand we show, in Proposition D.0.19, that the partition function associated with any Gaussian potential can be explicitly evaluated, and thus is amenable to a direct asymptotic analysis when $N \to \infty$. Note that when ω_1/ω_2 is rational, such Gaussian partition functions have been evaluated in [9] by using biorthogonal analogues of the Stieltjes–Wigert orthogonal polynomials.

Lemma D.0.18 *There exists a unique sequence of Gaussian potentials*

$$V_{G;N} = g_N \lambda^2 + t_N \lambda \tag{D.0.1}$$

such that their associated equilibrium measure has support $\sigma_{\mathrm{eq}}^{(N)} = [a_N ; b_N]$. *The coefficients* g_N, t_N *take the form*

$$g_N = \pi\beta(\omega_1+\omega_2)\left\{b_N - a_N + N^{-\alpha}\sum_{p=1}^{2}\frac{1}{\pi\omega_p}\ln\left(\frac{\omega_1\omega_2}{\omega_p(\omega_1+\omega_2)}\right)\right\}^{-1} + \mathrm{O}(N^{-\infty}) \tag{D.0.2}$$

and

$$t_N = -(a_N + b_N) g_N + \mathrm{O}\big(N^{-\infty}\big) . \tag{D.0.3}$$

Proof Let $V_G(\lambda) = g\lambda^2 + t\lambda$ be any Gaussian potential. Since it strictly convex, all previous results apply. Suppose that V_G gives rise to an equilibrium measure supported on $\sigma_{\mathrm{eq}}^{(N)} = [a_N ; b_N]$. This means that the potential V_G has to satisfy the system of two equations that are linear in V_G':

G. Borot et al., *Asymptotic Expansion of a Partition Function Related to the Sinh-model*, Mathematical Physics Studies,
DOI 10.1007/978-3-319-33379-3

$$\int_{a_N}^{b_N} \mathcal{W}_N[V'_G](\xi)\,\mathrm{d}\xi = 1 \quad \text{and} \quad \int_{\mathbb{R}+\mathrm{i}\epsilon'} \frac{\mathrm{d}\mu}{2\mathrm{i}\pi}\,\chi_{11}(\mu) \int_{a_N}^{b_N} V'_G(\eta)\mathrm{e}^{\mathrm{i}N^\alpha \mu(\eta-b_N)}\,\mathrm{d}\eta = 0\,. \tag{D.0.4}$$

It follows from the multi-linearity in (g, t) of V_G and from the evaluation of single integrals carried out in Lemma 6.1.2 and Proposition 6.1.6 that there exist two linear forms L_1, L_2 of (g, t) whose norm is a $\mathrm{O}(N^{-\infty})$ and such that

$$1 = \frac{g}{\pi\beta(\omega_1+\omega_2)}\bigg\{(b_N - a_N) + \frac{1}{N^\alpha}\cdot\sum_{p=1}^{2}\frac{1}{\omega_p\pi}\ln\Big(\frac{\omega_1\omega_2}{\omega_p(\omega_1+\omega_2)}\Big)\bigg\} + L_1(g,t) \tag{D.0.5}$$

where we have used that

$$\int_0^{+\infty} \mathfrak{b}_0(x)\,\mathrm{d}x = \frac{1}{2\pi\beta(\omega_1+\omega_2)}\cdot\sum_{p=1}^{2}\frac{1}{\omega_p\pi}\ln\Big(\frac{\omega_1\omega_2}{\omega_p(\omega_1+\omega_2)}\Big) \tag{D.0.6}$$

a formula that can be established with the help of (5.1.71) and (5.1.91). One also obtains that

$$0 = \frac{2}{N^\alpha\sqrt{\omega_1+\omega_2}}\Big(g(b_N+a_N)+t\Big) + L_2(g,t)\,. \tag{D.0.7}$$

In virtue of the unique solvability of perturbations of linear solvable systems, the existence and uniqueness of the potential $V_{G;N}$ follows. ∎

Proposition D.0.19 *The partition function $Z_N[V_G]$ at $\beta = 1$ associated with the Gaussian potential $V_G(\lambda) = g\lambda^2 + t\lambda$ can be explicitly computed as*

$$Z_N[V_G]_{|\beta=1} = \frac{N!}{2^{N(N-1)}}\Big(\frac{\pi}{gN^{1+\alpha}}\Big)^{N/2} \times \exp\bigg\{\frac{N^{2+\alpha}t^2}{4g} + \frac{\pi^2(\omega_1+\omega_2)^2}{12g}N^\alpha(N^2-1)\bigg\}\prod_{j=1}^{N}\Big(1-\mathrm{e}^{-\frac{2jN^\alpha}{gN}\pi^2\omega_1\omega_2}\Big)^{N-j}\,. \tag{D.0.8}$$

Proof We can get rid of the linear term in the potential by a translation of the integration variables. Then

$$Z_N[V_G]_{|\beta=1} = \exp\bigg\{\frac{N^{2+\alpha}t^2}{4g}\bigg\}\cdot Z_N[\widetilde{V}_G]_{|\beta=1} \quad \text{where} \quad \widetilde{V}_G(\lambda) = g\lambda^2\,. \tag{D.0.9}$$

Further, the products over hyperbolic sinh's can be recast as two Vandermonde determinants

$$\prod_{a<b}^{N}\left\{\sinh\left[\pi\omega_1 N^{\alpha}(\lambda_a-\lambda_b)\right]\sinh\left[\pi\omega_2 N^{\alpha}(\lambda_a-\lambda_b)\right]\right\}$$
$$=\prod_{a=1}^{N}\left\{\frac{\mathrm{e}^{-\pi(\omega_1+\omega_2)N^{\alpha}(N-1)\lambda_a}}{2^{N-1}}\right\}\cdot\prod_{p=1}^{2}\det_N\left[\mathrm{e}^{2\pi\omega_p N^{\alpha}\lambda_j(k-1)}\right]. \quad \text{(D.0.10)}$$

Inserting this formula into the multiple integral representation for $Z_N[V_G]_{|\beta=1}$ and using the symmetry of the integrand, one can replace one of the determinants by $N!$ times the product of its diagonal elements. Then, the integrals separate and one gets:

$$Z_N[\widetilde{V}_G]_{|\beta=1} = \frac{N!}{2^{N(N-1)}}\cdot\det_N\left[\int_{\mathbb{R}}\mathrm{e}^{-\pi(\omega_1+\omega_2)N^{\alpha}(N-1)\lambda}\mathrm{e}^{2\pi N^{\alpha}[\omega_1(k-1)+\omega_2(j-1)]\lambda}\cdot\mathrm{e}^{-gN^{1+\alpha}\lambda^2}\mathrm{d}\lambda\right]. \quad \text{(D.0.11)}$$

The integral defining the (k, j)th entry of the determinant is Gaussian and can thus be computed. This yields, upon factorising the trivial terms arising in the determinant,

$$Z_N[\widetilde{V}_G]_{|\beta=1} = \left(\frac{\pi}{gN^{1+\alpha}}\right)^{\frac{N}{2}}\frac{N!}{2^{N(N-1)}}\cdot\prod_{j=1}^{N}\mathrm{e}^{\frac{\pi^2}{4g}N^{\alpha-1}(\omega_1^2+\omega_2^2)(2j-1-N)^2}\cdot D_N\,, \quad \text{(D.0.12)}$$

where

$$D_N = \det_N\left[\exp\left\{\frac{\pi^2}{2g}N^{\alpha-1}\omega_1\omega_2(2k-N-1)(2j-1-N)\right\}\right]. \quad \text{(D.0.13)}$$

The last determinant can be reduced to a Vandermonde determinant. Indeed, we have:

$$D_N = \exp\left\{\frac{\pi^2}{2g}\omega_1\omega_2(N-1)^2N^{\alpha}\right\}\cdot\cdot\prod_{j=1}^{N}\left\{\mathrm{e}^{-2\frac{\pi^2}{g}\omega_1\omega_2N^{\alpha-1}(N-1)(j-1)}\right\}$$
$$\times\det_N\left[\exp\left\{2\frac{\pi^2}{g}N^{\alpha-1}\omega_1\omega_2(k-1)(j-1)\right\}\right]$$
$$=\exp\left\{-\frac{\pi^2}{2g}\omega_1\omega_2(N-1)^2N^{\alpha}\right\}\cdot\prod_{k>j}^{N}\left(\mathrm{e}^{2\frac{\pi^2}{g}N^{\alpha-1}\omega_1\omega_2(k-1)}-\mathrm{e}^{2\frac{\pi^2}{g}N^{\alpha-1}\omega_1\omega_2(j-1)}\right). \quad \text{(D.0.14)}$$

In order to present the last product into a convergent form, we factor out the largest exponential of each term. The product of these contributions is computable as

$$\prod_{k>l}^{N}\left(\mathrm{e}^{2\frac{\pi^2}{g}N^{\alpha-1}\omega_1\omega_2(k-1)}\right)=\prod_{k=1}^{N-1}\mathrm{e}^{2\frac{\pi^2}{g}N^{\alpha-1}\omega_1\omega_2k^2}=\exp\left\{\frac{\pi^2}{3g}\omega_1\omega_2N^{\alpha}(N-1)(2N-1)\right\}, \quad \text{(D.0.15)}$$

where we took advantage of

$$\sum_{p=1}^{N} p^2 = \frac{N(N+1)(2N+1)}{6} . \tag{D.0.16}$$

Putting all of the terms together leads to the claim. ■

The large-N asymptotic behaviour of the partition function at $\beta = 1$ and associated to a Gaussian potential can be extracted from (D.0.8).

Proposition D.0.20 *Assume* $0 < \alpha < 1$. *We have the asymptotic expansion:*

$$\begin{aligned}\ln Z_N[V_G]|_{\beta=1} = {} & N^{2+\alpha} \cdot \Big[\frac{t^2}{4g} + \frac{\pi^2(\omega_1+\omega_2)^2}{12g}\Big] - N^2 \cdot \ln 2 - N^{2-\alpha} \cdot \frac{g}{12\omega_1\omega_2} \\ & + N^{2-2\alpha} \cdot \frac{g^2\,\zeta(3)}{(2\pi^2\omega_1\omega_2)^2} + (1-\alpha)\,N\ln N + N \cdot \ln\Big(\frac{2/\mathrm{e}}{\sqrt{\omega_1\omega_2}}\Big) \\ & - N^{\alpha} \cdot \frac{\pi^2(\omega_1+\omega_2)^2}{12g} + \ln N \cdot \frac{\alpha+5}{12} + \frac{1}{12}\ln\Big(\frac{128\pi^8\omega_1\omega_2}{g}\Big) + \zeta'(-1) + \mathrm{o}(1) .\end{aligned} \tag{D.0.17}$$

Proof The sole problematic terms demanding some further analysis is the last product in (D.0.8). The latter can be recast as:

$$\begin{aligned}&\prod_{\ell=1}^{N}(1-\mathrm{e}^{-\tau_N\ell})^{N-\ell} \\ &= \Bigg[\frac{M_0\big(\mathrm{e}^{-N\tau_N};\mathrm{e}^{-\tau_N}\big)}{M_0(1;\mathrm{e}^{-\tau_N})}\Bigg]^N \cdot \frac{M_1(1;\mathrm{e}^{-\tau_N})}{M_1\big(\mathrm{e}^{-N\tau_N};\mathrm{e}^{-\tau_N}\big)} \qquad \text{where} \quad \tau_N = \frac{2N^{\alpha}}{gN}\pi^2\omega_1\omega_2\end{aligned} \tag{D.0.18}$$

and $M_r(a,q)$ corresponds to the infinite products $M_r(a;q) = \prod_{\ell=1}^{\infty}(1-aq^{\ell})^{-\ell^r}$.

We will exploit the fact that asymptotics of $M_r(a;\mathrm{e}^{-\tau})$ when $\tau \to 0^+$ up to $o(1)$ can be read-off from the singularities of the Mellin transform of its logarithm

$$\mathfrak{M}_r(a;s) = \int_0^{\infty} \ln M_r(a;\mathrm{e}^{-t})\,t^{s-1}\mathrm{d}t \qquad \text{where} \qquad \ln M_r(a;q) \equiv -\sum_{\ell=1}^{+\infty}\ell^r \ln\big(1-aq^{\ell}\big) . \tag{D.0.19}$$

The above Mellin transform is well-defined for $\mathrm{Re}(s) > r+1$ and can be easily computed. For any $|a| \le 1$, we have:

$$\mathfrak{M}_r(a;s) = \sum_{\ell=1}^{\infty}\sum_{m=1}^{\infty}\frac{\ell^r\,a^m}{m}\int_0^{\infty} t^{s-1}\,\mathrm{e}^{-t\ell m}\,\mathrm{d}t = \Gamma(s)\,\zeta(s-r)\,\mathrm{Li}_{s+1}(a) . \tag{D.0.20}$$

Above, ζ refers to the Riemann zeta function whereas $\mathrm{Li}_s(z)$ is the polylogarithm which is defined by its series expansion in a variable z inside the unit disk:

$$\mathrm{Li}_s(z) = \sum_{k \geq 1} \frac{z^k}{k^s} . \qquad \text{(D.0.21)}$$

Note that, when $\mathrm{Re}\, s > 1$, the series also converges uniformly up to the boundary of the unit disk. We remind that the first two polylogarithms can be expressed in terms of elementary functions:

$$\mathrm{Li}_0(z) = \frac{z}{1-z} \quad \text{and} \quad \mathrm{Li}_1(z) = -\ln(1-z) . \qquad \text{(D.0.22)}$$

In both cases $|a| < 1$ or $a = 1$, $\mathfrak{M}_r(a; s)$ admits a meromorphic extension from $\mathrm{Re}\, s > 1$ to $\mathbb{C}$. When $|a| < 1$ this is readily seen at the level of the series expansion of the polylogarithm whereas when $a = 1$, this follows from $\mathrm{Li}_{s+1}(1) = \zeta(s+1)$. Furthermore, this meromorphic continuation is such that $\mathfrak{M}_r(a; x+\mathrm{i}y) = \mathrm{O}(e^{-c|y|})$, $c > 0$, when $y \to \pm\infty$. This estimate is uniform for a in compact subsets of the open unit disk and for x belonging to compact subsets of $\mathbb{R}$. The same type of bounds also holds for $a = 1$, namely $\mathfrak{M}_r(1; x + \mathrm{i}y) = \mathrm{O}(e^{-c|y|})$, $c > 0$, when $y \to \pm\infty$ for x belonging to a compact subset of $\mathbb{R}$. This is a consequence of three facts:

- $\Gamma(x+\mathrm{i}y)$ decays exponentially fast when $|y| \to +\infty$ and x is bounded, as follows from the Stirling formula;
- $|\zeta(x+\mathrm{i}y)| \leq C|x+\mathrm{i}y|^c$ for some $c > 0$ valid provided that x is bounded [10];
- $\mathrm{Li}_{x+\mathrm{i}y}(a)$ is uniformly bounded for x in compact subsets of $\mathbb{R}$ and a in compact subsets of the open unit disk, as is readily inferred from the series representation (D.0.21).

Thanks to the inversion formula for the Mellin transform

$$\ln M_r(a, \mathrm{e}^{-\tau}) = \int_{c-\mathrm{i}\infty}^{c+\mathrm{i}\infty} \tau^{-s}\, \mathfrak{M}_r(a; s) \frac{\mathrm{d}s}{2\mathrm{i}\pi} \quad \text{with} \quad c > r+1 , \qquad \text{(D.0.23)}$$

we can compute the $\tau \to 0$ asymptotic expansion of $\ln M_r(a, \mathrm{e}^{-\tau})$—this principle is the basis of the transfer theorems of [11]. To do so, we deform the contour of integration to the region $\mathrm{Re}\, s < 0$. The residues at the poles of $\mathfrak{M}_r(a; s)$ are picked up in the process. There are two cases to distinguish since the polylogarithm factor in (D.0.20) is entire if $|a| < 1$, while for $a = 1$ one has $\mathrm{Li}_{s+1}(1) = \zeta(s+1)$ what generates an additional pole at $s = 0$. We remind that:

$$\Gamma(s) \underset{s \to 0}{=} \frac{1}{s} - \gamma_E + \mathrm{O}(s) , \qquad \zeta(s) = \frac{1}{s-1} + \gamma_E + \mathrm{O}(s) \qquad \text{(D.0.24)}$$

where γ_E is the Euler constant. For $a < 1$, $\mathfrak{M}_r(a;s)$ has simple poles at $s = 1+r$ and $s = 0$:

$$\mathfrak{M}_r(a;s) = \frac{\mathrm{Li}_{2+r}(a)\,r!}{s-(1+r)} + \mathrm{O}(1)\,, \qquad \mathfrak{M}_r(a;s) = \frac{-\zeta(-r)\ln(1-a)}{s} + \mathrm{O}(1)\,. \tag{D.0.25}$$

Notice that here $r \in \{0, 1\}$ and the Riemann zeta function has the special values $\zeta(0) = -1/2$ and $\zeta(-1) = -1/12$. Therefore,

$$\ln M_r(a; e^{-\tau}) = \frac{r!\,\mathrm{Li}_{2+r}(a)}{\tau^{1+r}} - \zeta(-r)\ln(1-a) + \mathrm{o}(1)\,, \qquad \tau \to 0^+ \tag{D.0.26}$$

and the remainder is uniform for a uniformly away from the boundary of the unit disk. For $a = 1$, $\mathfrak{M}_r(a;s)$ has the same simple pole at $s = 1+r$ with residue $r!\,\zeta(2+r)$, but now a double pole at $s = 0$:

$$\mathfrak{M}_r(1;s) = \frac{\zeta(-r)}{s^2} + \frac{\zeta'(-r)}{s} + \mathrm{O}(1)\,, \qquad \mathfrak{M}_r(1;s) = \frac{r!\,\zeta(2+r)}{s-(1+r)} + \mathrm{O}(1) \tag{D.0.27}$$

and we remind the special value $\zeta'(0) = -\ln(2\pi)/2$. In this case, we thus have:

$$M_r(1; e^{-\tau}) = \frac{r!\,\zeta(2+r)}{\tau^2} - \zeta(-r)\ln\tau + \zeta'(-r) + \mathrm{o}(1), \qquad \tau \to 0^+\,. \tag{D.0.28}$$

Collecting all the terms from (D.0.26)–(D.0.28), we obtain the asymptotics of the product (D.0.18) that are uniform in a belonging to compact subsets of the unit disk:

$$\begin{aligned}\ln\left[\prod_{\ell=1}^{N-1}(1-e^{-\tau_N})^{N-\ell}\right] &= \frac{\zeta(3) - \mathrm{Li}_2(\mathrm{e}^{-N\tau_N})}{\tau_N^2} + \frac{N}{\tau_N}\left(\mathrm{Li}_1(\mathrm{e}^{-N\tau_N}) - \frac{\pi^2}{6}\right) \\ &+ \left(\frac{N}{2} - \frac{1}{12}\right)\ln\left(\frac{1-\mathrm{e}^{-N\tau_N}}{\tau_N}\right) + \frac{N\ln(2\pi)}{2} + \zeta'(-1) + \mathrm{o}(1)\,.\end{aligned} \tag{D.0.29}$$

Here, we have used the special value $\zeta(2) = \pi^2/6$. It remains to insert in (D.0.29) the value of the parameter of interest $\tau_N = N^{\alpha-1}\,2\pi^2\omega_1\omega_2/g$, and return to the original formula. The announced result for the Gaussian partition function (D.0.17) follows, upon using the Stirling approximation $N! \sim \sqrt{2\pi}\,N^{N+1/2}e^{-N}$ for the factorial pre-factor.

We remark that for $\alpha \geq 1$, $\tau_N \geq 0$ is not anymore going to 0 when $N \to \infty$, therefore the asymptotic regime will be different.

■

Appendix E
Summary of Symbols

Empirical and Equilibrium Measures

$\mathcal{E}_{(\text{ply})}[\mu]$	(B.0.13)	Energy functional for the baby integral of Section 2.1
$E(\xi, \eta)$	(B.0.13)	Its kernel function
$\mu_{\text{eq}}^{(\text{ply})}$	Section B.0.4	Minimiser of $\mathcal{E}_{(\text{ply})}$
$\mathcal{E}_N[\mu]$	(2.4.7)	N-dependent energy functional
$\mathcal{E}_\infty[\mu]$	(2.4.1)	Same one at $N = \infty$
$\mathfrak{D}[\mu, \nu]$	Definition 3.1.1	Pseudo-distance between probability measures induced by $\mathcal{E}_N$
$\mu_{\text{eq}}^{(N)}$	(2.4.8) and (2.4.9)	N-dependent equilibrium measure (maximiser of $\mathcal{E}_N$)
$\rho_{\text{eq}}^{(N)}$	Theorem 2.4.2	Density of $\mu_{\text{eq}}^{(N)}$
$[a_N, b_N]$	Theorem 2.4.2	Support of $\mu_{\text{eq}}^{(N)}$
x_a^N	Definition 3.1.3	Classical positions for $\mu_{\text{eq}}^{(N)}$
$V_{N;\text{eff}}$	(3.1.16)	Effective potential
$L_N^{(\boldsymbol{\lambda})}$	(2.5.2)	Empirical measure
$\tilde{\boldsymbol{\lambda}}$	Definition 3.1.5	Deformation of $\boldsymbol{\lambda}$ enforcing a minimal spacing
$L_{N;u}^{(\boldsymbol{\lambda})}$	Definition 3.1.5	Convolution of $L_N^{(\boldsymbol{\lambda})}$ with uniform law of small support
$\mathcal{L}_N^{(\boldsymbol{\lambda})}$	Definition 3.1.8	Centred empirical measure with respect to $\mu_{\text{eq}}^{(N)}$
$\mathbb{M}_{N;\kappa}^{(n)}$	Definition 3.1.7	Probability measure including exponential regularization of n variables

Partition Functions

$Z_N[V]$	(1.5.28)	Partition function of the sinh model with potential V
$V_{G;N}$	Lemma D.0.18	Gaussian potential leading to support $[a_N; b_N]$

G. Borot et al., *Asymptotic Expansion of a Partition Function Related to the Sinh-model*, Mathematical Physics Studies,
DOI 10.1007/978-3-319-33379-3

Pairwise Interactions

$s_N(\xi)$	(3.1.1)	Pairwise interaction kernel
$S(\xi)$	(2.4.15)	Derivative of $\beta \ln \lvert\sinh(\pi\omega_1\xi)\sinh(\pi\omega_2\xi)\rvert$ *viz.* $\frac{1}{2}\partial_\xi s_N(N^{-\alpha}\xi)$
$S_{\mathrm{reg}}(\xi)$	(5.2.6)	S minus its pole at 0
$\mathcal{S}_N$	(2.4.15)	Integral operator with kernel $S(N^{\alpha}(\xi_1 - \xi_2))$
$\mathcal{S}_{N;\gamma}$	(4.0.1)	Same one with extended support
$\mathscr{S}_{N;\gamma}$	(4.1.2)	Same one in rescaled and centered variables

Operators

$\mathcal{K}_\kappa$	Definition 3.1.7	Multiplication by a decreasing exponential
$\Xi^{(p)}$	Definition 3.2.2	Operator inserting a copy of ξ_1 at position p
$\mathcal{U}_N$	(3.2.1)	Master operator
$\mathcal{D}_N$	(3.2.4)	Hyperbolic analog of the non-commutative derivative
$\mathcal{V}_N$	Proposition 5.2.1	Building block of $\mathcal{U}_N^{-1}$
$\mathcal{W}_N$	(2.4.17)	Inverse of $\mathcal{S}_N$
$\mathcal{X}_N$	Definition 3.3.5	Linear form related to $\mathscr{I}_{11}$
$\widetilde{\mathcal{X}}_N$	Definition 3.3.5	Projection to the hyperplane $\mathfrak{X}_s([a_N\,;\,b_N]) = \mathrm{Ker}\,\mathcal{X}_N$
$\widetilde{\mathcal{U}}_N^{-1}$, $\widetilde{\mathcal{W}}_N$	(3.3.33)	operators composed to the right with $\widetilde{\mathcal{X}}_N$
$\mathscr{W}_N$	(4.3.58)	Operator $\mathcal{W}_N$ in rescaled and centered variables
$\widetilde{\mathscr{W}}_{\vartheta;z_0}$	(4.3.15)	A pseudo-inverse of $\mathscr{S}_{N;\gamma}$
$\mathscr{I}_{11}$, $\mathscr{I}_{12}$	Proposition 4.3.4	Functionals appearing in the inversion of $\mathscr{S}_{N;\gamma}$
$\mathscr{J}_{1a}(\lambda)$	(4.3.34)	Related functionals
$w_{k;a}^{(1/2)}$, $w_{k;a}^{(1)}$	(4.3.36)	Functionals appearing in the large λ expansion of the latter
$H^{\wedge}$	Definition 5.1.2	Reflection of the function H (exchanging left and right boundary)
$\mathcal{G}_N$	(6.3.2)	2-variable operator related to $\mathcal{W}_N$
$\mathcal{T}_{\mathrm{even}}$, $\mathcal{T}_{\mathrm{odd}}$	(6.3.23)	Some even/odd averaging operator

Decomposition of Operators for Asymptotic Analysis

$\mathcal{W}^{(\infty)}$, $\delta\mathcal{W}$	(5.0.1)	Leading and subleading terms in $\mathcal{W}_N$ when $N \to \infty$
$\mathcal{W}_R$, $\mathcal{W}_L$	(5.1.3)	Contribution of the right/left boundary to $\mathcal{W}_N$
$\mathcal{W}_{R;k}$	Proposition 5.1.6	Terms contributing to the latter up to $O(N^{-k\alpha})$ ...
$\Delta_{[k]}\mathcal{W}_R$	Proposition 5.1.6	...and the remainder
$\mathcal{W}_{R;k}^{(\mathrm{as})}$, $\Delta_{[k]}\mathcal{W}_{R;k}^{(\mathrm{as})}$	Lemma 5.1.11	Putting aside exponentially small terms in $\mathcal{W}_{R;k}$
$\mathcal{W}_{\mathrm{bk}}$	(5.1.3)	Contribution of the bulk to $\mathcal{W}_N$
$\mathcal{W}_{\mathrm{bk};k}$	Proposition 5.1.6	The terms contributing to the latter up to $O(N^{-k\alpha})$ ...
$\Delta_{[k]}\mathcal{W}_{\mathrm{bk}}$	Proposition 5.1.6	...and the remainder
$\mathcal{W}_{\mathrm{bk};k}^{(\mathrm{as})}$, $\Delta_{[k]}\mathcal{W}_{\mathrm{bk};k}^{(\mathrm{as})}$	Lemma 5.1.11	Putting aside exponentially small terms in the bulk operator
$\mathcal{W}_{\mathrm{exp}}$	(5.1.5)	Exponentially small contribution
$(\Delta_{[k]}\mathcal{W}_N)_R$	(6.3.8)	Local right boundary remainder

Similar notations are used throughout Section 6.3 for the decompositions of $\mathcal{G}$ and the various $\mathfrak{J}$.

Riemann–Hilbert Problems

$R(\lambda)$	(4.1.18)	Reflection coefficient
κ_N	(4.1.17)	Coefficient of $1/\lambda$ term
$R_{\uparrow/\downarrow}(\lambda)$	(4.1.24) and (4.1.25)	Wiener–Hopf factors of $R(\lambda)$
$\upsilon(\lambda)$	(4.1.23)	Related, piecewise holomorphic function
Φ	Lemma 4.1.1	Two-dimensional vector in correspondence with solutions of $\mathscr{S}_{N;\gamma}[f]=g$
$\chi(\lambda)$	Proposition 4.2.1	2×2 matrix solution of the homogeneous Riemann–Hilbert problem with jump G_χ
$\chi^{(\mathrm{as})}_{\uparrow/\downarrow}(\lambda)$	(6.1.7)	Leading part of $\chi(\lambda)$ when $N\to\infty$
$\chi^{(\mathrm{exp})}_{\uparrow/\downarrow}(\lambda)$	(6.1.8)	Exponentially small part of $\chi(\lambda)$
χ_k	(4.2.30)	Matrix coefficients in the large λ expansion of $\chi(\lambda)$
G_χ	(4.1.7)	Jump matrix of the Riemann–Hilbert problem of Φ and χ
$\Psi(\lambda)$	(4.2.20) and Fig. 4.1	2×2 matrix related to $\chi(\lambda)$
$\Pi(\lambda)$	(4.2.14) and Fig. 4.2	Related 2×2 matrix
$\Delta\Pi(\lambda)$	(6.1.8)	Difference between $\Pi(\lambda)$ minus identity
G_Ψ	(4.2.4) and (4.2.5)	Jump matrix of the auxiliary Riemann–Hilbert problem
$\varkappa_\epsilon$	(4.2.7)	Rate of exponential decay of $G_\Psi - I_2$
$\mathcal{R}_{\uparrow/\downarrow}(\lambda)$	(4.1.30)	Some factors of the jump matrix
$\mathcal{R}^{(\infty)}_{\uparrow/\downarrow}$	(4.1.31)	Their non-oscillatory parts
$M_{\uparrow/\downarrow}(\lambda)$	(4.1.32)	Some factors of the jump matrix
$P_R(\lambda)$, $P_{L;\uparrow/\downarrow}(\lambda)$	(4.1.33)	Some factors in the auxiliary Riemann–Hilbert problem
θ_R	(4.2.20)	A constant involved in the auxiliary Riemann–Hilbert problem
$\Upsilon(\lambda)$	(4.3.5)–(4.3.13)	Polynomial remainder in the inhomogeneous Riemann–Hilbert problem
$\boldsymbol{H}(\lambda)$	(4.1.7)	Two-dimensional vector on the right-hand side of the inhomogeneous Riemann–Hilbert problem
$\widehat{\boldsymbol{H}}(\lambda)$	(4.3.14)	Related quantity

Auxiliary Functions, Contours, and Constants

$J_k(\lambda)$	(4.3.41)	Model integral appearing in the asymptotics of $\mathcal{J}_{1a}(\lambda)$
x_R, x_L	Definition 5.1.1	Reduced variables centered at the right and left boundary
$\Gamma_{\uparrow/\downarrow}$	Figure 5.1	Contours in the upper/lower half-plane
$\mathscr{C}_{\text{reg}}^{(\pm)}$	Definition 5.1.3 and Fig. 5.1	Contours between $\Gamma_{\uparrow/\downarrow}$ and $\mathbb{R}$
$J(x)$	Definition 5.1.3	Related to the Fourier transform of $1/R(\lambda)$
$\varrho_0(x)$	(5.1.33) and (5.1.34)	Proportional to a primitive of $J(x)$
$\varrho_\ell(x)$	Definition 5.1.5	Related to higher primitives
$\varpi_\ell(x)$	Definition 5.1.5	Integrals of $x^\ell J(x)$ from x to ∞
u_ℓ	Definition 5.1.5	Coefficients in the Taylor expansion of $1/R(\lambda)$ at $\lambda = 0$
$\mathfrak{u}_\ell(x)$	Definition 5.1.9, (5.2.48)	Related to the ℓth order truncation of the Taylor series of $1/R$
$\mathfrak{a}_\ell(x)$, $\mathfrak{b}_\ell(x)$	Definition 5.1.9	Combinations of the above, involved in asymptotics of $\mathcal{W}_N$
$\daleth_p$	Definition 6.1.1	Negative moments of $1/R_\downarrow$
$\daleth_{s,\ell}$	Definition 6.1.5	sth order moment of $\mathfrak{b}_\ell$
$\beth_\ell$	Definition 6.3.4	ℓth order moments related to J and S
$\mathcal{P}_\ell$, $\mathcal{Q}_\ell$	(6.3.73)	Some universal multivariable polynomial
$\mathfrak{g}_{R;\ell}$, $\mathfrak{g}_{\text{bk};\ell}$	(6.3.77)–(6.3.80)	A specialisation of the latter involving the functions above

Answer for the Partition Function

$\mathfrak{I}_{\text{s}}[H, G]$	Definition 6.1.4	Bilinear pairing induced by $\mathcal{W}_N$
$\mathfrak{I}_{\text{s};\beta}^{(1)}[H, G]$	(3.4.1)	Related expression appearing only for $\beta \neq 1$
$\mathfrak{I}_{\text{s};\beta}^{(1)}[H, G]$	(3.4.2)	Related expression appearing only for $\beta \neq 1$
$\mathfrak{I}_{\text{d}}[H, G]$	(6.3.1)	Related expression
$\mathfrak{I}_{\text{d};\beta}[H, G]$	(3.4.4)	Related expression appearing only for $\beta \neq 1$
$\mathfrak{Q}[V, V_0]$	(2.3.7)	A functional appearing in the interpolation
$\mathfrak{c}(x)$	Definition 6.3.9	A function involving the $\mathfrak{a}$'s and $\mathfrak{b}$'s appearing in expansion of $\mathfrak{I}_{\text{d}}$
$\aleph_0$	Definition 6.3.9	A constant involving integrals of J, S and $R_{\uparrow/\downarrow}$, appears in expansion of $\mathfrak{I}_{\text{d}}$

Norms

$\mathcal{N}_N^{(\ell)}[\phi]$	Definition 3.3.1	Weighted norms involving W_k^∞ norms for $k \in [\![0 ; \ell]\!]$
$\mathfrak{n}_\ell[V]$	Definition 3.3.2	Some estimates for the magnitude of potential

Miscellaneous

$q(z)$	(5.2.8)	Square root
$q_R(z)$	(5.2.23)	Square root at the right boundary

References

1. Hunt, R., Muckenhoupt, B., Wheeden, R.: Weighted norm inequalities for the conjugate function and Hilbert transform. Trans. Am. Math. Soc. **176**, 227–251 (1973)
2. Calderon, A.P.: Cauchy integrals on Lipschitz curves and related operators. Proc. Natl. Acad. Sci. USA **74**(4), 1324–1327 (1977)
3. Paley, R., Wiener, N.: Fourier Transforms in the Complex Domain, vol. 19. Colluquium Publications, AMS (1934)
4. Anderson, G.W., Guionnet, A., Zeitouni, O.: An introduction to random matrices. Cambridge studies in advances mathematics, vol. 118. Cambridge University Press (2010)
5. Dembo, A., Zeitouni, O.: Large deviation techniques and applications. Stochasting modelling and applied probabilities. Springer (1991)
6. Saff, E.B., Totik, V.: Logarithmic potentials with external fields. Grundlehren der mathematischen Wissenschaften, vol. 316. Springer, Berlin, Heidelberg (1997)
7. Landkof, N.S.: Foundations of modern potential theory (Translated from Osnovy sovremennoi teorii potenciala, Nauka, Moscow). Springer, Berlin (1972)
8. Borot, G., Guionnet, A., Kozlowski, K.K.: Large-N asymptotic expansion for mean field models with Coulomb gas interaction. Int. Math. Res. Not. (2015). math-ph/1312.6664
9. Dolivet, Y., Tierz, M.: Chern-Simons matrix models and Stieltjes-Wigert polynomials. J. Math. Phys. **48**, 023507 (2007). hep-th/0609167
10. Titchmarsh, E.C., Heath-Brown, D.R.: The theory of the riemann zeta function, 2nd edn. Oxford University Press (1986)
11. Flajolet, P., Gourdon, X., Dumas, P.: Mellin transforms and asymptotics: harmonic sums. Theor. Comput. Sci. **144**, 3–58 (1995)

Index

G. Borot et al., *Asymptotic Expansion of a Partition Function Related to the Sinh-model*, Mathematical Physics Studies,
DOI 10.1007/978-3-319-33379-3

Printed in the United States
By Bookmasters